ROUGH VOLATILITY

Financial Mathematics

Carmona, René, *Lectures on BSDEs, Stochastic Control, and Stochastic Differential Games with Financial Applications*

Bayer, Christian, Friz, Peter K., Fukasawa, Masaaki, Gatheral, Jim, Jacquier, Antoine, and Rosenbaum, Mathieu, editors, *Rough Volatility*

ROUGH VOLATILITY

Edited by

Christian Bayer
Weierstrass Institute for Applied Analysis and Stochastics
Berlin, Germany

Peter K. Friz
Technische Universität Berlin and
Weierstrass Institute for Applied Analysis and Stochastics
Berlin, Germany

Masaaki Fukasawa
Osaka University
Osaka, Japan

Jim Gatheral
Baruch College, CUNY
New York, USA

Antoine Jacquier
Imperial College London
London, England

Mathieu Rosenbaum
École Polytechnique
Paris, France

siam.

Society for Industrial and Applied Mathematics
Philadelphia

10 9 8 7 6 5 4 3 2

Publications Director	Kivmars H. Bowling
Executive Editor	Elizabeth Greenspan
Acquisitions Editor	Paula M. Callaghan
Developmental Editor	Rose Kolassiba
Managing Editor	Kelly Thomas
Production Editor	Lisa Briggeman
Copy Editor	Lisa Briggeman
Production Manager	Rachel Ginder
Production Coordinator	Cally A. Shrader
Compositor	Cheryl Hufnagle
Graphic Designer	Doug Smock

Library of Congress Control Number 2023946731

Cover: Detail crop from the following artwork by Oscar Murillo: *Disrupted Frequencies (Colombia, Brazil, Turkey, China)*, 2013–2019, ballpoint pen, fountain pen, graphite, felt tip pen, highlighter pen, permanent marker, paint, crayon, staples, natural pigments, debris, oil, oil stick, and other mixed media on canvas, 190 x 220 cm, OM 1078.

Contributors

Christian Bayer
Weierstrass Institute for Applied Analysis and Stochastics

Mikkel Bennedsen
Aarhus University

Peter K. Friz
TU-Berlin and Weierstrass Institute for Applied Analysis and Stochastics

Masaaki Fukasawa
Osaka University

Paul Gassiat
Université Paris Dauphine

Jim Gatheral
Baruch College, CUNY

Stefan Gerhold
TU Wien

Blanka Horvath
University of Oxford

Antoine Jacquier
Imperial College London and Alan Turing Institute

Thibault Jaisson
Banque Pictet & Cie, Geneva

Martin Keller-Ressel
TU Dresden

Paul Jusselin
Tower Research Capital, New York

Martin Larsson
Carnegie Mellon University

Asger Lunde
Copenhagen Economics

Jörg Martin
PTB, Braunschweig

Mikko Pakkanen
University of Waterloo and Imperial College London

Sergio Pulido
ENSIIE, Évry

Mathieu Rosenbaum
École Polytechnique, Paris

Benjamin Stemper
Squarepoint Capital, London

Peter Tankov
ENSAE, Paris

Contents

List of Figures

List of Tables

Foreword

Jean-Philippe Bouchaud
Capital Fund Management & Académie des Sciences

Since we will never really know *why* the prices of financial assets move, we should at least make a faithful model of *how* they move. This was the motivation of Bachelier in 1900, when he wrote in the very first page of his thesis that "contradictory opinions in regard to [price] fluctuations are so diverse that at the same instant buyers believe the market is rising and sellers that it is falling."[1] He went on to propose the first mathematical model of prices: the Brownian motion. He then built an option pricing theory that he compared to empirical data available to him—which already revealed, quite remarkably, what is now called the volatility smile.[2]

After 120 years of improvements and refinements, we are closing in on a remarkably realistic model, which reproduces almost all known stylized facts of financial price series. But are we there yet? As Mandelbrot once remarked, "in economics, there can never be a 'theory of everything.' But I believe each attempt comes closer to a proper understanding of how markets behave."[3] In order to close the gap and justify the modern mathematical apparatus that has slowly matured, we will need to understand the interactions between the behavior of zillions of traders—with individual investment styles, trading frequencies, risk limits, etc.—and the price process itself. Interestingly, recent research strongly suggests that markets self-organize in a subtle way so as to be poised at the border between stability and instability. This could be the missing link—or the holy grail—that researchers have been looking for to justify the host of empirical regularities, in particular various forms of scale invariance, that appear in price time series—and are indeed captured by the rough volatility models reviewed in this book.

For many years, the only modification to Bachelier's random walk model was to consider that log-prices, not prices themselves, are described by a Brownian motion. Apart from the fact that this modification prevents prices from becoming negative, none of the flaws of the Bachelier model were seriously tackled. Notwithstanding, the heyday of Brownian finance came when Black and Scholes published their famous 1973 paper, with the striking result that perfect delta-hedging is possible. But this is because, in the Black–Scholes world, price jumps are absent and crashes impossible. This is of course a very problematic assumption, especially because the fat-tailed distribution of returns had been highlighted as early as 1963 by Mandelbrot—who noted,

[1] https://www.investmenttheory.org/uploads/3/4/8/2/34825752/emhbachelier.pdf

[2] Looking at Bachelier's table on page 30, one clearly see a smile that flattens with the maturity of the options, as routinely observed nowadays! As we now understand, this flattening comes from the slow convergence of returns towards Gaussian random variables as the time-lag increases.

[3] B. Mandelbrot, "The variation of certain speculative prices," *The Journal of Business*, 36-4, (Oct., 1963), pp. 394–419.

in the same paper, that "large changes tend to be followed by large changes, of either sign, and small changes tend to be followed by small changes," an effect now commonly referred to as *volatility clustering* and captured by the extended family of GARCH models.

It took the violent crash of October 1987, exacerbated by the massive impact of Black–Scholes' delta-hedging, for new models to emerge. The Heston model, published in 1993, is among the most famous post-Black–Scholes models, encapsulating volatility clustering within a continuous time, Brownian motion formalism. However, like GARCH, the Heston model predicts that volatility fluctuations decay over a single time scale—in other words, that periods of high or low volatility have a rather well defined duration. This is not compatible with market data because volatility fluctuations have no clear characteristic time scale; volatility bursts can last anywhere from a few hours to a few years—suggesting that a scale invariant version of the Heston model was needed.

Mandelbrot had been mulling about this for a long while and actually proposed in 1974 a model to describe a very similar phenomenon in turbulent flows, called "multifractality." He adapted his theory in 1997 to describe currency exchange rates, before Bacry, Muzy, and Delour formulated in 2000 a more convincing version of the model, which they called the *Multifractal Random Walk* (MRW). With a single extra parameter (interpreted as a kind of volatility of volatility), the MRW captures satisfactorily many important empirical observations, such as fat-tailed distribution of returns and long memory of volatility fluctuations.

In 2014, Gatheral, Jaisson, and Rosenbaum introduced their now famous "Rough Volatility" model, which took off like wildfire and is the subject of the present book. I remember reading their paper (reproduced in Chapter 1) the day it came out on arXiv and feeling they were onto something.

The rough volatility model can be seen as an extension of the MRW with an extra parameter allowing one to tune at will the roughness of volatility. More precisely, it postulates that the log-volatility is a fractional Brownian motion with Hurst exponent H less than $1/2$ (i.e., "rougher" that the Brownian motion) while the MRW model formally corresponds to $H \to 0$. Indeed, empirical data reveal that volatility is slightly less rough than what the MRW posits, at least for stock indices. Calibration suggests $H \approx 0.1$ for the S&P500, whereas very recent (2022) work by Wu, Bacry, and Muzy suggests that single stock returns are better adjusted with $H = 0$.

The next episode of this long saga came in 2009 when Zumbach noticed a subtle, yet crucial, aspect of empirical financial time series: they are not statistically invariant upon time reversal. Past and future are not equivalent, whereas most models up to that date, including the MRW, did not distinguish past from future. More precisely, Zumbach noticed that past price trends, whether up or down, lead to higher future volatility but not the other way around (or at least more than the other way around). In 2019, following some work by Blanc, Donier, and myself on Quadratic Hawkes Processes, and some epic discussions with Jim Gatheral and Mathieu Rosenbaum (Which earned me a bottle of Dom Perignon!), Dandapani, Jusselin, and Rosenbaum proposed describing financial time series with what they called a "Quadratic Rough Heston Model," which is a synthesis of all the ideas reviewed above. It is probably the most realistic model of financial price series to date. In particular, it provides a natural solution to a long-standing puzzle, namely the joint calibration of the volatility smile of the S&P 500 and VIX options, which had eluded quants for many years. The missing ingredient was, I believe, the Zumbach effect—although Julien Guyon has proposed another, possibly complementary viewpoint.

Is this the end of the saga? From a purely engineering point of view, the latest version of the Rough Volatility model is probably hard to beat. But the remaining challenge is to justify how this particular model emerges from the underlying flow of buy and sell trades that interacts with market makers and high frequency traders. Parts of the story are already clear; in particular, as alluded to above and argued by Jaisson, Jusselin, and Rosenbaum in a remarkable series of papers, the Rough Volatility model is intimately related to the proximity of an instability that would justify the rough, multi-timescale nature of volatility.

But what is the self-organizing mechanism through which all markets appear to settle close to such a critical point? Could this scenario allow one to understand why financial time series all look so much alike? Stocks, futures, commodities, exchange rates, etc. all share very similar statistical features, in particular in the tails. Beyond being the denouement of a 120-year odyssey, we may allow ourselves to believe that the final model is not only a figment of our mathematical imagination but also a robust, trustworthy framework for risk management and derivative pricing.

In the meantime, I hope students and seasoned researchers alike will enjoy as much as I did the present collection of papers. This book embodies, in my view, quantitative finance at its best: cutting-edge mathematical tools developed not for their own sake but rather to describe, as best as possible, the real world.

Foreword: Roughness versus arbitrage

Walter Schachermayer
University of Vienna

The most fundamental pillar of Mathematical Finance is the *principle of no-arbitrage*. It states that, in a liquid financial market, agents should have no way of making a profit without risk. This sentence has to be interpreted under some technical provisions, e.g., the book-keeping should be done in discounted terms, the market should be frictionless, etc. But let us not focus on these technicalities but rather on the heart of the matter. The paradigm of *no-arbitrage* is highly convincing from an economic point of view, and surprisingly far-reaching from a mathematical one when combined with the concept of dynamic trading.

This brings us to another fundamental principle of Mathematical Finance: the modeling of price processes should be done in *continuous time*. As often encountered in mathematics, things become *easier* and more elegant when passing to a continuous-time limit rather than remaining in a discrete time setting.

At this point every reader of this book immediately thinks of the Black–Merton–Scholes model [2, 10]: in the setting of (geometric) Brownian motion these authors deduce unique option prices relying only on the principle of no-arbitrage. The Black–Merton–Scholes model is the multiplicative version of standard Brownian motion as was used by L. Bachelier [1] in his thesis from 1900, anticipating Einstein's and Smoluchowski's analyses of this process by five years. Bachelier's aim was to find a rational method for the pricing of options. He achieved this goal in a brilliant way.

Bachelier's model assumes—similarly to Black–Merton–Scholes—a constant diffusion parameter, which he gave the French name *coefficient de nervosité du marché*. Of course, this is a much more colorful expression than the corresponding modern term *volatility*. It is worth noting that Bachelier emphasized in 1900 that in the real world the "coefficient de nervosité du marché" does vary substantially in time.

Jumping a couple of decades ahead, Itô calculus made it clear—some 50+ years ago—that the replication arguments used by Black–Scholes–Merton hinge on the *quadratic variation* of the price process $(S_t)_{0\leq t\leq T}$. However, the concept of quadratic variation is a mathematical abstraction involving delicate limits in continuous time. It is hard to verify empirically the soundness of this concept for financial applications. This issue quickly raised serious doubts about whether the quadratic variation of the price process $(S_t)_{0\leq t\leq T}$ is a reliable concept for a financially relevant theory.

As early as 1963, Mandelbrot [8] proposed the use of fractional Brownian motion with a Hurst parameter $H \neq \frac{1}{2}$ for financial data (the case $H = \frac{1}{2}$ for the price process $(S_t)_{0 \le t \le T}$ corresponds to Brownian motion). The discussion of what the "true" value of H is has been raging for decades. Mandelbrot argued that it is "inescapable" to go beyond the case of a finite quadratic variation (which is nontrivial for $H = \frac{1}{2}$ only), the reason simply being that market data rejects this hypothesis.

According to Mandelbrot there is an "Inescapable Need for Fractal Tools in Finance." This is the title of a note in the *Annals of Finance* [9]. We quote the following from this paper:

> While my substitutes for the Brownian have long been resisted, my objections were widely heard and innumerable alternative models reacted by introducing "fixes", each specifically designed to avoid one of the "anomalies" I pioneered, such as discontinuity, divergent moments, and divergent dependence. Examined in the light of the fractal/multifractal approach, the fixes have a common feature: they automatically reset the local roughness to 1/2.

This point of view, fiercely defended by Mandelbrot and his followers, created an (almost) eternal, seemingly "inescapable" conflict between Mandelbrot's view and mainstream Mathematical Finance. Following the work by Harrison, Kreps, and Pliska [6, 7], this theory is built on the notion of an *equivalent martingale measure.* As everyone who has read up to this page knows, this requires the existence of a measure $\mathbb{Q}$, equivalent to the original measure $\mathbb{P}$, such that the price process $(S_t)_{0 \le t \le T}$ is a $\mathbb{Q}$-martingale (or local martingale, or sigma-martingale, once things become technical). This requirement has the mathematical consequence that $(S_t)_{0 \le t \le T}$ must be a semimartingale under the original measure $\mathbb{P}$. Indeed, the class of semimartingales (which includes, of course, the class of martingales) is invariant under equivalent changes of measures. To apply the martingale technology of Mathematical Finance we therefore *must* have a meaningful notion of quadratic variation for the price process $(S_t)_{0 \le t \le T}$. This is in flagrant contradiction to Mandelbrot's insistence on using fractional Brownian motion with $H \neq \frac{1}{2}$ as a driver for the price process $(S_t)_{0 \le t \le T}$.

There is economic support for the above argument in favor of martingale measures as well as a nontrivial quadratic variation of $(S_t)_{0 \le t \le T}$, namely the *Fundamental Theorem of Asset Pricing.* In fact, this is not a single theorem, but rather a collection of many (mathematically precise) versions of the following (mathematically not precise) metatheorem: *no-arbitrage implies the existence of an equivalent martingale measure.* For example, Theorem 7.6 of [4] states that, for any continuous stochastic process $(S_t)_{0 \le t \le T}$, the negation of the following condition ensures the existence of an equivalent local martingale measure: there is $\alpha > 0$ such that for $M > 0$ there exist *finitely many* points in time $0 = t_0 < t_1 < \cdots < t_n = T$ such that by trading at these points of time, an agent can obtain a gain of M euros, with probability at least α, while risking the loss of at most one cent. The latter property is termed a *free lunch with vanishing risk for simple integrands* in [4]. The idea is that, for practical purposes, such a "free lunch" is just as good as an arbitrage, which is defined as a totally riskless profit opportunity.

Summing up, we have a seemingly "inescapable" conflict between the economically convincing idea of no-arbitrage and Mandelbrot's claim that price processes $(S_t)_{0 \le t \le T}$ with nontrivial quadratic variation do not fit market data. The good news is that this conflict is completely solved by now, and the present book gives an excellent overview of the state of the art. The basic ingredient is the notion of a stochastic volatility process $(\sigma_t)_{0 \le t \le T}$. The idea is simple. Why not take Bachelier's observation that the "coefficient de nervosité du marché" varies in time seriously? It seems natural to model these dynamics using a stochastic process. This idea was taken up during the past decades by many authors. Typically, one loses the completeness of the market, a feature which makes Black–Merton–Scholes so appealing. But this could even be considered a plus, as it saves us from the "illusion of replication," to quote Nassim Taleb.

At this stage it becomes clear that there is no logical obstruction to using fractal models for the *underlying volatility process*, while the resulting price process $(S_t)_{0 \le t \le T}$ is driven by usual Brownian motion, thus becoming a continuous semimartingale. This idea was taken up in 1998 by Comte and Renault [3]. A breakthrough came in 2018 with the paper [5] by Gatheral, Jaisson, Rosenbaum. They showed that the choice of a very rough volatility process (with H in the order of 0.1) gives a perfect resolution to the above dilemma: on the one hand, this model has a remarkably good fit to market data, while on the other hand, it is a perfectly well-behaved and tractable semimartingale. This opened the door for a large number of impressive achievements in past years. The present book gives an excellent account of the recent developments of this beautiful and practically relevant line of research.

Bibliography

[1] L. Bachelier, *Théorie de la Spéculation,* Annales Scientifiques de l'Ecole Normale Supérieure 17 (1900), pp. 21–86. English translation in: *The Random Character of Stock Market Prices*, P. Cootner, ed., MIT Press, Cambridge, MA, 1964.

[2] F. Black and M. Scholes, *The pricing of options and corporate liabilities,* Journal of Political Economy, 81 (1973), pp. 637–659.

[3] F. Comte and E. Renault, *Long memory in continuous-time stochastic volatility models,* Mathematical Finance, 8 (1998), pp. 291–323.

[4] F. Delbaen and W. Schachermayer, *A general version of the fundamental theorem of asset pricing,* Mathematische Annalen, 300 (1994), pp. 463–520.

[5] J. Gatheral, T. Jaisson, and M. Rosenbaum, *Volatility is rough*, Quantitative Finance, 18 (2018), pp. 933–949.

[6] J.M. Harrison and D.M. Kreps, *Martingales and arbitrage in multiperiod securities markets,* Journal of Economic Theory, 20 (1979), pp. 381–408.

[7] J.M. Harrison and S.R. Pliska, *Martingales and stochastic integrals in the theory of continuous trading,* Stochastic Processes and Their Applications, 11 (1981), pp. 215–260.

[8] B.B. Mandelbrot, *The variation of certain speculative prices,* The Journal of Business, 36 (1963), pp. 394–419.

[9] ———, *The inescapable need for fractal tools in finance,* Annals of Finance, 1 (2005), pp. 193–195.

[10] R.C. Merton, *The theory of rational option pricing,* Bell Journal of Economics and Management Science, 4 (1973), pp. 141–183.

Preface

What can be found in this book

This book offers the reader a comprehensive and modern exploration of volatility processes in finance, shedding light on the argument that modeling volatility with processes whose paths are rougher than those of standard Brownian motion is essential for both fundamental reasons and practical applications. The chapters provide a unique blend of original content written specifically for this publication and well-known articles previously published in leading academic journals. This approach allows the reader to access seminal papers that have played a central role in the development of rough volatility, without getting lost in the vast literature they have generated; these influential papers are presented in their original published format, ensuring that the reader can fully grasp their essential contributions. Additionally, these papers are written in a manner that is easily understandable by not only researchers and quantitative analysts but also graduate students. Other chapters in the book aim to offer the reader a succinct compilation of established results, providing a comprehensive and accurate overview of the subject matter. These chapters serve to present a broad perspective on the entire subject. The final chapters give the reader a glimpse into cutting-edge advancements in the field of rough volatility, showcasing some of the most recent research contributions.

Who is this book for?

This book caters to a diverse range of readers who are interested in understanding and applying rough volatility models. The target audience includes the following:

1. Researchers and Academics: This book provides an extensive examination of rough volatility, presenting fundamental results and theoretical insights that contribute to the ongoing research in this field.

2. Quantitative Analysts: This book offers a practical toolbox for the computation and application of rough volatility models in finance. It equips analysts with the necessary knowledge and techniques to implement these models in real-world scenarios.

3. Graduate Students: This book is designed to be accessible to graduate students, allowing them to grasp the concepts and implications of rough volatility. It serves as an educational resource to expand their understanding of volatility modeling in finance.

4. Financial Professionals: Professionals working in the finance industry, such as traders, risk managers, and portfolio managers, can benefit from the insights provided in this book and develop a deeper understanding of volatility dynamics, enabling better decision-making.

Why this book is important

The importance of this book lies in its contribution to the understanding and application of rough volatility models. By exploring the motivation for rough volatility modeling and providing a toolbox for computation and practical implementation, the book equips the reader with the necessary knowledge and tools to work with rough volatility. The logical organization of the chapters reflects the development of rough volatility, allowing the reader to grasp the progression and evolution of the field. This book fills a gap by providing a comprehensive yet accessible resource on rough volatility, serving as an important reference for both academic researchers and practitioners in the field of finance.

Prerequisites

Since this book is about volatility, an obvious prerequisite is a basic understanding of stochastic volatility such as may be obtained, for example, from *Derivatives in Financial Markets with Stochastic Volatility* by Fouque, Papanicolaou and Sircar [139] or from *The Volatility Surface: A Practitioner's Guide* by one of the editors of this book [163]. A background in stochastic analysis for finance may be found, for example, in *Martingale Methods in Financial Modelling* by Musiela and Rutkowski [244] or in *Stochastic Calculus for Finance* II: *Continuous-Time Models* by Shreve [268].

On the contents

Aptly enough, the first chapter of this book is a reprint of the paper that started it all, "Volatility is rough" [166]. Econometric evidence for rough volatility is presented, and a simple model consistent with this evidence then follows. As an application, an efficient forecast of future realized variance is developed. Chapter 2 is a reprint of the paper "Pricing under rough volatility" [39]. Whereas the analysis of Chapter 1 is under $\mathbb{P}$, the physical or statistical measure, Chapter 2 explores the consequences of rough volatility for pricing, under the $\mathbb{Q}$ measure. Under the simplest possible deterministic change of measure, a rough version of the Bergomi model—dubbed (obviously) rough Bergomi—is obtained and is shown to fit the SPX implied volatility surface amazingly well with very few parameters.

The rough volatility property of financial prices appears to be a universal phenomenon. Chapter 3, a reprint of the paper "No-arbitrage implies power-law market impact and rough volatility" [217], addresses the fundamental question, why is volatility rough? It seems natural to suppose that prices move due to order flow and that the properties of order flow should affect the properties of the volatility process. Modeling order flow as a Hawkes process shows that rough volatility follows from the principle of no-arbitrage together with the existence of market impact. Specifically, the volatility process is a rough (or even super-rough) version of the Heston model. In addition, surprisingly enough, it connects the shape of market impact curves with the roughness of the volatility.

Chapter 4 further explores the properties of the rough Heston model together with its relatives and descendants. Computation of option prices is explicitly discussed. This chapter concludes with a presentation of the quadratic rough Heston model, which has many attractive properties; volatility is approximately lognormal rather than square-root, and it reproduces the Zumbach effect. This feature is key in the ability of the quadratic rough Heston model to fit both SPX and VIX implied volatility surfaces simultaneously, a long-standing research objective of volatility modelers. This chapter also introduces multifactor Markovian approximations of rough volatility models, which allow us to view rough volatility as a superposition of classical stochastic volatility models. Such approximations play a crucial role in practice as they provide a viable method for generating random paths of rough volatility models.

Chapter 5 presents an overview of recent work on rough affine models, of which the rough Heston model of Chapter 4 is probably the simplest and best known example. The rough Heston model was originally derived as the limit of a Hawkes process based model of order flow. This chapter explains the big picture behind this, showing that rough affine models are natural generalizations of classical affine models. Affine models are popular because their characteristic functions are expressible in closed form in terms of solutions of Riccati ODEs. This analytical tractability of classical affine models carries over to the rough case, with the exception that in the this case the characteristic function is expressed in terms of the solution of a convolution integral equation.

Chapter 6 addresses the issue of hedging under rough volatility. The inherent non-Markovian nature of rough volatility models suggests that hedging might be very hard or even impossible. Fortunately, rough volatility models turn out to be Markovian in the (infinite-dimensional) forward variance curve; the forward variance curve retains sufficient memory of the past to enable the computation of perfect hedges. This chapter concludes with a cute empirical analysis that showcases the reduction in hedging error when using a rough volatility model rather than a classical model to hedge.

Chapter 7, a reprint of the paper "Hybrid scheme for Brownian semistationary processes" [50], introduces a scheme for simulating Brownian semistationary processes, a particular example of which is the rough Bergomi volatility process. Exact simulation of the rough Bergomi model is very inefficient and impractical; this hybrid scheme offers a very efficient alternative to exact simulation. Indeed, the hybrid scheme is now the simulation method of choice for the rough Bergomi model, and variations on the scheme are increasingly seen as the method of choice for other rough volatility models, too. Efficient simulation of rough volatility models is no longer a challenge.

Chapter 8 concerns itself with the asymptotic behavior of option prices or, equivalently, of the implied volatility surface, under rough volatility. The asymptotic regimes considered include small volatility of volatility, small times, large deviations, and moderate deviations. This exhaustive compilation of existing results with consistent notation will surely be welcomed by many readers. Moreover, formulae are tested numerically, and the results of these tests are presented graphically. One of the various applications of these asymptotic formulae is to facilitate intelligent selection of starting points for calibrating rough volatility models to option prices, leading to more efficient and effective calibration processes.

Chapter 9 presents the diamond product and the forest expansion, whereby the cumulant generating function of any continuous semimartingale and its quadratic variation may be expressed as an infinite (in general) sum of diamond trees, each of which corresponds to an iterated integral. A number of model-free results are given, including a generalization of the Bergomi–Guyon expansion to arbitrary order. Unsurprisingly, computations are easiest in affine forward variance models—the rough Heston model in particular. Explicit computations are performed under rough Heston and also under rough Bergomi (which is not affine), where computations are possible but somewhat more complicated.

Finally, in Chapter 10, a reprint of the paper "A regularity structure for rough volatility" [38], Fields medalist Martin Hairer's theory of regularity structures, which is a major breakthrough in the field of stochastic partial differential equations, is applied to the strong approximation of stochastic integrals appearing in rough volatility model computations. The analysis in this chapter is not just theoretical; pseudo code is provided and numerical convergence results are given.

Why did we decide to write this book?

The theory of rough volatility, while rooted in data-driven observations, possesses a striking and inherent beauty. Historical time series are reconciled with the prices of options with very few stable parameters. Most importantly, the rough volatility paradigm has microstructural foundations, so we know that volatility is indeed rough and why it has to be rough.

To the delight of mathematicians, the study of rough volatility requires cutting-edge mathematical tools and theories. These include fractional Brownian motion, Volterra equations, Hawkes point processes, and affine processes as well as elements from infinite-dimensional analysis, numerical analysis, rough paths, regularity structures, stochastic partial differential equations, and other studies. This multifaceted nature of the study of rough volatility only adds to its scientific allure.

However, the most captivating aspect of rough volatility lies in its ability to translate scientific elegance into applied beauty. The editors firmly believe that the research results indicate a clear direction: rough volatility should be regarded as a fundamental and benchmark approach. In other words, regardless of the specific problem in financial engineering involving volatility, rough volatility is likely to provide a reasonable and robust solution. While simplified versions of rough volatility models may be more relevant for certain specific tasks, and although more complex versions tailored to precise objectives may be required in other applications, the core concept of rough volatility remains fundamental.

From a statistical testing perspective, these ideas can be framed as the statistical null hypothesis H0: "Volatility is rough." Remarkably, it is exceedingly challenging to statistically reject this null hypothesis. This reinforces rough volatility as the fundamental paradigm, setting it apart from other approaches. For instance, classical stochastic volatility models can be rejected based on historical data due to their inherent stationarity over longer time scales necessary to compensate for Brownian scaling. In contrast, the multifractal nature of rough volatility exhibits remarkable consistency with historical data.

In summary, our conviction may be succinctly stated as follows: any viable model describing the dynamics of volatility should exhibit a close resemblance to rough volatility. With this in mind, the objective of this book is to provide readers, whether they be graduate students, mathematicians, or quantitative analysts, with the necessary insights to embrace this new paradigm. We firmly believe that adopting this point of view will empower readers to develop a more comprehensive and fundamental understanding of volatility—its nature, its behavior, and the art of modeling it effectively.

Berlin, London, New York,
Osaka, Paris, June 2023

Christian Bayer, Peter K. Friz, Masaaki Fukasawa,
Jim Gatheral, Antoine Jacquier, Mathieu Rosenbaum

Chapter 1

Volatility is rough[4]

Jim Gatheral, Thibault Jaisson, Mathieu Rosenbaum

1.1 ▪ Introduction

1.1.1 ▪ Volatility modeling

In the derivatives world, log-prices are often modeled as continuous semimartingales. For a given asset with log-price Y_t, such a process takes the form

$$dY_t = \mu_t dt + v_t dW_t,$$

where μ_t is a drift term and W_t is a one-dimensional Brownian motion. The term v_t denotes the volatility process and is the most important ingredient of the model. In the Black–Scholes framework, the volatility function is either constant or a deterministic function of time. In Dupire's local volatility model (see [124]) the local volatility $\sigma(Y_t, t)$ is a deterministic function of the underlying price and time, chosen to match observed European option prices exactly. Such a model is by definition time-inhomogeneous; its dynamics are highly unrealistic, typically generating future volatility surfaces (see Section 1.1.3) completely unlike those we observe. A corollary of this is that prices of exotic options under local volatility can be substantially off-market. On the other hand, in so-called stochastic volatility models, the volatility v_t is modeled as a continuous Brownian semimartingale. Notable among such stochastic volatility models are the Hull and White model [206], the Heston model [199], and the SABR model [188]. While stochastic volatility dynamics are more realistic than local volatility dynamics, generated option prices are not consistent with observed European option prices. We refer the reader to [163] and [244] for more detailed reviews of the different approaches to volatility modeling. More recent market practice is to use local-stochastic-volatility (LSV) models which both fit the market exactly and generate reasonable dynamics.

Consistent with our focus on derivatives, our goal in this work is to replicate features of the observed time series of volatility over time scales from one day to ten years say. Indeed, volatility modeling is probably only relevant at time scales of order one day or more. Below this, at

[4]This chapter is derived from "Volatility is rough," *Quantitative Finance*, 18 (2018), pp. 933–949, https://doi.org/10.1080/14697688.2017.1393551.

the subsecond time scale, for example, it is not even clear what the meaning of volatility is (independently of a specific model). Nevertheless, in order to get accurate volatility measurements, we will rely on high frequency methods in our estimation procedures.

1.1.2 ▪ Fractional volatility

In terms of the smoothness of the volatility process, the preceding models offer two possibilities: very regular sample paths in the case of Black–Scholes, and volatility trajectories with regularity close to that of Brownian motion for the local and stochastic volatility models. Starting from the stylized fact that volatility is a long-memory process, various authors have proposed models that allow for a wider range of regularity for the volatility. In a pioneering paper, Comte and Renault [99] proposed modeling log-volatility using fractional Brownian motion (fBm for short), ensuring long memory by choosing the Hurst parameter $H > \frac{1}{2}$. A large literature has subsequently developed around such fractional volatility models; see, for example, [92, 97, 262].

The fBm $(W_t^H)_{t\in\mathbb{R}}$ with Hurst parameter $H \in (0,1)$, introduced in [236], is a centered self-similar Gaussian process with stationary increments satisfying, for any $t \in \mathbb{R}$, $\Delta \geq 0$, $q > 0$,

$$\mathbb{E}[|W_{t+\Delta}^H - W_t^H|^q] = K_q \Delta^{qH}, \tag{1.1}$$

with K_q the moment of order q of the absolute value of a standard Gaussian variable. For $H = \frac{1}{2}$, we retrieve the classical Brownian motion. The sample paths of W^H are Hölder continuous with exponent r, for any $r < H$.[5] Finally, when $H > \frac{1}{2}$, the increments of the fBm are positively correlated and exhibit long memory in the sense that

$$\sum_{k=0}^{+\infty} \mathrm{Cov}[W_1^H, W_k^H - W_{k-1}^H] = +\infty.$$

Indeed, $\mathrm{Cov}[W_1^H, W_k^H - W_{k-1}^H]$ is of order k^{2H-2} as $k \to \infty$. Note that in the case of the fBm, there is a one-to-one correspondence between regularity and long memory through the Hurst parameter H.

As mentioned earlier, the long-memory property of the volatility process has been widely accepted as a stylized fact since the seminal analyses of Ding, Granger, and Engle [119], Andersen and Bollerslev [16], and Andersen et al. [18]. Initially, it appears that the term *long memory* referred to the slow decay of the autocorrelation function (of absolute returns, for example), anything slower than exponential. Over time, however, it seems that this term has acquired the more precise meaning that the autocorrelation function is not integrable (see [54]), and even more precisely, that it decays as a power law with exponent less than 1. Much of the more recent literature, for example, [52, 91, 94], assumes long memory in volatility in this more technical sense. Indeed, meaningful results probably can only be obtained under such a specification, since it is not possible to estimate the asymptotic behavior of the covariance function without assuming a specific form. Nevertheless, analyses such as that of Andersen et al. [18] use data that predate the advent of high frequency electronic trading, and the evidence for long memory has never been sufficient to satisfy remaining doubters, such as Mikosch and Starica in [242]. To quote Cont [100, p. 309], "... the econometric debate on the short range or long range nature of dependence in volatility still goes on (and may probably never be resolved)...".

In the spirit of the above quote, in our view, the question as to whether or not the volatility time series exhibits long memory (in a technical sense) is not a very useful or fruitful one. Indeed, we believe that in practice, the concept of long memory is too fragile to be applicable to an

[5] Actually, H corresponds to the regularity of the process in a more accurate way: in terms of Besov smoothness spaces. See Section 1.2.1.

analysis involving ultra high frequency data (for example, any seasonality may give rise to spurious long memory). Therefore we do not focus on long memory in this work. Still, we do show that the autocorrelation function of volatility does not behave as a power law, at least at usual time scales of observation. In particular, we are able to provide explicit expressions enabling us to analyze thoroughly the dependence structure of the volatility process.

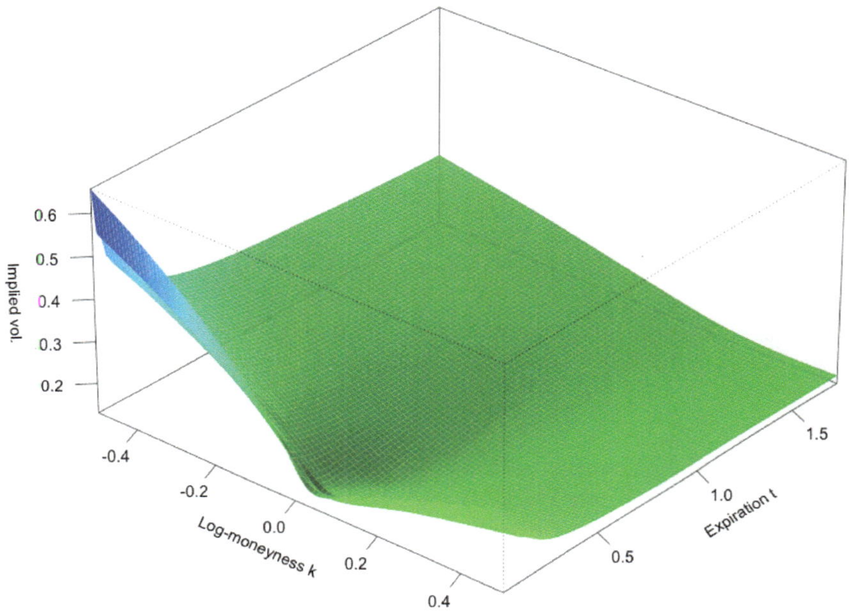

Figure 1.1: The S&P volatility surface as of June 20, 2013.

1.1.3 ▪ The shape of the implied volatility surface

As is well known, the implied volatility $\sigma_{\text{imp}}(k,\tau)$ of an option (with log-moneyness k and time to expiration τ) is the value of the volatility parameter in the Black–Scholes formula required to match the market price of that option. Plotting implied volatility as a function of strike price and time to expiry generates the *volatility surface*, explored in detail in, for example, [163]. A typical volatility surface generated from a "stochastic volatility inspired" (SVI) [165] fit to closing SPX option prices as of June 20, 2013[6] is shown in Figure 1.1. It is a stylized fact that, at least in equity markets, although the level and orientation of the volatility surface do change over time, the general overall shape of the volatility surface does not change, at least to a first approximation. This suggests that it is desirable to model volatility as a time-homogeneous process, i.e., a process whose parameters are independent of price and time.

However, conventional time-homogeneous models of volatility, such as the Hull and White, Heston, and SABR models, do not fit the volatility surface. In particular, as shown in Figure 1.2, the observed term structure of at-the-money ($k = 0$) volatility skew

$$\psi(\tau) := \left|\frac{\partial}{\partial k}\sigma_{\text{imp}}(k,\tau)\right|_{k=0}$$

[6]Closing prices of SPX options for all available strikes and expirations as of June 20, 2013 were sourced from OptionMetrics (www.optionmetrics.com) via Wharton Research Data Services (WRDS).

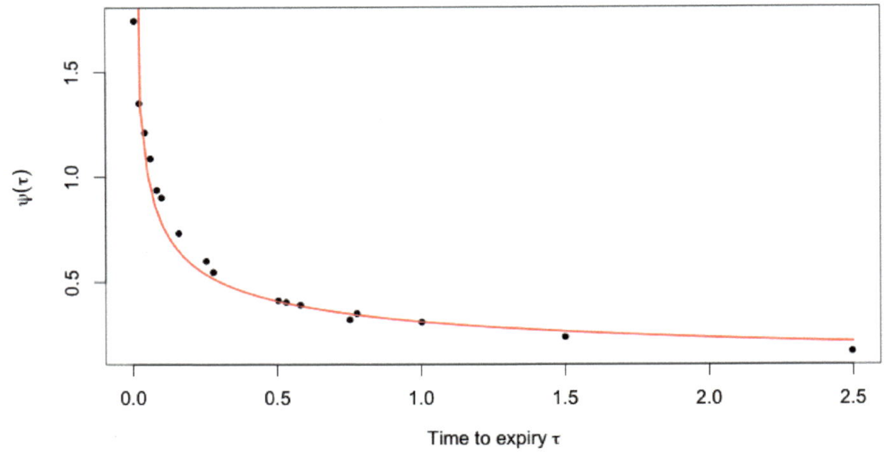

Figure 1.2: The black dots are nonparametric estimates of the S&P ATM volatility skews as of June 20, 2013; the red curve is the power-law fit $\psi(\tau) = A\,\tau^{-0.4}$.

is well-approximated by a power-law function of time to expiry τ. In contrast, conventional stochastic volatility models generate a term structure of at-the-money (ATM) skew that is *constant* for small τ and behaves as a sum of decaying exponentials for larger τ.

In Section 3.3 of [151], as an example of the application of his martingale expansion, Fukasawa shows that a stochastic volatility model where the volatility is driven by fractional Brownian motion with Hurst exponent H generates an ATM volatility skew of the form $\psi(\tau) \sim \tau^{H-\frac{1}{2}}$, at least for small τ. This is interesting in and of itself in that it provides a counterexample to the widespread belief that the explosion of the volatility smile as $\tau \to 0$ (as clearly seen in Figures 1.1 and 1.2) implies the presence of jumps [87]. The main point here is that for a model of the sort analyzed by Fukasawa to generate a volatility surface with a reasonable shape, we would need to have a value of H close to zero. As we will see in Section 1.2, our empirical estimates of H from time series data are in fact very small.

The volatility model that we will specify in Section 1.3.1, driven by fBm with $H < \frac{1}{2}$, therefore has the potential to be not only consistent with the empirically observed properties of the volatility time series but also consistent with the shape of the volatility surface. In this chapter, we focus on the modeling of the volatility time series. A more detailed analysis of the consistency of our model with option prices is provided in [39].

1.1.4 ▪ Main results and organization of the chapter

In Section 1.2, we report our estimates of the smoothness of the log-volatility for selected assets. This smoothness parameter lies systematically between 0.08 and 0.2 (in the sense of Hölder regularity, for example). Furthermore, we find that increments of the log-volatility are approximately normally distributed and that their moments enjoy a remarkable monofractal scaling property. This leads us to model the log of volatility using an fBm with Hurst parameter $H < \frac{1}{2}$ in Section 1.3. Specifically, we adopt the fractional stochastic volatility (FSV) model of Comte and Renault [99]. We call our model rough FSV (RFSV) to underline that, in contrast to FSV, we take $H < \frac{1}{2}$. We also show in the same section that the RFSV model is remarkably consistent with volatility time series data. The issue of volatility persistence is considered through the lens of the RFSV model in Section 1.4. Our main finding is that although the RFSV model does not have any long-memory property, classical statistical procedures aiming at detecting volatility persistence tend to conclude the presence of long memory in data generated from it. This sheds new light on the supposed long memory in the volatility of financial data. In Section 1.5, we finally apply our model to volatility forecasting. In particular, we show that RFSV volatility forecasts outperform conventional AR and HAR volatility forecasts. Some proofs are relegated to the appendix (Section 1.7).

1.2 ▪ Smoothness of the volatility: Empirical results

In this section we report estimates of the smoothness of the volatility process for the following four assets:

- The DAX and Bund futures contracts, for which we estimate integrated variance directly from high frequency data using an estimator based on the model with uncertainty zones; see [260, 261]. This model enables us to safely use all the ultra high frequency price data in order to perform our estimation and thus obtain accurate estimates over short time windows.
- The S&P and NASDAQ indices, for which we use precomputed realized variance estimates from the Oxford-Man Institute of Quantitative Finance Realized Library.[7]

1.2.1 ▪ Estimating the smoothness of the volatility process

Let us first pretend that we have access to discrete observations of the volatility process on a time grid with mesh Δ on $[0,T]$: $v_0,\ v_\Delta,\dots,v_{k\Delta},\dots,\ k\in\{0,\lfloor T/\Delta\rfloor\}$. Set $N=\lfloor T/\Delta\rfloor$; then for $q\geq 0$, we define

$$m(q,\Delta)=\frac{1}{N}\sum_{k=1}^{N}|\log(v_{k\Delta})-\log(v_{(k-1)\Delta})|^q.$$

In the spirit of [264], our main assumption is that for some $s_q>0$ and $b_q>0$, as Δ tends to zero,

$$N^{qs_q}m(q,\Delta)\to b_q. \tag{1.2}$$

Under additional technical conditions, equation (1.2) essentially says that the volatility process belongs to the Besov smoothness space $\mathcal{B}^{s_q}_{q,\infty}$ and does not belong to $\mathcal{B}^{s'_q}_{q,\infty}$, for $s'_q>s_q$; see [263]. Hence s_q can really be viewed as the regularity of the volatility when measured in l_q norm. In particular, functions in $\mathcal{B}^{s}_{q,\infty}$ for every $q>0$ enjoy the Hölder property with parameter h for any $h<s$. For example, if $\log(v_t)$ is an fBm with Hurst parameter H, then for any $q\geq 0$, equation (1.2) holds in probability with $s_q=H$, and it can be shown that the sample paths of the process indeed belong to $\mathcal{B}^{H}_{q,\infty}$ almost surely. Assuming the increments of the log-volatility process are stationary and that a law of large numbers can be applied, $m(q,\Delta)$ can also be seen as the empirical counterpart of

$$\mathbb{E}[|\log(v_\Delta)-\log(v_0)|^q].$$

Of course the volatility process is not directly observable, and an exact computation of $m(q,\Delta)$ is not possible in practice. We must therefore proxy spot volatility values by appropriate estimated

[7] http://realized.oxford-man.ox.ac.uk/data/download. The Oxford-Man Institute's Realized Library contains a selection of daily nonparametric estimates of volatility of financial assets, including realized variance (rv) and realized kernel (rk) estimates. A selection of such estimators is described and their performances compared in, for example, [169].

values. Since the minimal Δ will be equal to one day in the sequel, we proxy the (true) spot volatility daily at a fixed given time of the day (11 a.m., for example). Two daily spot volatility proxies will be considered:

- For our ultra high frequency intraday data (DAX future contracts and Bund future contracts,[8] 1248 days from 13/05/2010 to 01/08/2014[9]), we use the estimator of the integrated variance from 10 a.m. to 11 a.m. London time obtained from the model with uncertainty zones; see [260, 261]. After renormalization, the resulting estimates of integrated variance over very short time intervals can be considered as good proxies for the unobservable spot variance. In particular, the hour-long window on which they are computed is small compared to the extra day time scales that will be of interest here.

- For the S&P and NASDAQ indices,[10] we proxy daily spot variances by daily realized variance estimates from the Oxford-Man Institute of Quantitative Finance Realized Library (3540 trading days from January 3, 2000 to March 31, 2014). Since these estimates of integrated variance are for the whole trading day, we expect estimates of the smoothness of the volatility process to be biased upwards, integration being a regularizing operation. We compute the extent of this bias by simulation in Section 1.3.4 and more quantitatively in Appendix 1.7.3.

In the following, we retain the notation $m(q, \Delta)$ with the understanding that we are only proxying the (true) spot volatility as explained above. We now proceed to estimate the smoothness parameter s_q for each q by computing the $m(q, \Delta)$ for different values of Δ and regressing $\log m(q, \Delta)$ against $\log \Delta$. Note that for a given Δ, several $m(q, \Delta)$ can be computed depending on the starting point. Our final measure of $m(q, \Delta)$ is the average of these values.

1.2.2 ▪ DAX and Bund futures contracts

DAX and Bund futures are among the most liquid assets in the world, and, moreover, the model with uncertainty zones used to estimate volatility is known to apply well to them; see [111]. So we can be confident in the reliability of our volatility proxy. Nevertheless, as an extra check, we will confirm the quality of our volatility proxy by Monte Carlo simulation in Section 1.3.4. Plots of $\log m(q, \Delta)$ vs. $\log \Delta$ for different values of q are displayed for the DAX in Figure 1.3 and for the Bund in Figure 1.4.

For both DAX and Bund, for a given q, the points essentially lie on a straight line. Under stationarity assumptions, this implies that the log-volatility increments enjoy the following scaling property in expectation:

$$\mathbb{E}[|\log(\nu_\Delta) - \log(\nu_0)|^q] = b_q \Delta^{\zeta_q},$$

where $\zeta_q = q s_q > 0$ is the slope of the line associated to q. Moreover, the smoothness parameter s_q does not seem to depend on q. Indeed, plotting ζ_q against q, we obtain that $\zeta_q \sim H q$ with H equal 0.125 for the DAX and 0.082 for the Bund; see Figure 1.5.

We remark that the graphs for ζ_q are actually very slightly concave. However, we observe the same small concavity effect when we replace the log-volatility by simulations of an fBm with the same number of points. We conclude that this effect relates to finite sample size and is thus not significant.

[8]For every day, we only consider the future contract corresponding to the most liquid maturity.

[9]Data kindly provided by QuantHouse EUROPE/ASIA, https://www.iress.com/.

[10]And also the CAC40, Nikkei, and FTSE indices in some specific parts of the chapter.

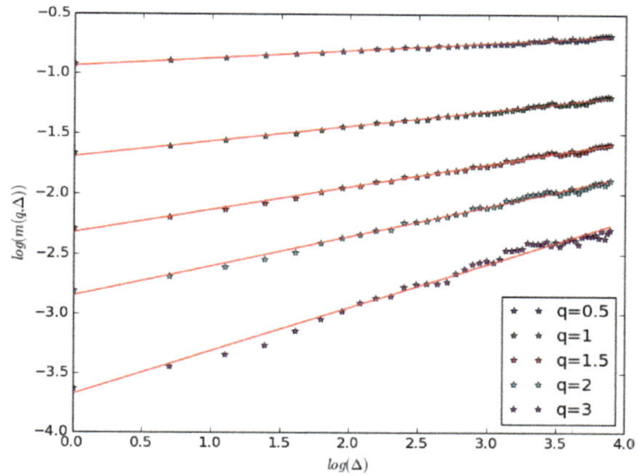

Figure 1.3: $\log m(q,\Delta)$ as a function of $\log \Delta$, DAX.

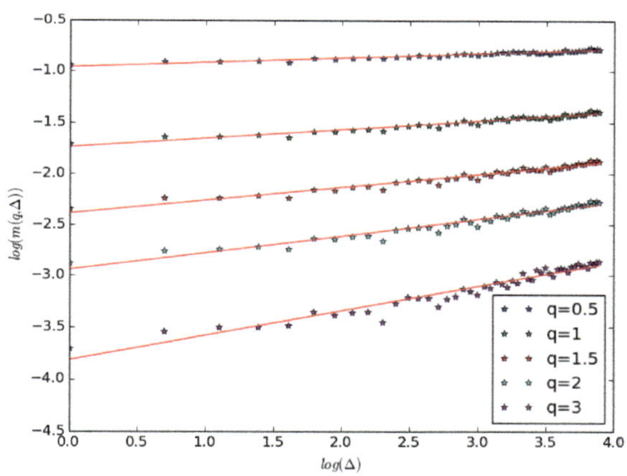

Figure 1.4: $\log m(q,\Delta)$ as a function of $\log \Delta$, Bund.

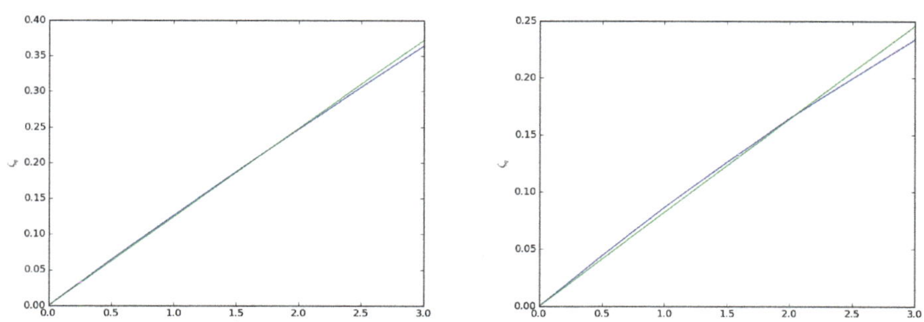

Figure 1.5: ζ_q (blue) and $0.125 \times q$ (green), DAX (left); ζ_q (blue) and $0.082 \times q$ (green), Bund (right).

1.2.3 ▪ S&P and NASDAQ indices

We report in Figures 1.6 and 1.7 similar results for the S&P and NASDAQ indices. The variance proxies used here are the precomputed five-minute realized variance estimates for the whole trading day made publicly available by the Oxford-Man Institute of Quantitative Finance.

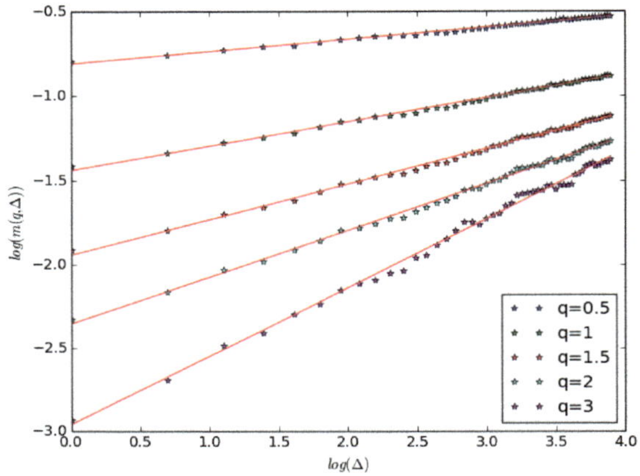

Figure 1.6: $\log m(q,\Delta)$ as a function of $\log \Delta$, S&P.

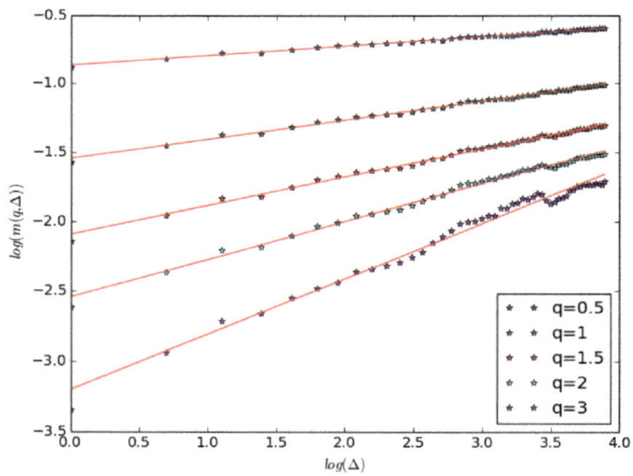

Figure 1.7: $\log m(q,\Delta)$ as a function of $\log(\Delta)$, NASDAQ.

We observe the same scaling property for the S&P and NASDAQ indices as we observed for DAX and Bund futures; and, again, the s_q do not depend on q. However, the estimated smoothnesses are slightly higher here: $H = 0.142$ for the S&P and $H = 0.139$ for the NASDAQ; see Figure 1.8.

Once again, we do expect these smoothness estimates to be biased high because we are using whole-day realized variance estimates, as explained earlier in Section 1.2. Finally, we remark that as for DAX and Bund futures, the graphs for ζ_q are slightly concave.

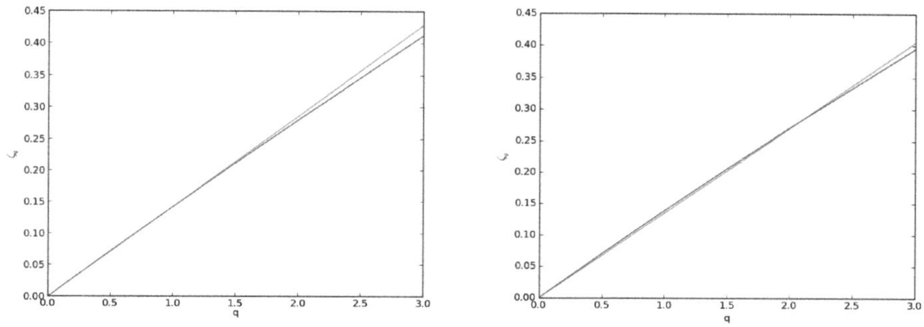

Figure 1.8: ζ_q (blue) and $0.142 \times q$ (green), S&P (left); ζ_q (blue) and $0.139 \times q$ (green), NASDAQ (right).

1.2.4 ▪ Other indices

Repeating the analysis of Section 1.2.3 for each index in the Oxford-Man dataset, we find that the $m(q, \Delta)$ present a universal scaling behavior. For each index and for $q = 0.5,\ 1,\ 1.5,\ 2,\ 3$, by doing a linear regression of $\log(m(q, \Delta))$ on $\log(\Delta)$ for $\Delta = 1, \ldots, 30$, we obtain estimates of ζ_q that we summarize in Table 1.3 in Appendix 1.7.2.

1.2.5 ▪ Distribution of the increments of the log-volatility

Having established that all our underlying assets exhibit essentially the same scaling behavior,[11] we focus in the rest of the chapter only on the S&P index, unless specified otherwise. That the distribution of increments of log-volatility is close to Gaussian is a well-established stylized fact reported, for example, in the papers [17] and [18] of Andersen et al. Looking now at the histograms of the increments of the log-volatility in Figure 1.9, with the fitted normal density superimposed in red, we see that, for any Δ, the empirical distribution of log-volatility increments is verified as being close to Gaussian. More impressive still is that rescaling the 1-day fit of the normal density by Δ^H generates (blue dashed) curves that are very close to the red fits of the normal density, consistent with the observed scaling.

The slight deviations from the normal distribution observed in Figure 1.9 are again consistent with the computation of the empirical distribution of the increments of a fractional Brownian motion on a similar number of points.

1.2.6 ▪ Does H vary over time?

In order to check whether our estimations of H depends on the time interval, we split the Oxford-Man realized variance dataset into two halves and re-estimate H for each half separately. The results are presented in Table 1.4 in Appendix 1.7.2. We note that although the estimated H all lie between 0.06 and 0.20, they seem to be higher in the second period which includes the financial crisis.

[11] We have also verified that this scaling relationship holds for crude oil and gold futures, with similar smoothness estimates.

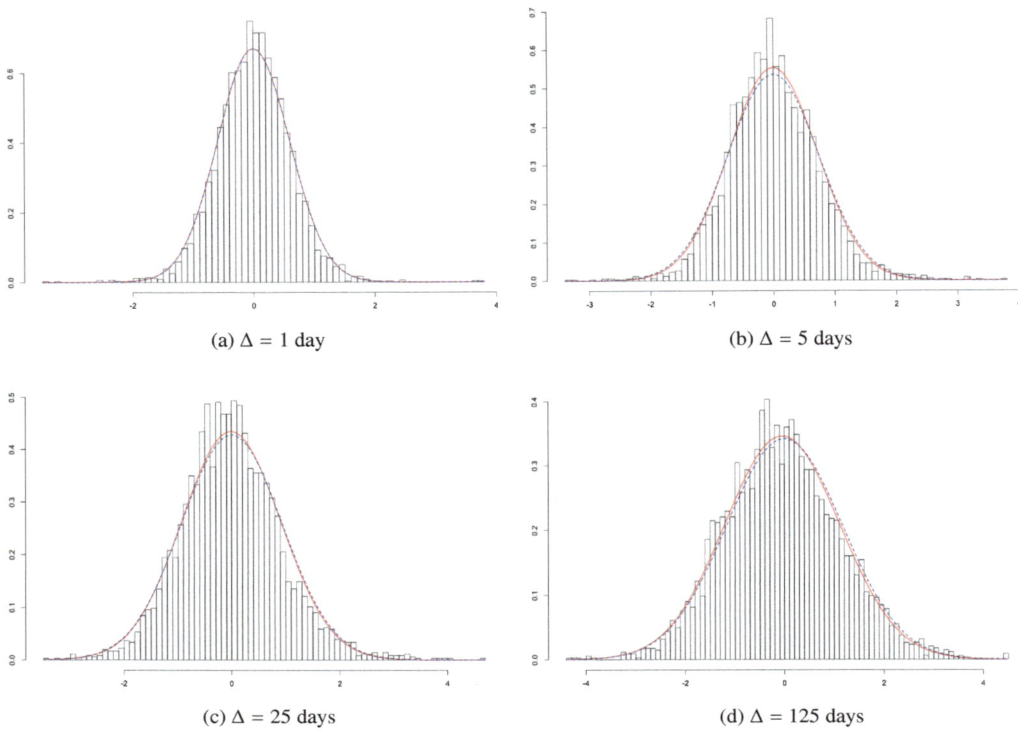

(a) $\Delta = 1$ day

(b) $\Delta = 5$ days

(c) $\Delta = 25$ days

(d) $\Delta = 125$ days

Figure 1.9: Histograms for various lags Δ of the (overlapping) increments $\log v_{t+\Delta} - \log v_t$ of the S&P log-volatility; normal fits in red; normal fit for $\Delta = 1$ day rescaled by Δ^H in blue.

1.3 ▪ A simple model compatible with the empirical scaling of the volatility

In this section, we specify the rough fractional stochastic volatility (RFSV) model and demonstrate that it reproduces the empirical facts presented in Section 1.2.

1.3.1 ▪ Specification of the RFSV model

In the previous section, we showed that, empirically, the increments of the log-volatility of various assets enjoy a scaling property with constant smoothness parameter and that their distribution is close to Gaussian. This naturally suggests the following simple model:

$$\log v_{t+\Delta} - \log v_t = \nu\left(W^H_{t+\Delta} - W^H_t\right), \tag{1.3}$$

where W^H is a fractional Brownian motion with Hurst parameter equal to the measured smoothness of the volatility, and ν is a positive constant. We may of course write (1.3) under the form

$$v_t = \sigma \exp\left(\nu W^H_t\right), \tag{1.4}$$

where σ is another positive constant.

However, this model is not stationary, and stationarity is desirable both for mathematical tractability and to ensure reasonableness of the model at very large times. This leads us to impose stationarity by modeling the log-volatility as a fractional Ornstein–Uhlenbeck process with a very long reversion time scale.

A stationary fractional Ornstein–Uhlenbeck process (X_t) is defined as the stationary solution of the stochastic differential equation

$$\mathrm{d}X_t = \nu\,\mathrm{d}W_t^H - \alpha(X_t - m)\mathrm{d}t,$$

where $m \in \mathbb{R}$, and ν and α are positive parameters; see [92]. As for usual Ornstein–Uhlenbeck processes, there is an explicit form for the solution which is given by

$$X_t = \nu \int_{-\infty}^{t} \mathrm{e}^{-\alpha(t-s)}\mathrm{d}W_s^H + m. \tag{1.5}$$

Here the stochastic integral with respect to fBm is simply a pathwise Riemann–Stieltjes integral; see again [92].

We thus arrive at the following final specification of our RFSV model for the volatility on the time interval of interest $[0,T]$:

$$\nu_t = \exp(X_t),\ t \in [0,T], \tag{1.6}$$

where (X_t) satisfies equation (1.5) for some $\nu > 0$, $\alpha > 0$, $m \in \mathbb{R}$, and $H < \frac{1}{2}$ the measured smoothness of the volatility. This model provides a very parsimonious description of the volatility process with only four parameters (and in practice only three; see below). Moreover, statistical inference methods for fractional Brownian motion and the fractional Ornstein–Uhlenbeck process are well known; see [70, 96, 207].

Such a model is indeed stationary. However, if $\alpha \ll 1/T$, the log-volatility behaves locally (at time scales smaller than T) as an fBm. This observation is formalized in the following proposition.

Proposition 1.1. *Let W^H be an fBm and X^α be defined by* (1.5) *for a given $\alpha > 0$. As α tends to zero,*

$$\mathbb{E}\left[\sup_{t\in[0,T]} \left|X_t^\alpha - X_0^\alpha - \nu W_t^H\right|\right] \to 0.$$

The proof is given in Appendix 1.7.1. Proposition 1.1 implies that in the RFSV model, if $\alpha \ll 1/T$, and we confine ourselves to the interval $[0,T]$, we can proceed as if the log-volatility process were an fBm. Indeed, simply setting $\alpha = 0$ in (1.5) gives (at least formally) $X_t - X_s = \nu(W_t^H - W_s^H)$, and we immediately recover our simple nonstationary fBm model (1.3). Consequently, although the RFSV model is technically stationary, its ergodic behavior is of no interest to us; for example, estimation of the mean of the volatility is not possible in practice. Indeed, at any time scale of practical interest (from one day to several years), we see no evidence of ergodicity in the data; see Figure 1.13.

The following corollary implies that the (exact) scaling property of the fBm is approximately reproduced by the fractional Ornstein–Uhlenbeck process when α is small.

Corollary 1.2. *Let $q > 0$, $t > 0$, $\Delta > 0$. As α tends to zero, we have*

$$\mathbb{E}[|X_{t+\Delta}^\alpha - X_t^\alpha|^q] \to \nu^q\, K_q\, \Delta^{qH}.$$

The proof is given in Appendix 1.7.1.

RFSV vs. FSV

We recognize our RFSV model (1.6) as a particular case of the classical FSV model of Comte and Renault [99]. The key difference is that here we take $H < \frac{1}{2}$ and $\alpha \ll 1/T$, whereas to accommodate the assumption of long memory, Comte and Renault have to choose $H > \frac{1}{2}$. The analysis of Fukasawa referred to earlier in Section 1.1.3 implies in particular that if $H > \frac{1}{2}$, the volatility skew function $\psi(\tau)$ is *increasing* in time to expiration τ (at least for small τ), which is obviously completely inconsistent with the approximately $1/\sqrt{\tau}$ skew term structure that is observed. To generate a decreasing term structure of volatility skew for longer expirations, Comte and Renault are then forced to choose $\alpha \gg 1/T$. Consequently, for very short expirations ($\tau \ll 1/\alpha$), models of the Comte–Renault type with $H > \frac{1}{2}$ still generate a term structure of volatility skew that is inconsistent with the observed one, as explained, for example, in Section 4 of [97].

In contrast, the choice $H < \frac{1}{2}$ enables us to reproduce the observed smoothness and scaling of the volatility process and generate a term structure of volatility skew in agreement with the observed one. The choice $H < \frac{1}{2}$ is also consistent with what is improperly called mean reversion by practitioners, which in fact corresponds to strong oscillations in the volatility process. Finally, taking α very small implies that the dynamics of our process is close to that of an fBm; see Proposition 1.1. This last point is particularly important. Indeed, recall that at the time scales we are interested in, the important feature we have in mind is really this fBm-like behavior of the log-volatility.

Finally, note that we could no doubt have considered other stationary models satisfying Proposition 1.1 and Corollary 1.2, where log-volatility behaves as an fBm at reasonable time scales; choosing the fractional Ornstein–Uhlenbeck process is probably the simplest way to accommodate this local behavior together with the stationarity property.

1.3.2 ▪ RFSV model autocovariance functions

From Proposition 1.1 and Corollary 1.2, we easily deduce the following corollary, where $o(1)$ tends to zero as α tends to zero.

Corollary 1.3. *Let $q > 0,\ t > 0,\ \Delta > 0$. As α tends to zero,*

$$\mathrm{Cov}[X_t^\alpha, X_{t+\Delta}^\alpha] = \mathbb{V}[X_t^\alpha] - \frac{1}{2}\nu^2\Delta^{2H} + o(1).$$

Consequently, in the RFSV model, for fixed t, the covariance between X_t and $X_{t+\Delta}$ is linear with respect to Δ^{2H}. This result is very well satisfied empirically. For example, in Figure 1.10, we see that for the S&P, the empirical autocovariance function of the log-volatility is indeed linear with respect to Δ^{2H}. Note in passing that at the time scales we consider, the term $\mathbb{V}[X_t^\alpha]$ is higher than $\frac{1}{2}\nu^2\,\Delta^{2H}$ in the expression for $\mathrm{Cov}[X_t^\alpha, X_{t+\Delta}^\alpha]$.

Having computed the autocovariance function of the log-volatility, we now turn our attention to the volatility itself. We have

$$\mathbb{E}[v_{t+\Delta}v_t] = \mathbb{E}[e^{X_t^\alpha + X_{t+\Delta}^\alpha}],$$

with X^α defined by (1.5). Since X^α is a Gaussian process, we deduce that

$$\mathbb{E}[v_{t+\Delta}v_t] = \exp\left\{\mathbb{E}[X_t^\alpha] + \mathbb{E}[X_{t+\Delta}^\alpha] + \frac{1}{2}\mathbb{V}[X_t^\alpha] + \frac{1}{2}\mathbb{V}[X_{t+\Delta}^\alpha] + \mathrm{Cov}[X_t^\alpha, X_{t+\Delta}^\alpha]\right\}.$$

Applying Corollary 1.3, we obtain that when α is small, $\mathbb{E}[v_{t+\Delta}v_t]$ is approximately equal to

$$\exp\left\{2\mathbb{E}[X_t^\alpha] + 2\mathbb{V}[X_t^\alpha] - \nu^2\frac{\Delta^{2H}}{2}\right\}. \tag{1.7}$$

It follows that in the RFSV model, $\log(\mathbb{E}[v_{t+\Delta}v_t])$ is also linear in Δ^{2H}. This property is again very well satisfied on data, as shown by Figure 1.11, where we plot the logarithm of the empirical counterpart of $\mathbb{E}[v_{t+\Delta}v_t]$ against Δ^{2H}, for the S&P with $H = 0.14$.

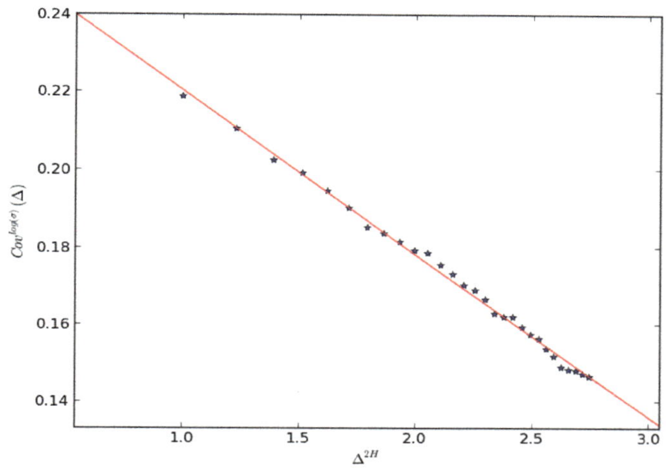

Figure 1.10: Autocovariance of the log-volatility as a function of Δ^{2H} for $H = 0.14$, S&P.

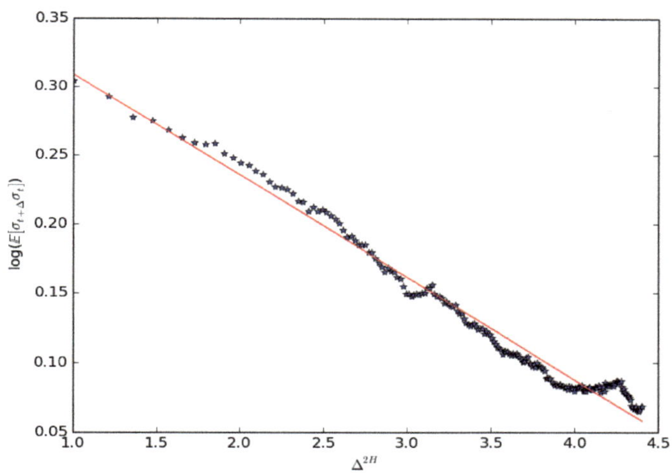

Figure 1.11: Empirical counterpart of $\log \mathbb{E}[v_{t+\Delta}v_t]$ as a function of Δ^{2H}, S&P.

We note that putting Δ^{2H} on the x-axis of Figure 1.11 is really crucial in order to retrieve linearity. In particular, a corollary of (1.7) is that the autocovariance function of the volatility does not decay as a power law as widely believed; see Figure 1.12, where we show that a log-log plot of the autocovariance function does not yield a straight line.

1.3.3 ▪ RFSV vs. FSV again

To further demonstrate the incompatibility of the classical long-memory FSV model with volatility data, we consider the quantity $m(2, \Delta)$. Recall that in the data (see Section 1.2) we observe

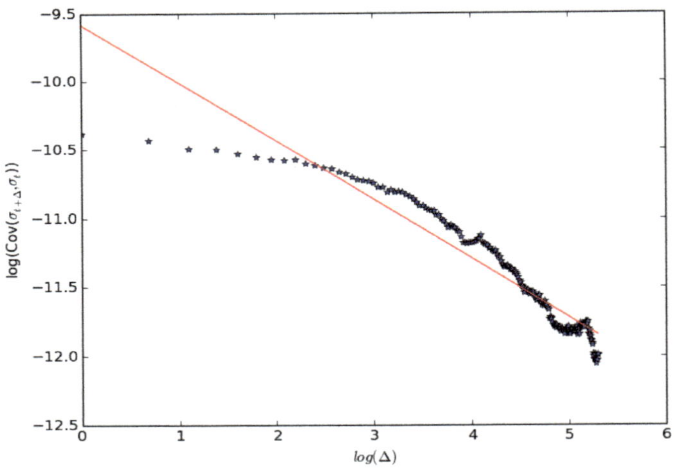

Figure 1.12: Empirical counterpart of $\log(\mathrm{Cov}[v_{t+\Delta}, v_t])$ as a function of $\log(\Delta)$, S&P.

the linear relationship $\log m(2,\Delta) \approx \zeta_2 \log\Delta + k$ for some constant k. Also, in both FSV and RFSV, we can consider

$$\begin{aligned} m(2,\Delta) &= \mathbb{E}\left[(X_{t+\Delta} - X_t)^2\right] \\ &= 2\,(\mathbb{V}[X_t] - \mathrm{Cov}[X_t, X_{t+\Delta}]). \end{aligned}$$

In Figure 1.13, we plot $m(2,\Delta)$ with the parameter $H = 0.53$, corresponding to the FSV model parameter estimate of Chronopoulou and Viens in [95], and the parameter $\alpha = 0.5$ to ensure some visible decay of the volatility skew. The slope of $m(2,\Delta)$ in the FSV model for small lags is driven by the value of H; the lag at which $m(2,\Delta)$ begins to flatten and stationarity kicks in corresponds to a time scale of order $1/\alpha$. It is clear from the figure that to fit the data, we must have $\alpha \ll 1/T$, and the value of H must be set by the initial slope of the regression line, which, as reported earlier in Section 1.2, is $\zeta_2 = 2 \times 0.14$.

1.3.4 ▪ Simulation-based analysis of the RFSV model

Our goal in this section is to show that in terms of smoothness measures, one obtains on simulated data from the RFSV model the same behaviors as those observed on empirical data. In particular, we would like to be able to quantify the positive bias associated with estimating H from whole-day realized variance data as in Section 1.2.3, relative to using data from a one-hour window as in Section 1.2.2.

We simulate the RFSV model for 2000 days (chosen to be between the lengths of our two datasets). In order to account for the overnight effect, we simulate the volatility v_t[12] and efficient price P_t[13] over the whole day. The parameters $H = 0.14$, $\nu = 0.3$, $m = X_0 = -5$, and $\alpha = 5 \times 10^{-4}$ are chosen to be consistent with our empirical estimates from Section 1.2. To model microstructure effects such as the discreteness of the price grid, we consider that the observed

[12] To simulate the fBm, we use a spectral method with 40,000,000 points (20,000 points per day). We then simulate X taking $X_{(n+1)\delta} - X_{n\delta} = \nu(W^H_{(n+1)\delta} - W^H_{n\delta}) + \alpha\delta(m - X_{n\delta})$ (with $\delta = 1/20000$).

[13] $P_{(n+1)\delta} - P_{n\delta} = P_{n\delta} v_{n\delta} \sqrt{\delta}\, U_n$ where the U_n are independent and identically distributed (i.i.d.) standard Gaussian variables.

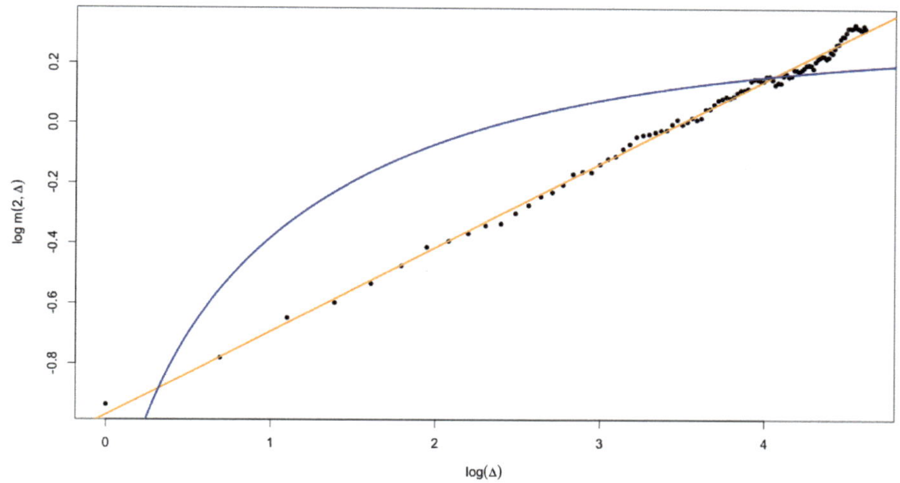

Figure 1.13: Long-memory models such as the FSV model of Comte and Renault are not compatible with S&P volatility data. Black points are empirical estimates of $m(2, \Delta)$; the blue line is the FSV model with $\alpha = 0.5$ and $H = 0.53$; the orange line is the RFSV model with $\alpha = 0$ and $H = 0.14$.

price process is generated from P_t using the uncertainty zones model of [260] with tick value 5×10^{-4} and parameter $\eta = 0.25$.

Exactly as in Section 1.2, for each of the 2000 days, we consider two volatility proxies obtained from the observed price and based on the following:

- The integrated variance estimator using the model with uncertainty zones over one-hour windows, from 10 a.m. to 11 a.m.

- The five-minute realized variance estimator, over eight-hour windows (the trading day).

We now repeat our analysis of Section 1.2, generating graphs analogous to Figures 1.3, 1.4, 1.6, and 1.7 obtained on empirical data. Figure 1.14 compares smoothness measures obtained using the uncertainty zones estimator on one-hour windows with those obtained using the realized variance estimator on eight-hour windows.

When the uncertainty zones estimator is applied on a one-hour window (1/24 of a simulated day) as in Section 1.2.2, we estimate $H = 0.16$, which is close to the true value $H = 0.14$ used in the simulation. The results obtained with the realized variance estimator over daily eight-hour windows (1/3 of a simulated day) do exhibit the same scaling properties as those we see in the empirical data with a smoothness parameter that does not depend on q. However, the estimated H is biased slightly higher at around 0.18. As discussed in Section 1.2.1, this extra positive bias is no surprise and is due to the regularizing effect of the integral operator over the longer window. We note also that the estimated values of ν ("volatility of volatility" in some sense), obtained from the intercepts of the regressions, are lower with the longer time windows, again as expected. A detailed computation of the bias in the estimated H associated with the choice of window length in an analogous but more tractable model is presented in Appendix 1.7.3.

We end this section by presenting in Figure 1.15 a sample path of the model-generated volatility (spot volatility, directly from the simulation rather than estimated from the simulated price series) together with a graph of S&P volatility over 3500 days.

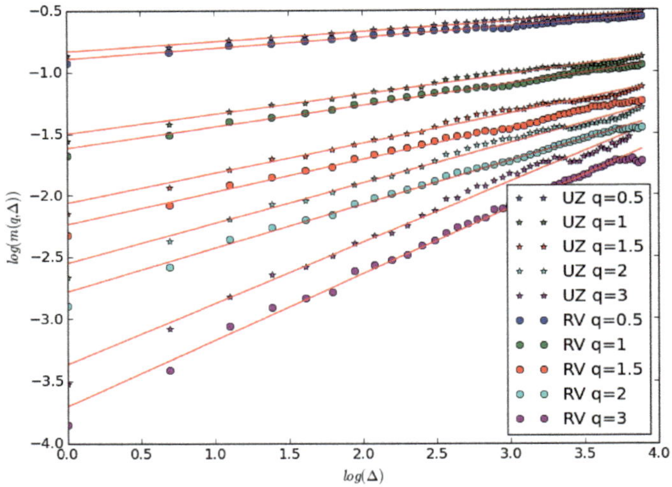

Figure 1.14: $\log(m(q,\Delta))$ as a function of $\log(\Delta)$, simulated data, with realized variance and uncertainty zones estimators.

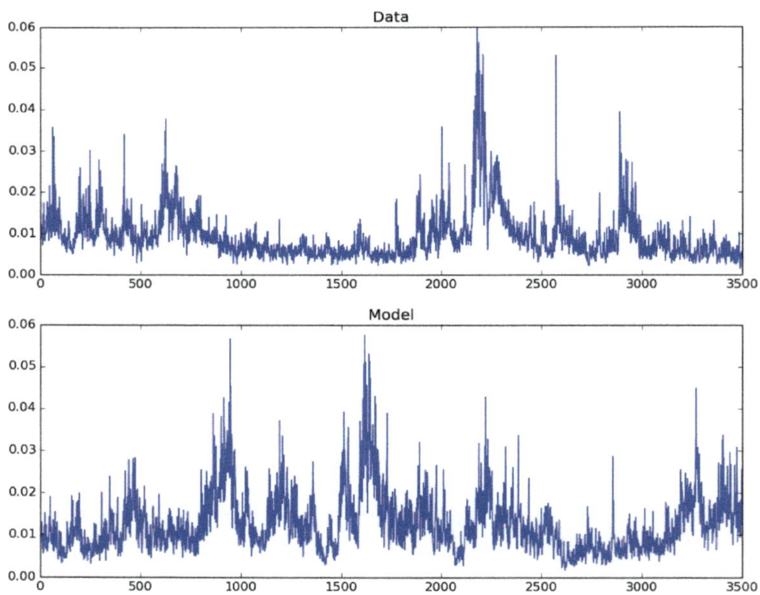

Figure 1.15: Volatility of the S&P (top) and generated by the model (bottom).

A first reaction to Figure 1.15 is that the simulated and actual graphs look very alike. In particular, in both of them, persistent periods of high volatility alternate with low volatility periods. On closer inspection of the empirical volatility series, we observe that the sample path of the volatility on a restricted time window seems to exhibit the same kind of qualitative properties as those of the global sample path (for example, periods of high and low activity). This fractal-type behavior of the volatility has been investigated both empirically and theoretically in, for example, [27, 67, 237].

At the visual level, we observe that this fractal-type behavior is also reproduced in our model, as we now explain. Denote by $L^{x,H}$ the law of the geometric fractional Brownian motion with Hurst exponent H and volatility x on $[0,1]$, that is, $(e^{xW_t^H})_{t\in[0,1]}$. Then, when α is very small, the rescaled volatility process on $[0,\Delta]$, $(v_{t\Delta}/v_0)_{t\in[0,1]}$, has approximately the law $L^{\nu\Delta^H,H}$. Now note that for H small, the function u^H increases very slowly. Thus, over a large range of observation scales Δ, the rescaled volatility processes on $[0,\Delta]$ have approximately the same law. For example, between an observation scale of one day and five years (1250 open days), the coefficient x characterizing the law of the volatility process is "only" multiplied by $1250^{0.14} = 2.7$. It follows that in the RFSV model, the volatility process over one day resembles the volatility process over a decade.

1.4 ▪ Spurious long memory of volatility?

We revisit in this section the issue of long memory of volatility through the lens of our model. Recall that a stationary time series is said to exhibit long memory if the autocovariance function $\mathrm{Cov}[\log(v_t), \log(v_{t+\Delta})]$ (or sometimes $\mathrm{Cov}[v_t, v_{t+\Delta}]$) goes slowly to zero as $\Delta \to \infty$, and often, even more precisely, that it behaves as $\Delta^{-\gamma}$,[14] with $\gamma < 1$ as $\Delta \to \infty$.

Thus, the classical approach to long memory is to consider a parametric class of models and to estimate within this class the parameter γ, typically based on empirical autocovariances; see [19] and Figure 1.12. As mentioned earlier in the introduction, the long memory of volatility is widely accepted as a stylized fact.

Specifically, in the RFSV model, we have from Corollary 1.3 that

$$\mathrm{Cov}[\log(v_t), \log(v_{t+\Delta})] \approx A - B\Delta^{2H}$$

and from equation (1.7) that

$$\mathrm{Cov}[v_t, v_{t+\Delta}] \approx C\, e^{-B\Delta^{2H}} - D,$$

for some constants A, B, C, and D. Moreover, we demonstrated in Figures 1.10 and 1.11 that these relations are consistent with the data. Thus the autocovariance function does not decay as a power law in the RFSV model nor appears to decay as a power law in the data.

Nevertheless, as an experiment, we can apply to both the data and the sample paths of the RFSV model some standard statistical procedures aimed at identifying long memory that have been used in the financial econometrics literature. Such procedures are of course designed to identify long memory under rather strict modeling assumptions. Consequently, spurious results may obviously then be obtained if the model underlying the estimation procedure is misspecified, which is the case with the RFSV model.[15]

With the same model parameters as in Section 1.3.4, we simulate our model over 3500 days, which corresponds to the size of our dataset. Consider first the procedure in [18], where in the context of a fractional Gaussian noise (FGN) model with Hurst parameter $\widehat{H}$, the authors test for long memory in the volatility by studying the scaling behavior of the quantity

$$V(\Delta) := \mathbb{V}\left[\int_0^\Delta v_s^2 ds\right]$$

[14]Indeed, the notion of empirical long memory does not make much sense outside the power law case; the empirical values of covariances at very large time scales are never measurable, and thus one cannot conclude whether the series of covariances converges in general.

[15]Recall in particular that the RFSV model is only formally stationary.

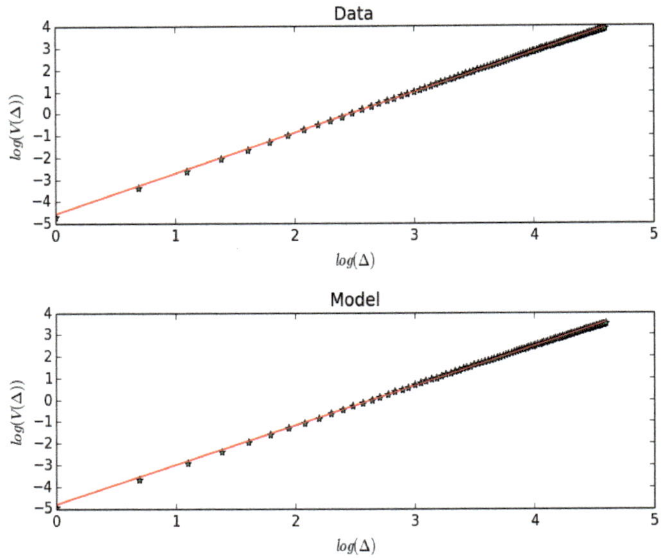

Figure 1.16: Empirical counterpart of $\log(V(\Delta))$ as a function of $\log(\Delta)$ on S&P (top) and simulation (bottom).

with respect to Δ. In the FGN model, as $\Delta \to \infty$, the autocorrelation function $\varrho(\Delta)$ behaves asymptotically as $\Delta^{2\widehat{H}-2}$, and $V(\Delta)$ behaves asymptotically as $\Delta^{2\widehat{H}}$ as $\Delta \to \infty$. Figure 1.16 presents the graph of the logarithm of the empirical counterpart of $V(\Delta)$ against the logarithm of Δ on the S&P data and within our simulation framework.

We note from Figure 1.16 that both our simulated model and market data lead to very similar graphs, close to straight lines with slope 1.86, giving $\widehat{H} = 0.93$.[16] Accordingly, in the setting of [18], we would deduce power law behavior of the autocorrelation function with exponent 0.14 and therefore long memory. Thus, if the data are generated by a model like the RFSV model, one can easily be wrongly convinced that the volatility time series exhibits long memory.

In [19], in the context of an ARFIMA$(0, d, 0)$ model, the authors deduce long memory in the volatility by showing that the process ε_t obtained by fractional differentiation of the log-volatility $\varepsilon_t = (1 - L)^d \log(v_t)$, with $d = 0.401$[17] (which is obtained by regression of the log-periodogram using the GPH estimator [173]) and L the lag operator, behaves as a white noise. To check for this, they compute the autocorrelation function of ε_t. We give in Figure 1.17 the autocorrelation functions of the logarithm of v_t and ε_t, again both on the data and on the simulated path.

Once again, the data and the simulation generate very similar plots. We conclude that this procedure for estimating long memory is just as fragile as the first, and it is easy to wrongly deduce volatility long memory when applying it.

In conclusion, the RFSV model is yet another model in which classical estimation procedures identify spurious long memory; see [15, 117, 176, 177] for various other such examples. Moreover, these procedures estimate the same long-memory parameter from data generated from a suitably calibrated RFSV model as they estimate from empirical data. Once again, although the

[16]Note that there is no reason to expect that there should be any direct connection between $\widehat{H}$ estimated for the FGN model and the H we estimated for the RFSV model.

[17]It is shown in [173] that the autocorrelation functions of the ARFIMA$(0, d, 0)$ and the FGN model with Hurst parameter $\widehat{H}$ have the same asymptotic behavior as $\Delta \to \infty$ if $d = \widehat{H} - \frac{1}{2}$.

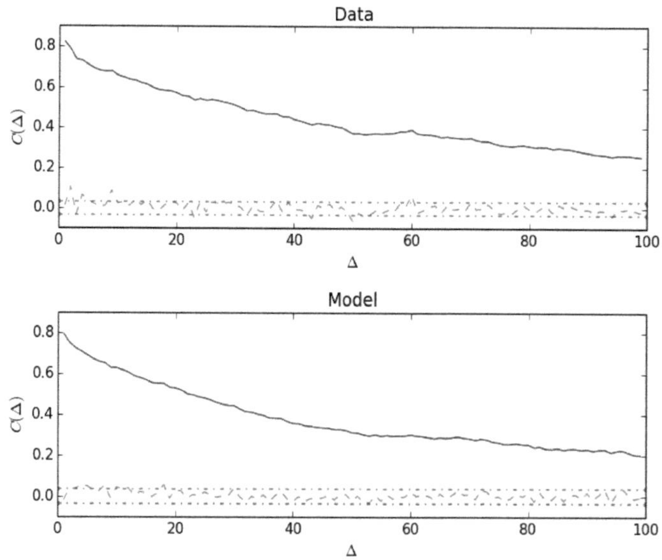

Figure 1.17: Autocorrelation functions of $\log(v_t)$ (in blue) and ε_t (in green) and the Bartlett standard error bands (in red), for S&P data (top) and for simulated data (bottom).

(near) nonstationarity of the RFSV model induces long swings in volatility, mirroring long-range dependence, it does not exhibit long memory in the classical power-law sense.

1.5 ▪ Forecasting using the RFSV model

In this section, we present an application of our model: forecasting the log-volatility and the variance.

1.5.1 ▪ Forecasting log-volatility

The key formula on which our prediction method is based is the following:

$$\mathbb{E}[W^H_{t+\Delta}|\mathcal{F}_t] = \frac{\cos(H\pi)}{\pi}\Delta^{H+\frac{1}{2}}\int_{-\infty}^{t}\frac{W^H_s}{(t-s+\Delta)(t-s)^{H+\frac{1}{2}}}\mathrm{d}s,$$

where W^H is an fBm with $H < \frac{1}{2}$ and $\mathcal{F}_t$ is the filtration it generates; see Theorem 4.2 of [254]. By construction, over any reasonable time scale of interest, as formalized in Corollary 1.2, we may approximate the fractional Ornstein–Uhlenbeck volatility process in the RFSV model as $\log\sigma^2_t \approx 2\nu W^H_t + C$, for some constants ν and C. Our prediction formula for log-variance then follows:[18]

$$\mathbb{E}\left[\log\sigma^2_{t+\Delta}|\mathcal{F}_t\right] = \frac{\cos(H\pi)}{\pi}\Delta^{H+\frac{1}{2}}\int_{-\infty}^{t}\frac{\log v^2_s}{(t-s+\Delta)(t-s)^{H+\frac{1}{2}}}\mathrm{d}s. \tag{1.8}$$

This formula, or rather its approximation through a Riemann sum (we assume in this section that the volatilities are perfectly observed, although they are in fact estimated), is used to forecast the log-volatility 1, 5, and 20 days ahead ($\Delta = 1,\ 5,\ 20$).

[18]The constants 2ν and C cancel when deriving the expression.

We now compare the predictive power of formula (1.8) with that of autoregressive (AR for short) and heterogeneous autoregressive (HAR for short) forecasts, in the spirit of [104].[19] Recall that for a given integer $p > 0$, the AR(p) and HAR predictors take the following form (where the index i runs over the series of daily volatility estimates):

- AR(p):

$$\widehat{\log(\sigma^2_{t+\Delta})} = K_0^\Delta + \sum_{i=0}^{p} C_i^\Delta \log(\sigma^2_{t-i}).$$

- HAR:

$$\widehat{\log(\sigma^2_{t+\Delta})} = K_0^\Delta + C_0^\Delta \log(\sigma_t^2) + C_5^\Delta \frac{1}{5}\sum_{i=0}^{4} \log(\sigma^2_{t-i}) + C_{20}^\Delta \frac{1}{20}\sum_{i=0}^{19} \log(\sigma^2_{t-i}).$$

We estimate AR coefficients using the R `stats` library[20] on a rolling time window of 500 days. In the HAR case, we use standard linear regression to estimate the coefficients as explained in [104]. In the sequel, we consider $p = 5$ and $p = 10$ in the AR formula. Indeed, these parameters essentially give the best results for the horizons at which we wish to forecast the volatility (1, 5, and 20 days). For each day, we forecast volatility for five different indices.[21]

We then assess the quality of the various forecasts by computing the ratio P between the mean squared error of our predictor and the (approximate) variance of the log-variance:

$$P = \frac{\sum_{k=500}^{N-\Delta} \left(\log(\sigma^2_{k+\Delta}) - \widehat{\log(\sigma^2_{k+\Delta})}\right)^2}{\sum_{k=500}^{N-\Delta} \left(\log(\sigma^2_{k+\Delta}) - \mathbb{E}[\log(\sigma^2_{t+\Delta})]\right)^2},$$

where $\mathbb{E}[\log(\sigma^2_{t+\Delta})]$ denotes the empirical mean of the log-variance over the whole time period.

We note from Table 1.1 that the RFSV forecast consistently outperforms the AR and HAR forecasts, especially at longer horizons. Moreover, our forecasting method is more parsimonious since it only requires the parameter H to forecast the log-variance. Compare this with the AR and HAR methods, for which coefficients depend on the forecast time horizon and must be recomputed if this horizon changes.

Note that our predictor can be linked to that of [121], where the issue of the prediction of the log-volatility in the multifractal random walk model of [27] is tackled. In this model,

$$\mathbb{E}[\log(\nu^2_{t+\Delta})|\mathcal{F}_t] = \frac{1}{\pi}\sqrt{\Delta}\int_{-\infty}^{t} \frac{\log(\nu_s^2)}{(t-s+\Delta)\sqrt{t-s}}ds,$$

which is the limit of our predictor when H tends to zero.

Note also that our prediction formula may be rewritten as

$$\mathbb{E}[\log(\sigma^2_{t+\Delta})|\mathcal{F}_t] = \frac{\cos(H\pi)}{\pi}\int_0^{+\infty} \frac{\log(\nu^2_{t-\Delta u})}{(u+1)\,u^{H+\frac{1}{2}}}du.$$

[19]Note that we do not consider GARCH models here since we have access to high frequency volatility estimates and not only to daily returns. Indeed, it is shown in [19] that forecasts based on the time series of realized variances outperform GARCH forecasts based on daily returns.

[20]More precisely, we use the default Yule–Walker method.

[21]In addition to S&P and NASDAQ, we also investigate CAC40, FTSE, and Nikkei, over the same time period as S&P and NASDAQ. For simplicity, the parameter H used in our predictor is computed only once for each asset, using the whole time period. This yields similar results to using a moving time window adapted in time.

Table 1.1: Ratio P for the AR, HAR, and RFSV predictors.

	AR(5)	AR(10)	HAR(3)	RFSV
SPX2.rv $\Delta = 1$	0.317	0.318	0.314	**0.313**
SPX2.rv $\Delta = 5$	0.459	0.449	0.437	**0.426**
SPX2.rv $\Delta = 20$	0.764	0.694	0.656	**0.606**
FTSE2.rv $\Delta = 1$	0.230	0.229	0.225	**0.223**
FTSE2.rv $\Delta = 5$	0.357	0.344	0.337	**0.320**
FTSE2.rv $\Delta = 20$	0.651	0.571	0.541	**0.472**
N2252.rv $\Delta = 1$	0.357	0.358	0.351	**0.345**
N2252.rv $\Delta = 5$	0.553	0.533	0.513	**0.504**
N2252.rv $\Delta = 20$	0.875	0.795	0.746	**0.714**
GDAXI2.rv $\Delta = 1$	0.237	0.238	0.234	**0.231**
GDAXI2.rv $\Delta = 5$	0.372	0.362	0.350	**0.339**
GDAXI2.rv $\Delta = 20$	0.661	0.590	0.550	**0.498**
FCHI2.rv $\Delta = 1$	0.244	0.244	0.241	**0.238**
FCHI2.rv $\Delta = 5$	0.378	0.373	0.366	**0.350**
FCHI2.rv $\Delta = 20$	0.669	0.613	0.598	**0.522**

For a given small $\varepsilon > 0$, let r be the smallest real number such that

$$\int_r^{+\infty} \frac{1}{(u+1)\,u^{H+\frac{1}{2}}} \mathrm{d}u \leq \varepsilon.$$

Then we have, with an error of order ε,

$$\mathbb{E}[\log(\sigma_{t+\Delta}^2)|\mathcal{F}_t] \approx \frac{\cos(H\pi)}{\pi} \int_0^r \frac{\log(v_{t-\Delta u}^2)}{(u+1)\,u^{H+\frac{1}{2}}} du.$$

Consequently, the volatility process needs to be considered (roughly) down to time $t - \Delta\, r$ if one wants to forecast up to time Δ in the future. The relevant regression window is thus linear in the forecasting horizon. For example, for $r = 1$, $\varepsilon = 0.35$ which is not so unreasonable. In this case, as is well known to practitioners, to predict log-volatility one week ahead, one should essentially look at the volatility over the last week. If trying to predict log-volatility one month ahead, one should look at the volatility over the last month.

1.5.2 ▪ Variance prediction

Recall that $\log v_t^2 \approx 2\,\nu\, W_t^H + C$ for some constant C. In [254], it is shown that $W_{t+\Delta}^H$ is conditionally Gaussian with conditional variance,

$$\mathbb{V}[W_{t+\Delta}^H|\mathcal{F}_t] = c\,\Delta^{2H},$$

where

$$c = \frac{\Gamma(\frac{3}{2} - H)}{\Gamma(H + \frac{1}{2})\,\Gamma(2 - 2H)}.$$

Thus, we obtain the following form for the RFSV predictor of the variance:

$$\widehat{\sigma_{t+\Delta}^2} = \exp\left(\widehat{\log \sigma_{t+\Delta}^2} + 2\,c\,\nu^2 \Delta^{2H}\right), \tag{1.9}$$

where $\widehat{\log(\sigma^2_{t+\Delta})}$ is the predictor from Section 1.5.1, and ν^2 is estimated as the exponential of the intercept in the linear regression of $\log(m(2,\Delta))$ on $\log(\Delta)$.

As previously, we compare in Table 1.2 the performances of the RFSV forecast with those of AR and HAR forecasts (constructed on variance rather than log-variance this time).

Table 1.2: Ratio P for the AR, HAR, and RFSV predictors.

	AR(5)	AR(10)	HAR(3)	RFSV
SPX2.rv $\Delta = 1$	0.520	0.566	0.489	**0.475**
SPX2.rv $\Delta = 5$	0.750	0.745	0.723	**0.672**
SPX2.rv $\Delta = 20$	1.070	1.010	1.036	**0.903**
FTSE2.rv $\Delta = 1$	0.612	0.621	0.582	**0.567**
FTSE2.rv $\Delta = 5$	0.797	0.770	0.756	**0.707**
FTSE2.rv $\Delta = 20$	1.046	0.984	0.935	**0.874**
N2252.rv $\Delta = 1$	0.554	0.579	**0.504**	0.505
N2252.rv $\Delta = 5$	0.857	0.807	0.761	**0.729**
N2252.rv $\Delta = 20$	1.097	1.046	1.011	**0.964**
GDAXI2.rv $\Delta = 1$	0.439	0.448	0.399	**0.386**
GDAXI2.rv $\Delta = 5$	0.675	0.650	0.616	**0.566**
GDAXI2.rv $\Delta = 20$	0.931	0.850	0.816	**0.746**
FCHI2.rv $\Delta = 1$	0.533	0.542	0.470	**0.465**
FCHI2.rv $\Delta = 5$	0.705	0.707	0.691	**0.631**
FCHI2.rv $\Delta = 20$	0.982	0.952	0.912	**0.828**

We find again that the RFSV forecast typically outperforms AR and HAR, although it is worth noting that the HAR forecast is already visibly superior to the AR forecast.

Since our working paper first appeared, much work has been done to estimate the roughness of volatility of other assets, notably by Bennedsen, Lunde, and Pakkanen in [51]. In an analysis of E-mini S&P 500 futures data, at all time scales over 15 minutes, the out-of-sample forecasting performance of the estimator (1.9) is shown to be very similar to the performance of the other more highly parameterized estimators proposed. Interestingly, at daily and higher time scales, the simple estimator (1.9) is actually shown to outperform the more complicated estimators. It is also notable that rough volatility was confirmed for more than five thousand equities in [51]; we are not yet aware of any asset for which rough volatility has not been confirmed.

We emphasise that our point in this forecasting exercise is not to show that our rough volatility based approach is really superior to others. For example, we could have considered alternative competitors to the RFSV formula or refined the HAR procedure. We only wish to demonstrate that our forecast is probably at least as good as other predictors while being simpler, requiring only the estimation of H.

1.6 ▪ Conclusion

Using daily realized variance estimates as proxies for daily spot (squared) volatilities, we uncovered two startlingly simple regularities in the resulting time series. First, we found that the distributions of increments of log-volatility are approximately Gaussian, consistent with many prior studies. Second, we established the monofractal scaling relationship

$$\mathbb{E}\left[|\log(v_\Delta) - \log(v_0)|^q\right] = K_q\, \nu^q\, \Delta^{qH}, \tag{1.10}$$

where H can be seen as a measure of smoothness characteristic of the underlying volatility process, typically, $0.06 < H < 0.2$. The simple scaling relationship (1.10) naturally suggests that log-volatility may be modeled using fractional Brownian motion. The resulting rough fractional stochastic volatility (RFSV) model turns out to be formally almost identical to the FSV model of Comte and Renault [99], with one major difference: in the FSV model, $H > \frac{1}{2}$ to ensure long memory, whereas in the RFSV model, $H < \frac{1}{2}$, typically, $H \approx 0.1$. Moreover, in the FSV model, the mean reversion coefficient α has to be large compared to $1/T$ to ensure a decaying volatility skew; in the RFSV model, the volatility skew decays naturally just like the observed volatility skew, $\alpha \ll 1/T$ and indeed for time scales of practical interest, we may proceed as if α were exactly zero.

We further showed that applying standard test procedures to volatility time series simulated with the RFSV model would lead us to erroneously deduce the presence of long memory, with parameters similar to those found in prior studies. Despite the fact that volatility in the RFSV model (or in the data) is not a long-memory process, we can therefore explain why long memory of volatility is widely accepted as a stylized fact. Thus the RFSV model is able to replicate the stylized facts of the time series, i.e., *volatility is rough*. More precisely, we have shown that within a specific class of models (which we strongly argue to be relevant), empirical daily realized variance values are much more likely to be sampled from a rough volatility process than from a smooth process. Whether or not instantaneous variance is rough cannot of course be ultimately determined since instantaneous variance is latent and not observable; as mentioned in the introduction, it is not even clear that such an object exists independently of a specific model.

It is of course plausible that other models are compatible with many of our observations. In fact, there are probably many ways to design a process so that most of our empirical results are reproduced (for example, estimation errors when estimating volatility can be quite significant for some models, leading to downward biases in the measurement of the smoothness). However, what we show here is that we cannot find any evidence against the RFSV model. In statistical terms, the null hypothesis that the data generating process of the volatility is an RFSV model cannot be rejected based on our analysis. Even more, it is likely that the RFSV model is simpler, more parsimonious, and more tractable than any other such model. In particular we do not address the question of volatility jumps; neither do we insist that there are no jumps in volatility. Rather, one of the main messages of our work is that our model is able to replicate the stylized facts of the time series without having to appeal to jumps; see [62] for another example where a continuous process can mimic the properties obtained from data generated by a process with jumps.

As an application of the RFSV model, we showed how to forecast volatility at various time scales, at least as well as when using Fulvio Corsi's impressive HAR estimator, but with only one parameter—H!

We focus in this work on the statistical properties of the RFSV model. In [39], the authors explore the implications of the RFSV model (written under the physical measure $\mathbb{P}$) for option pricing (under the pricing measure $\mathbb{Q}$). In particular, following Mandelbrot and Van Ness, the fBm that appears in the definition (1.6) of the RFSV model may be represented as a fractional integral of a standard Brownian motion as follows [236]:

$$W_t^H = \int_0^t \frac{dW_s}{(t-s)^\gamma} + \int_{-\infty}^0 \left[\frac{1}{(t-s)^\gamma} - \frac{1}{(-s)^\gamma}\right] dW_s, \tag{1.11}$$

with $\gamma = \frac{1}{2} - H$. The observed anticorrelation between price moves and volatility moves may then be modeled naturally by anticorrelating the Brownian motion W that drives the volatility process with the Brownian motion driving the price process. As already proved by Fukasawa [151], such a model with a small H reproduces the observed decay of at-the-money volatility skew with respect to time to expiry, asymptotically for short times. It is shown that an appropriate extension of Fukasawa's model, consistent with the RFSV model, fits the entire implied volatility surface remarkably well. In particular, this model accurately reproduces the extreme short-dated smiles, with no jumps. Moreover, despite that it would seem from (1.11) that knowledge of the entire path $\{W_s : s < t\}$ of the Brownian motion would be required, it turns out that the statistics of this path necessary for option pricing are traded and thus easily observed. Remarkably, Heston-type formulas can also be obtained in the rough volatility framework; see [129, 131].

Finally, note that there are microstructural foundations to rough volatility models. Indeed, it is explained in [126] how rough volatility emerges as the scaling limit of a Hawkes process based description of the order flow in the context of high frequency trading and metaorder splitting.

1.7 ▪ Appendix

1.7.1 ▪ Proofs

Proof of Proposition 1.1

Starting from equation (1.5) and applying integration by parts, we get

$$X_t^\alpha = \nu W_t^H - \int_{-\infty}^t \nu\alpha e^{-\alpha(t-s)} W_s^H ds + m.$$

Therefore,

$$(X_t^\alpha - X_0^\alpha) - \nu W_t^H = -\int_0^t \nu\alpha e^{-\alpha(t-s)} W_s^H ds - \int_{-\infty}^0 \nu\alpha(e^{-\alpha(t-s)} - e^{\alpha s}) W_s^H ds.$$

Consequently,

$$\sup_{t\in[0,T]} |(X_t^\alpha - X_0^\alpha) - \nu W_t^H| \le \nu\alpha T \widehat{W}_T^H + \int_{-\infty}^0 \nu\alpha(e^{\alpha s} - e^{-\alpha(T-s)}) \widehat{W}_s^H ds,$$

where $\widehat{W}_t^H = \sup_{s\in[0,t]} |W_s^H|$. Using the maximum inequality of [250], we get

$$\mathbb{E}\left[\sup_{t\in[0,T]} \left|(X_t^\alpha - X_0^\alpha) - \nu W_t^H\right|\right] \le c\left(\nu\alpha T^H + \int_{-\infty}^0 \nu\alpha(T\alpha e^{\alpha s})|s|^H ds\right),$$

with c some constant. The term on the right-hand side is easily seen to go to zero as α tends to zero.

Proof of Corollary 1.2

We first recall equation (2.2) in [92] which reads:

$$\mathrm{Cov}[X_{t+\Delta}^\alpha, X_t^\alpha] = K \int_{\mathbb{R}} e^{i\Delta x} \frac{|x|^{1-2H}}{\alpha^2 + x^2} dx,$$

with $K = \nu^2\Gamma(2H+1)\sin(\pi H)/(2\pi)$.[22] Now note that

$$\mathbb{E}[(X_{t+\Delta}^\alpha - X_t^\alpha)^2] = 2\mathbb{V}[X_t^\alpha] - 2\mathrm{Cov}[X_{t+\Delta}^\alpha, X_t^\alpha].$$

[22] This covariance is real because it is the Fourier transform of an even function.

Therefore,

$$\mathbb{E}[(X^\alpha_{t+\Delta} - X^\alpha_t)^2] = 2K \int_{\mathbb{R}} (1 - \mathrm{e}^{\mathrm{i}\Delta x}) \frac{|x|^{1-2H}}{\alpha^2 + x^2} \mathrm{d}x.$$

This implies that for fixed Δ, $\mathbb{E}[|X^\alpha_{t+\Delta} - X^\alpha_t|^2]$ is uniformly bounded by

$$2K \int_{\mathbb{R}} (1 - \mathrm{e}^{\mathrm{i}\Delta x}) \frac{|x|^{1-2H}}{x^2} \mathrm{d}x.$$

Moreover, $X^\alpha_{t+\Delta} - X^\alpha_t$ is a Gaussian random variable, and thus for every q, its moment of order $(q+1)$ is uniformly bounded (in α) so that the family $|X^\alpha_{t+\Delta} - X^\alpha_t|^q$ is uniformly integrable. Therefore, since by Proposition 1.1,

$$|X^\alpha_{t+\Delta} - X^\alpha_t|^q \to \nu^q |W^H_{t+\Delta} - W^H_t|^q, \text{ in law,}$$

we get the convergence of the sequence of expectations.

1.7.2 ▪ Estimations of H

On different indices

Table 1.3: Estimates of ζ_q for all indices in the Oxford-Man dataset.

Index	$\zeta_{0.5}/0.5$	ζ_1	$\zeta_{1.5}/1.5$	$\zeta_2/2$	$\zeta_3/3$
SPX2.rv	0.128	0.126	0.125	0.124	0.124
FTSE2.rv	0.132	0.132	0.132	0.131	0.127
N2252.rv	0.131	0.131	0.132	0.132	0.133
GDAXI2.rv	0.141	0.139	0.138	0.136	0.132
RUT2.rv	0.117	0.115	0.113	0.111	0.108
AORD2.rv	0.072	0.073	0.074	0.075	0.077
DJI2.rv	0.117	0.116	0.115	0.114	0.113
IXIC2.rv	0.131	0.133	0.134	0.135	0.137
FCHI2.rv	0.143	0.143	0.142	0.141	0.138
HSI2.rv	0.079	0.079	0.079	0.080	0.082
KS11.rv	0.133	0.133	0.134	0.134	0.132
AEX.rv	0.145	0.147	0.149	0.149	0.149
SSMI.rv	0.149	0.153	0.156	0.158	0.158
IBEX2.rv	0.138	0.138	0.137	0.136	0.133
NSEI.rv	0.119	0.117	0.114	0.111	0.102
MXX.rv	0.077	0.077	0.076	0.075	0.071
BVSP.rv	0.118	0.118	0.119	0.120	0.120
GSPTSE.rv	0.106	0.104	0.103	0.102	0.101
STOXX50E.rv	0.139	0.135	0.130	0.123	0.101
FTSTI.rv	0.111	0.112	0.113	0.113	0.112
FTSEMIB.rv	0.130	0.132	0.133	0.134	0.134

On different time intervals[23]

Table 1.4: Estimates of H over two different time intervals for all indices in the Oxford-Man dataset.

Index	H (first half)	H (second half)
SPX2.rk	0.115	0.158
FTSE2.rk	0.140	0.156
N2252.rk	0.083	0.134
GDAXI2.rk	0.154	0.168
RUT2.rk	0.098	0.149
AORD2.rk	0.059	0.114
DJI2.rk	0.123	0.151
IXIC2.rk	0.094	0.156
FCHI2.rk	0.140	0.146
HSI2.rk	0.072	0.129
KS11.rk	0.109	0.147
AEX.rk	0.168	0.151
SSMI.rk	0.206	0.183
IBEX2.rk	0.122	0.149
NSEI.rk	0.112	0.124
MXX.rk	0.068	0.118
BVSP.rk	0.074	0.134
GSPTSE.rk	0.075	0.147
STOXX50E.rk	0.138	0.132
FTSTI.rk	0.080	0.171
FTSEMIB.rk	0.133	0.140

1.7.3 ▪ The effect of smoothing

Although we are really interested in the model

$$\log v_{t+\Delta} - \log v_t = \nu \left(W^H_{t+\Delta} - W^H_t \right),$$

consider the more tractable fractional Stein and Stein (fSS) model

$$\mathfrak{v}_{t+\Delta} - \mathfrak{v}_t = \alpha \left(W^H_{t+\Delta} - W^H_t \right),$$

where $\mathfrak{v}_t = \sigma^2$. We cannot observe $\mathfrak{v}_t$, but suppose we can proxy it by the average

$$\widehat{\mathfrak{v}}^\delta_t = \frac{1}{\delta} \int_0^\delta \mathfrak{v}_u du.$$

[23]Note that we used realized kernel rather than realized variance estimates to generate Table 1.4. Results obtained using different estimators are almost indistinguishable.

We would like to estimate, for example, $m(2,\Delta) = \mathbb{E}\left[(\mathfrak{v}_{t+\Delta} - \mathfrak{v}_t)^2\right]$. However, we need to proxy spot variance with integrated variance, so instead we have the estimate

$$
\begin{aligned}
m^\delta(2,\Delta) &= \mathbb{E}\left[(\widehat{\mathfrak{v}}^\delta_{t+\Delta} - \widehat{\mathfrak{v}}^\delta_t)^2\right] \\
&= \frac{1}{\delta^2}\,\mathbb{E}\left[\left(\int_0^\delta (\mathfrak{v}_{u+\Delta} - \mathfrak{v}_u)\mathrm{d}u\right)^2\right] \\
&= \frac{\alpha^2}{\delta^2}\int_0^\delta\int_0^\delta \mathbb{E}\left[(W^H_{u+\Delta} - W^H_u)(W^H_{s+\Delta} - W^H_s)\right]\mathrm{d}u\,\mathrm{d}s \\
&= \int_0^\delta\int_0^\delta \left(|u-s+\Delta|^{2H} - |u-s|^{2H}\right)\mathrm{d}u\,\mathrm{d}s,
\end{aligned} \tag{1.12}
$$

where the last step uses the fact that

$$
\mathbb{E}\left[W^H_u W^H_s\right] = \frac{1}{2}\left(u^{2H} + s^{2H} - |u-s|^{2H}\right),
$$

and the symmetry of the integral.

We assume that the length δ of the smoothing window is less than one day, so $\Delta > \delta$. Then easy computations give

$$
\begin{aligned}
&\int_0^\delta\int_0^\delta |u-s+\Delta|^{2H}\,\mathrm{d}u\,\mathrm{d}s \\
&= \frac{1}{2H+1}\frac{1}{2H+2}\left((\Delta+\delta)^{2H+2} - 2\Delta^{2H+2} + (\Delta-\delta)^{2H+2}\right)
\end{aligned}
$$

and

$$
\int_0^\delta\int_0^\delta |u-s|^{2H}\,\mathrm{d}u\,\mathrm{d}s = \frac{2}{2H+1}\frac{1}{2H+2}\delta^{2H+2}.
$$

Substituting back into (1.12) gives

$$
\begin{aligned}
m^\delta(2,\Delta) &= \alpha^2\Delta^{2H}\,\frac{1}{2H+1}\frac{1}{2H+2}\frac{1}{\theta^2}\left[(1+\theta)^{2H+2} - 2 - 2\theta^{2H+2} + (1-\theta)^{2H+2}\right] \\
&=: \alpha^2\Delta^{2H} f(\theta).
\end{aligned}
$$

where $\theta = \delta/\Delta$.

Figure 1.18 shows the effect of smoothing on the estimated variance in the fSS model. Keeping δ fixed, as Δ increases, $f(\theta) = f(\delta/\Delta)$ increases towards one. Thus, in a linear regression of $\log m^\delta(2,\Delta)$ against $\log\Delta$, we will obtain a higher effective H (from the higher slope) and a lower effective ("volatility of volatility") α, exactly as we observed in the RFSV model simulations in Section 1.3.4.

Numerical example

In the simulation of the RFSV model in Section 1.3.4, we have $H = 0.14$, $\delta_1 = 1/24$ for the UZ estimate and $\delta_2 = 1/3$ for the RV estimate. We now reproduce an fSS analogue of the RFSV simulation plots of $m(2,\Delta)$ in Figure 1.14. Specifically, for each $\Delta \in \{1,2,\ldots,100\}$, with $\alpha = 0.3$ and $\delta = \delta_1$ or $\delta = \delta_2$, we compute $m^\delta(2,\Delta)$ and regress $\log m^\delta(2,\Delta)$ against $\log\Delta$. The regressions are shown in Figure 1.19 and results are tabulated in Table 1.5.

In Figure 1.19 and Table 1.5, we observe similar qualitative and quantitative biases from our fSS model simulation as we observe in our simulation of the RFSV model with equivalent parameters in Section 1.3.4.

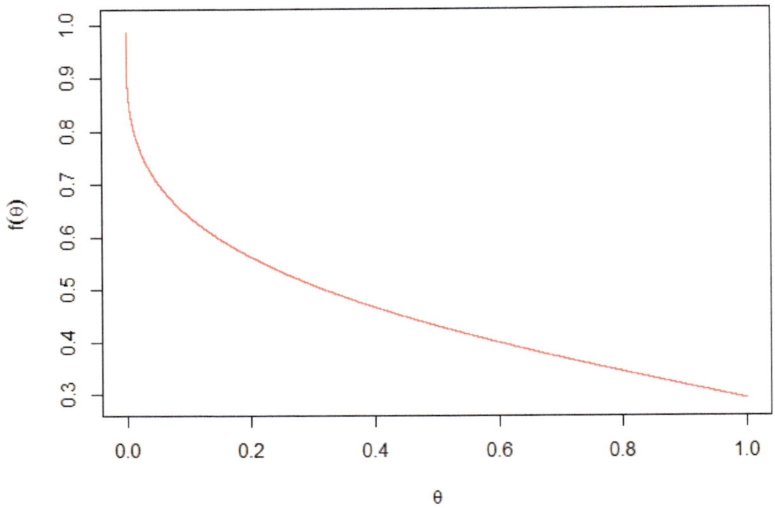

Figure 1.18: $f(\theta)$ vs. $\theta = \delta/\Delta$ with $H = 0.14$.

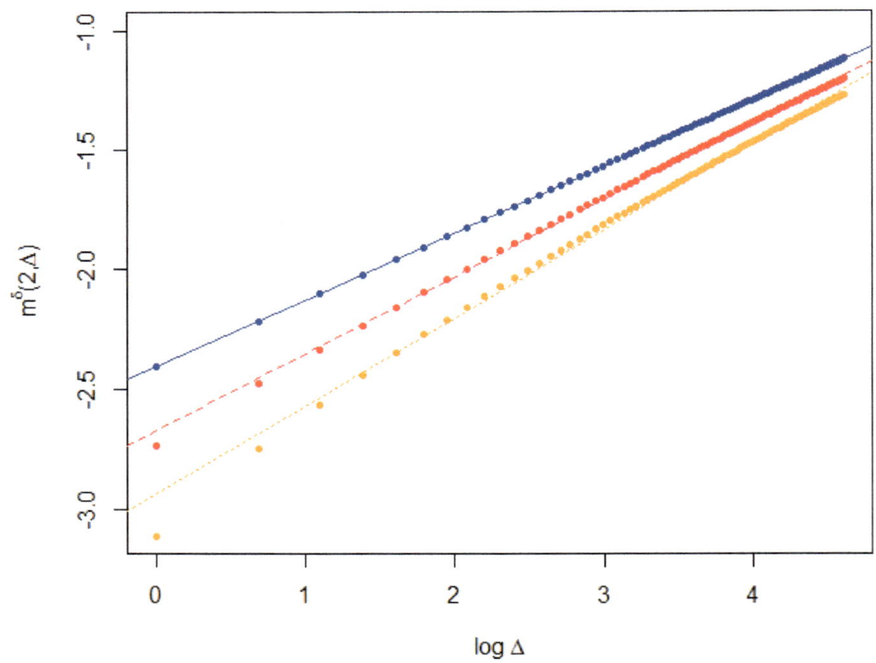

Figure 1.19: Analogue of Figure 1.14 in the fSS model: the blue solid line is the true $m(2,\Delta)$; the red long-dashed line is the UZ estimate $m^{\delta_1}(2,\Delta)$; the orange short-dashed line is the RV estimate $m^{\delta_2}(2,\Delta)$.

Table 1.5: Estimated model parameters from the regressions in Figure 1.19.

Estimate	Estimated α	Estimated H
Exact ($\delta = 0$)	0.300	0.140
UZ ($\delta = 1/24$)	0.263	0.161
RV ($\delta = 1/3$)	0.230	0.184

Acknowledgments

TJ gratefully acknowledges financial support from the chair "Risques Financiers" of the Risk Foundation and the chair "Marchés en Mutation" of the French Banking Federation. MR gratefully acknowledges financial support from the ERC under grant 679836 STAQAMOF. We are very grateful to the referees of *Econometrica*, the *Journal of the American Statistical Association*, the *Journal of Business and Economic Statistics*, and *Quantitative Finance* whose careful reading and valuable comments have helped us improve substantially our presentation of this work. All three authors thank Masaaki Fukasawa for several interesting discussions.

Chapter 2

Pricing under rough volatility[24]

Christian Bayer, Peter K. Friz, Jim Gatheral

2.1 ▪ Introduction

From an analysis of the time series of realized variance using recent high frequency data, Gatheral, Jaisson, and Rosenbaum [166] showed that the logarithm of realized variance behaves essentially as a fractional Brownian motion (fBm) with Hurst exponent H of order 0.1, at any reasonable time scale. The following stationary rough fractional stochastic volatility (RFSV) model was proposed:

$$\begin{aligned} \frac{\mathrm{d}S_t}{S_t} &= \mu_t \mathrm{d}t + V_t \mathrm{d}Z_t, \\ V_t &= \exp\{X_t\}, \ t \in [0, T], \end{aligned} \tag{2.1}$$

where μ_t is a suitable drift term, Z_t is a standard Brownian motion, X_t is a fractional Ornstein–Uhlenbeck process (fOU process for short) satisfying

$$\mathrm{d}X_t = \nu \mathrm{d}W_t^H - \alpha(X_t - m)\mathrm{d}t,$$

where $m \in \mathbb{R}$, and ν and α are positive parameters (see [92]), with Z and W^H correlated in general. Recall that sample paths of fractional Brownian motion W^H are $(H - \varepsilon)$-Hölder (but not H-Hölder and hence "rougher" than Brownian motion whenever $H < \frac{1}{2}$). The reversion time scale is understood to be very long so that $\alpha T \ll 1$ for any reasonable time scale T of practical interest (from one day to ten years say), in which case the log-variance behaves locally (at time scales smaller than T) as an fBm. The RFSV model is remarkably consistent with financial time series data. Moreover, the RFSV model has a quantitative market microstructure-based foundation based on the modeling of order flow using Hawkes processes; see [166, Section 6] and [215].

On the other hand, from the perspective of options pricing, it is well known that conventional low-dimensional Markovian stochastic volatility models, such as the Hull and White, Heston,

[24]This chapter is derived from "Pricing under rough volatility," *Quantitative Finance*, 16 (2016), pp. 887–904, https://doi.org/10.1080/14697688.2015.1099717.

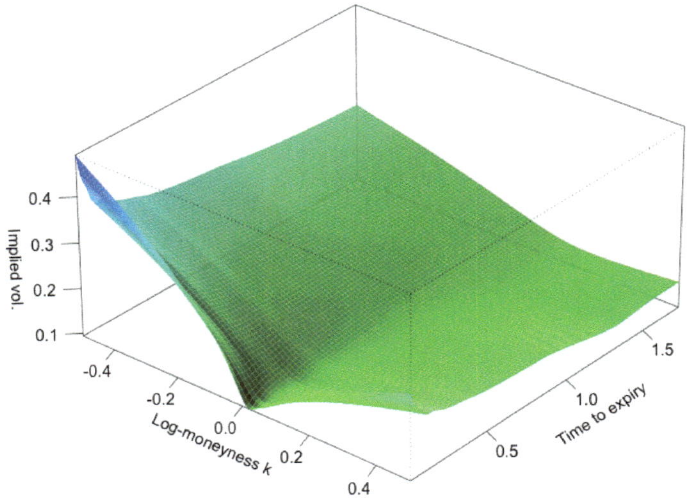

Figure 2.1: The SPX volatility surface as of August 14, 2013. Time measured in years.

and SABR generate implied volatility surfaces whose shapes differ substantially from that of the empirically observed volatility surface. A typical such volatility surface generated from a "stochastic volatility inspired" (SVI) [165] fit to closing SPX option prices as of August 14, 2013[25] is shown in Figure 2.1. It is a stylized fact that, at least in equity markets, although the level and orientation of the volatility surface do change over time, the general overall shape of the volatility surface does not change, at least to a first approximation. This suggests that it is desirable to model volatility as a time-homogeneous process, i.e., a process whose parameters are independent of price and time.

Given an implied volatility smile for a single expiration, little can be said about the process generating it; any process that generates uncertain integrated variance from inception to expiration and with some correlation between changes in volatility and returns of the underlying might suffice. To say more about the underlying process, the scaling of smiles with respect to time to expiration needs to be examined. In particular, one feature of the volatility surface that really does distinguish between models is the term structure of at-the-money (ATM) volatility skew defined as

$$\psi(\tau) := \left| \frac{\partial}{\partial k} \sigma_{\text{imp}}(k, \tau) \right|_{k=0}$$

where $\tau = T - t$ denotes time to expiration; and $k = \log K/S_t$ is the log-strike. In conventional stochastic volatility models, the ATM volatility skew $\psi(\tau)$ is constant for short dates and inversely proportional to τ for long dates. Empirically, as shown in Figure 2.2, we observe that $\psi(\tau)$ is proportional to $1/\tau^\alpha$ for some $0 < \alpha < \frac{1}{2}$ over a very wide range of expirations.

Let $V_u = v_u^2$ denote the instantaneous variance at time $u > t$ (we refer the reader to Appendix 2.10.1 for a reminder of the various notions of variance and volatility used in this work). Then the forward variance curve is given by

$$\xi_t(u) = \mathbb{E}[\, V_u | \mathcal{F}_t], \qquad \text{for } u \geq t.$$

[25]Closing prices of SPX options for all available strikes and expirations were sourced from OptionMetrics (www.optionmetrics.com) via Wharton Research Data Services (WRDS).

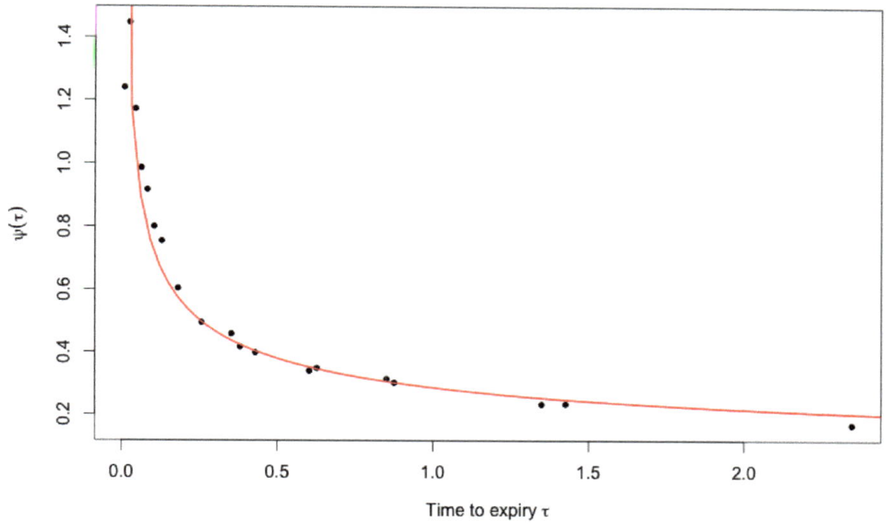

Figure 2.2: The black dots are nonparametric estimates of the S&P at-the-money (ATM) volatility skews as of August 14, 2013; the red curve is the power-law fit $\psi(\tau) = A\,\tau^{-0.407}$, τ measured in years.

Bergomi and Guyon [58] derive a small noise expansion for the smile in a stochastic volatility model written in the following forward variance curve form:

$$\begin{aligned} \frac{dS_t}{S_t} &= \sqrt{\xi_t(t)}\,dZ_t, \\ d\xi_t(u) &= \varepsilon\lambda(t,u,\xi_t)\,dW_t \end{aligned} \tag{2.2}$$

where Z_t is a Brownian motion driving the asset price S_t, W_t is a (suitably correlated) d-dimensional Brownian motion driving the evolution of the forward variance curve, and ε is a dimensionless expansion parameter. To first order in ε, The Bergomi–Guyon expansion takes the form[26]

$$\sigma_{\rm imp}(k,T) = \widehat{\sigma}_T + \varepsilon\sqrt{\frac{w}{T}}\frac{1}{2w^2}C^{x\xi}k + O(\varepsilon^2) \tag{2.3}$$

where the log-strike $k = \log K/S_0$, $w = \int_0^T \xi_0(s)ds$ is total variance to expiration T, and the autocorrelation functional is

$$C^{x\xi} = \int_0^T dt \int_t^T du \frac{\mathbb{E}\left[dx_t d\xi_t(u)\right]}{dt} \tag{2.4}$$

where $x_t = \log S_t/S_0$. Thus, given a stochastic model written in the forward variance curve form (2.2), we can easily (at least in principle) compute the term structure of ATM skew $\psi(\tau)$ to first order in ε (which corresponds to first order in volatility of volatility).

One well-known example of a stochastic volatility model expressed in forward variance curve form is the Bergomi model [57]. The n-factor Bergomi variance curve model may be written in

[26] The at-the-money (ATM) volatility term $\widehat{\sigma}_T$ in the Bergomi–Guyon expansion can also be expressed explicitly to order ε in terms of w and $C^{x\xi}$. See, for example, equation (2.21).

the form

$$\xi_t(u) = \xi_0(u)\mathcal{E}\left(\sum_{i=1}^{n} \eta_i \int_0^t e^{-\kappa_i(u-s)} dW_s^{(i)}\right) \tag{2.5}$$

where $\mathcal{E}(\cdot)$ denotes the stochastic exponential.

Remark 2.1.1. *For a continuous (semi)martingale Z, the stochastic exponential is classically defined as* $\mathcal{E}(Z)_t = \exp(Z_t - Z_0 - \frac{1}{2}[Z, Z]_{0,t})$. *If Z is a local martingale, then so is* $\mathcal{E}(Z)$. *On the other hand, for a zero-mean Gaussian random variable* Ψ, *it will be convenient to introduce the "Wick" exponential as*

$$\mathcal{E}(\Psi) := \exp\left(\Psi - \frac{1}{2}\mathbb{E}\left[|\Psi|^2\right]\right).$$

When Ψ *is the increment of a Gaussian martingale—such as* $\int_0^t f(s)dW_s$ *with deterministic integrand—the two notions coincide. Note, however, that we only ever use the notation* $\mathcal{E}$ *as shorthand; Wick calculus is not required for any of our arguments.*

$\xi_t(u)$ is thus a martingale in t, consistent with the fact that forward variances are tradable. As was pointed out by Bergomi, the entire forward variance curve $\xi_t(\cdot) = \{\xi_t(u) : u > t\}$ is determined by n-factors, each of OU type. Indeed, in the case $n = 1$ (for notational simplicity only) one has

$$\xi_t(u) = \xi_0(u)\exp\left\{\eta e^{-\kappa(u-t)} Y_t - \frac{1}{2}\eta^2 e^{-2\kappa(u-t)}\mathbb{E}[Y_t^2]\right\}$$

in terms of a scalar OU process,

$$dY_t = -\kappa Y_t dt + dW_t.$$

To achieve a decent fit to the observed volatility surface, and to control the forward smile, we need at least two factors. In the two-factor case, there are seven parameters, $\eta_1, \eta_2, \kappa_1, \kappa_2$, and the (constant) correlations $\varrho_{Z,W^{(1)}}, \varrho_{Z,W^{(2)}}, \varrho_{W^{(1)},W^{(2)}}$, in addition to the initial forward variance curve $\xi_0(u)$. When calibrating the two-factor Bergomi model to option prices, we find that it is already overparameterized. Any combination of the parameters η_i, κ_i and the various correlation parameters that gives a roughly $1/\sqrt{T}$ term structure of ATM skew fits well enough. Moreover, the calibrated correlations between the Brownian increments $dW_s^{(i)}$ tend to be high.

The Bergomi model generates a term structure of volatility skew $\psi(\tau)$ that has the qualitative form

$$\psi(\tau) \sim \sum_i \frac{\eta_i}{\kappa_i \tau}\left\{1 - \frac{1 - e^{-\kappa_i \tau}}{\kappa_i \tau}\right\}.$$

Indeed, it can be seen from the Bergomi–Guyon expansion that this functional form is related to the term structure of the autocorrelation functional $C^{x\xi}$, as defined in (2.4), which in turn is driven by the exponential kernels in the exponent in (2.5). To generate the empirically observed $\psi(\tau) \sim \tau^{-\alpha}$ for some α, it would be tempting to replace the exponential kernels in (2.5) with a power-law kernel. This would give a model of the form

$$\xi_t(u) = \xi_0(u)\mathcal{E}\left(\eta \int_0^t \frac{dW_s}{(u-s)^\gamma}\right) \tag{2.6}$$

with $\xi_t(u)$ again a martingale in t. Assuming constant $\xi_0(u) \equiv \sigma_0^2$, and with the Wick interpretation of the stochastic exponential, we would have instantaneous stochastic volatility

$$V_t = \sqrt{\xi_t(t)} = V_0\sqrt{\mathcal{E}\left(\eta \int_0^t \frac{dW_s}{(t-s)^\gamma}\right)} = V_0 \exp\left\{\frac{\eta}{2}\widehat{W}_t - \frac{\eta^2}{4}\mathbb{E}\left[\widehat{W}_t^2\right]\right\},$$

where $\widehat{W}_t = \int_0^t \frac{dW_s}{(t-s)^\gamma}$ is known as "Volterra" fBm with Hurst parameter $H = \frac{1}{2} - \gamma$ and has, similarly to classical fBm, $(H - \varepsilon)$-Hölder sample paths. We note a striking resemblance to the RFSV model (2.1). Moreover, by applying his martingale expansion to a special case of a model originally proposed by Alòs, León, and Vives [11], Fukasawa [151] shows formally that the volatility skew generated by such models has the form

$$\psi(\tau) \sim \frac{1}{\tau^\gamma}$$

for small τ.

In this chapter, we show that the RFSV model does indeed lead naturally to a non-Markovian generalization of the Bergomi model, which we call the rough Bergomi (rBergomi) model. This model fits the observed volatility surface markedly better than conventional Markovian stochastic volatility models, and with fewer parameters.

2.1.1 ▪ Main results and organization of the chapter

In Section 2.2, we show how the RFSV model leads naturally to an options pricing model. In Section 2.3, we analyze a special case of this model, the rBergomi model, where the change of measure from $\mathbb{P}$ to $\mathbb{Q}$ is deterministic. In Section 2.4, we show how to simulate the rBergomi model, and in Section 2.5 we show that volatility surfaces generated using the rBergomi model simulation are remarkably consistent with observed ones (at least on the two specific days presented). In Section 2.6, we examine consistency between the rBergomi model and the VIX options market, finding that in general, the rBergomi model is not consistent with the VIX options market. In Section 2.7, we compute coefficients of the Bergomi–Guyon expansion of the rBergomi model up to second order in volatility of volatility; sadly, we find that this asymptotic expansion does not converge with parameters of practical interest. In Section 2.8, we show that the evolution of market variance swap curves is consistent with forecasts obtained from the historical realized variance time series; we examine the cases of the collapse of Lehman Brothers and the Flash Crash in detail. Finally, in Section 2.9 we summarize and conclude. Additional detailed computations are relegated to the appendix.

2.2 ▪ Change of measure

In [166], using estimates of realized variance as proxies for daily instantaneous variances, the authors uncovered two startlingly simple regularities. First, consistent with many prior studies, distributions of increments of the logarithm of realized variance were found to be close to Gaussian. Second, and more interestingly, for reasonable time scales of practical interest (from one day to a few years say), the time series of realized variance was found to be consistent with the simple model

$$\log V_{t+\Delta} - \log V_t = \nu\left(W^H_{t+\Delta} - W^H_t\right) \tag{2.7}$$

where W^H is fBm, which is simply the RFSV model (2.1) with $\alpha = 0$. This relationship was found to hold for all 21 equity indices in the Oxford-Man database, Bund futures, Crude Oil futures, and Gold futures. Perhaps this feature of the time series of volatility is universal?

Consider the Mandelbrot–Van Ness representation of fBm W^H in terms of Wiener integrals:[27]

$$W^H_t = C_H \left\{ \int_{-\infty}^t \frac{dW^{\mathbb{P}}_s}{(t-s)^\gamma} - \int_{-\infty}^0 \frac{dW^{\mathbb{P}}_s}{(-s)^\gamma} \right\}$$

[27] Strictly speaking, this expression is only formal. The rigorous form, as used in the computations that follow, exploits the cancellation between the integrands as $s \to -\infty$.

where $\gamma = \frac{1}{2} - H$, and the choice $C_H = \sqrt{\frac{2H\Gamma(\frac{3}{2}-H)}{\Gamma(H+\frac{1}{2})\Gamma(2-2H)}}$ ensures that

$$\mathbb{E}\left[W_t^H W_s^H\right] = \frac{1}{2}\left\{t^{2H} + s^{2H} - |t-s|^{2H}\right\}.$$

Substituting into (2.7) (and in terms of $V_t = v_t^2$), we obtain the following model for the evolution of V_u under the physical measure $\mathbb{P}$:

$$\begin{aligned}
&\log V_u - \log V_t \\
&= 2\nu C_H \left\{ \int_{-\infty}^{u} \frac{dW_s^{\mathbb{P}}}{(u-s)^\gamma} - \int_{-\infty}^{t} \frac{dW_s^{\mathbb{P}}}{(t-s)^\gamma} \right\} \\
&= 2\nu C_H \left\{ \int_{t}^{u} \frac{1}{(u-s)^\gamma} dW_s^{\mathbb{P}} + \int_{-\infty}^{t} \left[\frac{1}{(u-s)^\gamma} - \frac{1}{(t-s)^\gamma} \right] dW_s^{\mathbb{P}} \right\} \\
&=: 2\nu C_H \{ M_t(u) + Z_t(u) \}.
\end{aligned} \tag{2.8}$$

Note that $Z_t(u)$ is $\mathcal{F}_t$-measurable, whereas $M_t(u)$ is independent of $\mathcal{F}_t$, and Gaussian with mean zero, and variance $(u-t)^{2H}/(2H)$. We introduce

$$\widehat{W}_t^{\mathbb{P}}(u) := \sqrt{2H} \int_t^u \frac{dW_s^{\mathbb{P}}}{(u-s)^\gamma} \tag{2.9}$$

which has the same properties as $M_t(u)$, only with variance $(u-t)^{2H}$. With $\eta := 2\nu C_H / \sqrt{2H}$ we have $2\nu C_H\, M_t(u) = \eta\, \widehat{W}_t^{\mathbb{P}}(u)$, and so

$$\mathbb{E}^{\mathbb{P}}[\,V_u | \mathcal{F}_t] = V_t \exp\left\{ 2\nu C_H\, Z_t(u) + \frac{1}{2}\eta^2 \mathbb{E}\left[\widehat{W}_t^{\mathbb{P}}(u)^2\right] \right\}.$$

As a consequence, in terms of the (Wick) stochastic exponential

$$V_u = V_t \exp\left\{ \eta \widehat{W}_t^{\mathbb{P}}(u) + 2\nu C_H\, Z_t(u) \right\} \tag{2.10a}$$

$$= \mathbb{E}^{\mathbb{P}}[\,V_u | \mathcal{F}_t]\, \mathcal{E}\left(\eta\, \widehat{W}_t^{\mathbb{P}}(u) \right). \tag{2.10b}$$

At this stage, two comments are in order: First, note that the non-Markovian nature of V_u is hidden in (2.10a) in the dependence of $Z_t(u)$ on the full history of $W^{\mathbb{P}}$. However, (2.10b) reveals that the conditional distribution of V_u depends on $\mathcal{F}_t$ only through the instantaneous variance forecasts $\mathbb{E}^{\mathbb{P}}[\,V_u | \mathcal{F}_t]$, $u > t$.[28]

Summarizing, we propose the following model under the physical probability measure $\mathbb{P}$:

$$\frac{dS_u}{S_u} = \mu_u du + \sqrt{V_u} dZ_u^{\mathbb{P}}, \tag{2.11a}$$

$$V_u = V_t \exp\left\{ \eta \widehat{W}_t^{\mathbb{P}}(u) + 2\nu C_H\, Z_t(u) \right\}, \tag{2.11b}$$

defined in terms of two standard Brownian motions $Z^{\mathbb{P}}$ and $W^{\mathbb{P}}$ with (constant) correlation ϱ.

Remark 2.2.1. *The process $\widehat{W}$ is an example of a* Brownian semistationary process *(see* [103] *for a nice introduction) that is a process of the form*

$$\int_{-\infty}^{t} g(t-s)\sigma_s dW_s$$

[28] This is analogous to what happens in Comte, Coutin, and Renault [97] in the context of their fractionally integrated square-root model.

for some deterministic weight function g and an adapted intermittency process σ. Such processes have been studied extensively; the application of numerical techniques developed in that context to our model remains to be explored.

2.2.1 ▪ Pricing under $\mathbb{Q}$

We have specified dynamics (2.11) for the underlying under the physical probability measure $\mathbb{P}$, defined in terms of two standard Brownian motions $Z^{\mathbb{P}}$ and $W^{\mathbb{P}}$ with constant correlation ϱ and a suitable drift term $\mu = \mu(u, \omega)$. Pricing of options on the underlying (for example, SPX in the foregoing) at (fixed) time t will be based on an equivalent martingale measure; that is, $\mathbb{Q} \sim \mathbb{P}$ on $[t, T]$ such that the discounted[29] asset price process S_t is a martingale under $\mathbb{Q}$. In our context,

$$\frac{dS_u}{S_u} = \sqrt{V_u}\, dZ_u^{\mathbb{Q}}, \qquad \text{for } t \le u \le T,$$

which is achieved, on some fixed time horizon $[t, T]$, by a Girsanov change of measure,

$$dZ_u^{\mathbb{Q}} = dZ_u^{\mathbb{P}} + \frac{\mu_u}{\sqrt{V_u}} du, \qquad t \le u \le T.$$

On the other hand, from (2.9), the Volterra-type process $\widehat{W}_t^{\mathbb{P}}$ has been built from $W^{\mathbb{P}}$, a Brownian motion correlated with $Z_u^{\mathbb{P}}$ in the sense that

$$W^{\mathbb{P}} = \varrho Z^{\mathbb{P}} + \bar{\varrho} \overline{Z}^{\mathbb{P}}, \qquad \varrho^2 + \bar{\varrho}^2 = 1$$

where $(Z^{\mathbb{P}}, \overline{Z}^{\mathbb{P}})$ is a pair of independent standard Brownian motions. A general change of measure for the second component is then of the form

$$d\overline{Z}_u^{\mathbb{Q}} = d\overline{Z}_u^{\mathbb{P}} + \gamma_u du,$$

where $\gamma = \gamma(u, \omega)$, for $u \in [t, T]$, is a suitable adapted process, called the *market price of volatility risk*. As a result,

$$dW_u^{\mathbb{Q}} = dW_u^{\mathbb{P}} + \left(\frac{\varrho \mu_u}{\sqrt{V_u}} + \bar{\varrho} \gamma_u \right) du, \qquad \text{for } t \le u \le T,$$

which we may rewrite as

$$dW_s^{\mathbb{P}} = dW_s^{\mathbb{Q}} + \lambda_s ds. \tag{2.12}$$

To recap, we have a model (2.10b) that accurately mimics the behavior of realized variance time series data, written under $\mathbb{P}$:

$$V_u = \mathbb{E}^{\mathbb{P}}[V_u | \mathcal{F}_t] \, \mathcal{E}\left(\eta \, \widehat{W}_t^{\mathbb{P}}(u) \right). \tag{2.13}$$

In particular, $\mathbb{E}^{\mathbb{P}}[V_u | \mathcal{F}_t]$ is adapted to the filtration generated by $W^{\mathbb{P}}$ (which we assume coincides with the filtration generated by $W^{\mathbb{Q}}$). We may then rewrite (2.13) as

$$\begin{aligned} V_u &= \mathbb{E}^{\mathbb{P}}[V_u | \mathcal{F}_t] \exp\left\{ \eta \sqrt{2H} \int_t^u \frac{1}{(u-s)^\gamma} dW_s^{\mathbb{P}} - \frac{\eta^2}{2} (u-t)^{2H} \right\} \\ &= \mathbb{E}^{\mathbb{P}}[V_u | \mathcal{F}_t] \, \mathcal{E}\left(\eta \, \widehat{W}_t^{\mathbb{Q}}(u) \right) \exp\left\{ \eta \sqrt{2H} \int_t^u \frac{\lambda_s}{(u-s)^\gamma} ds \right\}. \end{aligned} \tag{2.14}$$

The last term in the exponent obviously changes the marginal distribution of the V_u; although the conditional distribution of V_u under $\mathbb{P}$ is lognormal, it will not be lognormal in general under $\mathbb{Q}$.

[29] We set interest rates to zero which is, without loss of generality, assuming deterministic interest rates.

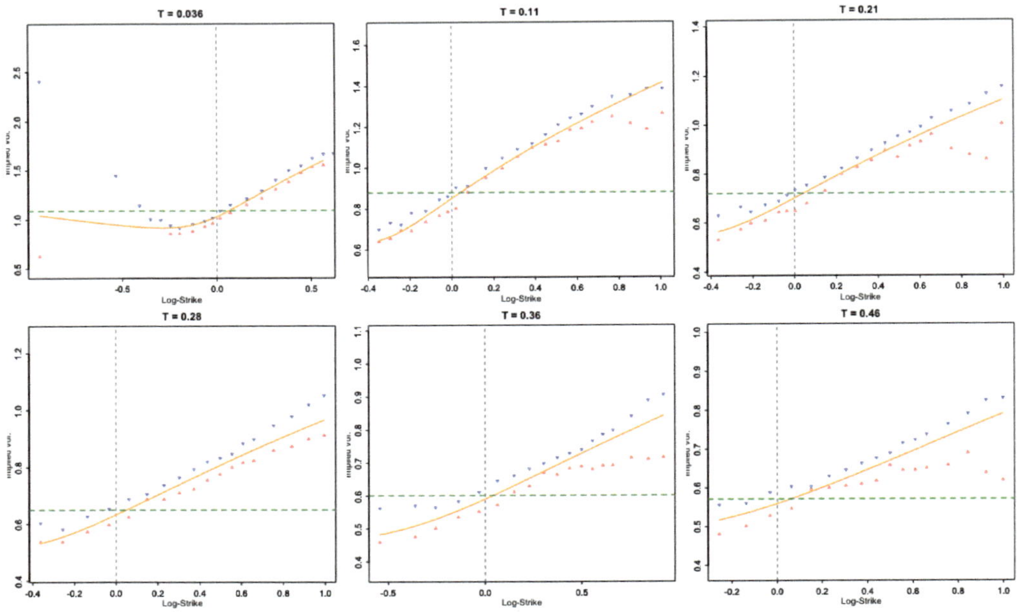

Figure 2.3: VIX implied volatility smiles as of February 4, 2010. Blue points are ask volatilities, red points are bid volatilities, orange lines are SVI fits, and green dashed lines represent the VIX log-strip (VVIX).

VIX smiles and the change of measure

In the case of SPX, it is obvious from the shape of VIX implied volatility smiles that the change of measure cannot be deterministic. If the change of measure were deterministic, it follows from (2.14) that V_u would be conditionally lognormal, VIX would also be approximately lognormal, and so the VIX implied volatility smiles would be approximately flat, a well-known problem with the conventional Bergomi model. In contrast, we observe VIX smiles that are strongly upward sloping (see Figure 2.3, for example), reflecting the intuition that high volatility scenarios are priced more highly by the market than low volatility scenarios. Specifically, we conclude that the change of measure λ must be positively correlated with $W^{\mathbb{Q}}$.

2.3 ▪ The rough Bergomi (rBergomi) model

Despite the inconsistency with VIX smiles pointed out above, let's nevertheless consider the simplest change of measure

$$dW_s^{\mathbb{P}} = dW_s^{\mathbb{Q}} + \lambda(s)ds,$$

assuming for simplicity (resp., as a first approximation) that $\lambda(s)$ is a deterministic function of s. Then from (2.14), we would have

$$\begin{aligned} V_u &= \mathbb{E}^{\mathbb{P}}[V_u|\mathcal{F}_t]\,\mathcal{E}\left(\eta\,\widehat{W}_t^{\mathbb{Q}}(u)\right)\exp\left\{\eta\,\sqrt{2H}\int_t^u \frac{\lambda(s)}{(u-s)^{\gamma}}ds\right\} \\ &= \xi_t(u)\mathcal{E}\left(\eta\widehat{W}_t^{\mathbb{Q}}(u)\right) \end{aligned} \tag{2.15}$$

where, by definition, $\xi_t(u) = \mathbb{E}^{\mathbb{Q}}[V_u|\mathcal{F}_t]$. Moreover, the forward variance curve

$$\xi_t(u) = \mathbb{E}^{\mathbb{P}}[V_u|\mathcal{F}_t] \exp\left\{\eta\sqrt{2H}\int_t^u \frac{\lambda(s)}{(u-s)^\gamma}\mathrm{d}s\right\}$$

is the product of two terms: $\mathbb{E}^{\mathbb{P}}[V_u|\mathcal{F}_t]$ which depends on the history of the driving Brownian motion as explained earlier, and another term which depends on the price of risk $\lambda(s)$. The model (2.15) is a non-Markovian generalization of the Bergomi model (2.5) which we might dub a *rough Bergomi* (or *rBergomi*) model. Specifically, this rBergomi model is non-Markovian in the instantaneous variance V_t: $\mathbb{E}^{\mathbb{Q}}[V_u|\mathcal{F}_t] \neq \mathbb{E}^{\mathbb{Q}}[V_u|V_t]$ but is Markovian in the (infinite-dimensional) state vector $\mathbb{E}^{\mathbb{Q}}[V_u|\mathcal{F}_t] = \xi_t(u)$.

Note also that with (2.15), we have achieved the goal we set in the introduction by replacing the exponential kernels in the Bergomi model (2.5) with a power-law kernel. We may therefore expect that the rBergomi model will generate a realistic term structure of ATM volatility skew.

The observed anticorrelation between price moves and volatility moves may be modeled naturally, just as in the conventional Bergomi model, by anticorrelating the Brownian motion W that drives the volatility process with the Brownian motion driving the price process. Thus,

$$\frac{\mathrm{d}S_t}{S_t} = \sqrt{V_t}\,\mathrm{d}Z_t$$

with

$$\mathrm{d}Z_t = \varrho\mathrm{d}W_t + \sqrt{1-\varrho^2}\mathrm{d}W_t^\perp$$

where ϱ is the correlation between volatility moves and price moves.

Remark 2.3.1. *Note that in the rBergomi model, option prices (and in fact the statistics of the process) as seen from time t depend only on the forward variance curve $\xi_t(u)$ and the three parameters H, η, and ϱ. In this sense there is no explicit dependence on t which is essentially just a label.*

2.3.1 ▪ Re-interpretation of the conventional Bergomi model

According to [57], the conventional Bergomi model is a *market model*, which means that $\xi_t(u)$ can be any given initial forward variance swap curve consistent with market prices. However, for the Bergomi model to properly describe the evolution of this curve, $\xi_t(u) = \mathbb{E}[V_u|\mathcal{F}_t]$ should be consistent with the assumed dynamics; in this sense, a conventional n-factor Bergomi model is not self-consistent in general. Viewed from the perspective of the rBergomi model, however, the initial curve $\xi_t(u)$ reflects the history $\{W_s; s < t\}$ of the driving Brownian motion up to time t. The exponential kernels in the exponent of (2.5) approximate more realistic power-law kernels. The conventional two-factor Bergomi model is then justified in practice as a tractable Markovian engineering approximation to a more realistic rBergomi model.

2.4 ▪ Simulation of the rBergomi model

To simplify notation, we set the origin of the simulation to be $t = 0$ and drop the explicit reference to the pricing measure $\mathbb{Q}$. From (2.15), the model to be simulated is

$$S_t = S_0\,\mathcal{E}\left(\int_0^t \sqrt{V_u}\mathrm{d}Z_u\right),$$
$$V_u = \xi_0(u)\mathcal{E}\left(\eta\sqrt{2H}\int_0^u \frac{1}{(u-s)^\gamma}\mathrm{d}W_s\right) = \xi_0(u)\mathcal{E}\left(\eta\,\widehat{W}_u\right),$$

where $\widehat{W}$ is a Volterra process[30] with the scaling property $\mathbb{V}[\widehat{W}_u] = u^{2H}$. So far, $\widehat{W}$ behaves just like fBm. However, the dependence structure is different. Specifically, for $v > u$,

$$\mathbb{E}\left[\widehat{W}_v\,\widehat{W}_u\right] = u^{2H}\,G\left(\frac{v}{u}\right)$$

where, for $x \geq 1$ and with $\gamma = \frac{1}{2} - H$,

$$G(x) = 2H\int_0^1 \frac{\mathrm{d}s}{(1-s)^{\gamma}\,(x-s)^{\gamma}} = \frac{1-2\gamma}{1-\gamma}x^{\gamma}{}_2F_1(1,\gamma,2-\gamma,x), \tag{2.16}$$

where ${}_2F_1(\cdot)$ denotes the hypergeometric function.

Remark 2.4.1. *The dependence structure of the Volterra process $\widehat{W}$ is markedly different from that of fBm, with the Molchan–Golosov kernel $K(u,s)$ given by*

$$K(u,s) = c_H\frac{1}{(u-s)^{\gamma}}{}_2F_1\left(\frac{1}{2}-H, H-\frac{1}{2}, H+\frac{1}{2}, \frac{s-u}{s}\right), \quad 0 < s < t,$$

for some constant c_H. In particular, for small H, correlations drop precipitously as the ratio u/v moves away from 1.

We also need covariances of the Brownian motion Z with the Volterra process $\widehat{W}$. With $v \geq u$, these are given by

$$\mathbb{E}\left[\widehat{W}_v Z_u\right] = \varrho D_H\left\{v^{H+\frac{1}{2}} - (v-u)^{H+\frac{1}{2}}\right\}$$

and

$$\mathbb{E}\left[Z_v \widehat{W}_u\right] = \varrho\, D_H u^{H+\frac{1}{2}}$$

where, for future convenience, we have defined the constant

$$D_H = \frac{\sqrt{2H}}{H+\frac{1}{2}}.$$

These two formulae may be conveniently combined as

$$\mathbb{E}\left[\widehat{W}_v Z_u\right] = \varrho D_H\left\{v^{H+\frac{1}{2}} - (v-\min(u,v))^{H+\frac{1}{2}}\right\}.$$

Lastly, of course, for $v \geq u$, $\mathbb{E}[Z_v Z_u] = u$. With m the number of time steps and n the number of simulations, our rBergomi model simulation algorithm may then be summarized as follows.

- Construct the joint covariance matrix for the Volterra process $\widehat{W}$ and the Brownian motion Z, and compute its Cholesky decomposition.
- For each time, generate independent and identically distributed (i.i.d.) normal random vectors, and multiply them by the lower-triangular matrix obtained by the Cholesky decomposition to get an $m \times 2n$ matrix of paths of $\widehat{W}$ and Z with the correct joint marginals.
- With these paths held in memory, we may evaluate the expectation under $\mathbb{Q}$ of any payoff of interest.

The simulation procedure we have described is, unsurprisingly, very slow because of the high number of matrix-vector multiplications with a lower-triangular but otherwise dense matrix. We leave for future research the search for faster simulation techniques based on the specific structure of the problem, including the specific choice of the correlation structure between Z and $\widehat{W}$.

[30] This is identical up to a constant factor to the definition of [98].

2.5 ▪ Consistency of the rBergomi model with the SPX volatility surface

As mentioned earlier, our simulation of the rBergomi model is very slow. However, the model has only three parameters: H, η, and ϱ. Although there is a little bit of art involved, these parameters may be identified in a relatively straightforward way from the observed implied volatility surface by exploiting the following observations:

- H controls the decay of the term structure of the ATM volatility skew $\psi(\tau)$ for small τ.
- Fixing H, the product $\varrho\,\eta$ is fixed by the level of $\psi(\tau)$ for larger τ.
- Focusing on a single expiration and fixing the product $\varrho\,\eta$, the minimum of the smile moves lower, and the position of the minimum moves to the right as ϱ is made more negative.

The rBergomi parameters can thus be guessed in practice.

We will now show that on two particular days in history, the rBergomi model was surprisingly consistent with the observed volatility surface. Fits for other days we chose are not always as impressive as these two but nevertheless visibly superior to fits of conventional Markovian stochastic volatility models.

2.5.1 ▪ Estimation of the variance swap curve

Variance swaps are actively traded, so in principle, computation of the forward variance swap curve should be straightforward. In practice, however, it is not easy to obtain high quality variance swap quote data, and in any case, the bid/ask spread is wide. We thus choose to proxy the value of a τ-maturity variance swap by the value of a τ-expiration log contract as explained, for example, in [163, Chapter 11]. To price the log contract for a particular expiration τ requires knowing the prices of τ-expiration options for all strikes; of course prices are only quoted for a finite number of strikes. We therefore choose to interpolate and extrapolate observed implied volatilities using the arbitrage-free SVI parameterization of the volatility surface as explained in [165]. For any given day, we obtain the closing prices of SPX options for all available strikes and expirations from OptionMetrics (www.optionmetrics.com) via Wharton Research Data Services (WRDS). Having estimated variance swaps to each expiration, we interpolate total variances using a monotonic spline to estimate variance swaps for intermediate dates. This allows us in turn to estimate the forward variance swap curve.

One subtlety is that by choosing SVI to interpolate and extrapolate, we may be assuming a smile that is inconsistent with the one generated by the rBergomi model, and therefore that the forward variance curve may not be accurate. The practical effect of this is that ATM implied volatilities are not matched in the first pass, with good agreement for very short expirations but rather less good agreement as time to expiry increases. A simple iteration on the forward variance curve soon reaches a fixed point that achieves consistency between model ATM volatilities and market ATM volatilities.

To summarize, we use the following procedure to compute parameters for the rBergomi model from market data.

(1) Use SVI to interpolate and extrapolate market prices of put and call options at any strikes and expirations;

(2) compute the price of the log-strip contract (for any expiration) and extract the forward variance curve by differentiation;

(3) plug the forward variance curve and guessed values of H, ϱ, and η into the rBergomi model;

(4) compute option prices and ATM implied volatilities in the rBergomi model;

(5) compare with observed ATM implied volatilities and adjust the forward variance curve (and possibly the parameters H, ϱ, and η) and go back to (3).

We follow the procedure until we reach a fixed point in forward variance. In (5), we essentially adjust the forward variance curve by shifting and scaling, but this is a purely ad hoc choice.

2.5.2 ▪ Fits to two specific days in history

February 4, 2010

For our first comparison of the model to SPX options data, we choose February 4, 2010, a day when the ATM volatility term structure happened to be pretty flat. With guessed parameters $H = 0.07$, $\eta = 1.9$, $\varrho = -0.9$, we obtain the impressive fit shown in Figure 2.4. Only three parameters is enough to get a very good fit to the whole SPX volatility surface, including the shortest dated smile (Figure 2.5).

In Figures 2.6 and 2.7, we see just how well the rBergomi model can match empirical skews and volatilities. Recall also that the parameters we used are just guesses!

August 14, 2013

For our second comparison, we choose a date just prior to an expiration date for SPX options. Rather than choosing the last Thursday of trading, we examine the volatility surface as of the close on the final Wednesday prior to expiration so that the shortest expiration smile is more meaningful; the latest such date available to us in our OptionMetrics dataset is Wednesday, August 14, 2013. With guessed parameters $H = 0.05$, $\eta = 2.3$, $\varrho = -0.9$, we obtain the fit shown in Figure 2.8. Once again, only three parameters is enough to get a very good fit to the whole SPX volatility surface, including the shortest dated smile (from options with only one day of trading left).

Jump-like behavior of the rBergomi price process

It has often been claimed that jumps are required to explain the observed extreme short-dated smile in SPX. In particular, in [87], Carr and Wu determine whether or not there are jumps in the asset process, and if so, whether such jumps are finite or infinite activity. They determine based on their analysis that jumps are required to generate the smiles observed in SPX. However, the class of processes that Carr and Wu consider is too restrictive, excluding models like rBergomi where the out-of-the-money smile explodes as time to expiration $\tau \to 0$. It is apparent from Figures 2.4 and 2.8 that the rBergomi model (where the price process is continuous) generates smiles consistent with those observed empirically *even for very short expirations*; there is no need for jumps.

2.6 ▪ The rBergomi model and VIX options

We pointed out earlier in Section 2.2.1 that observed VIX smiles are inconsistent with the rBergomi model. Nevertheless, even if the rBergomi model is misspecified, it may be possible to

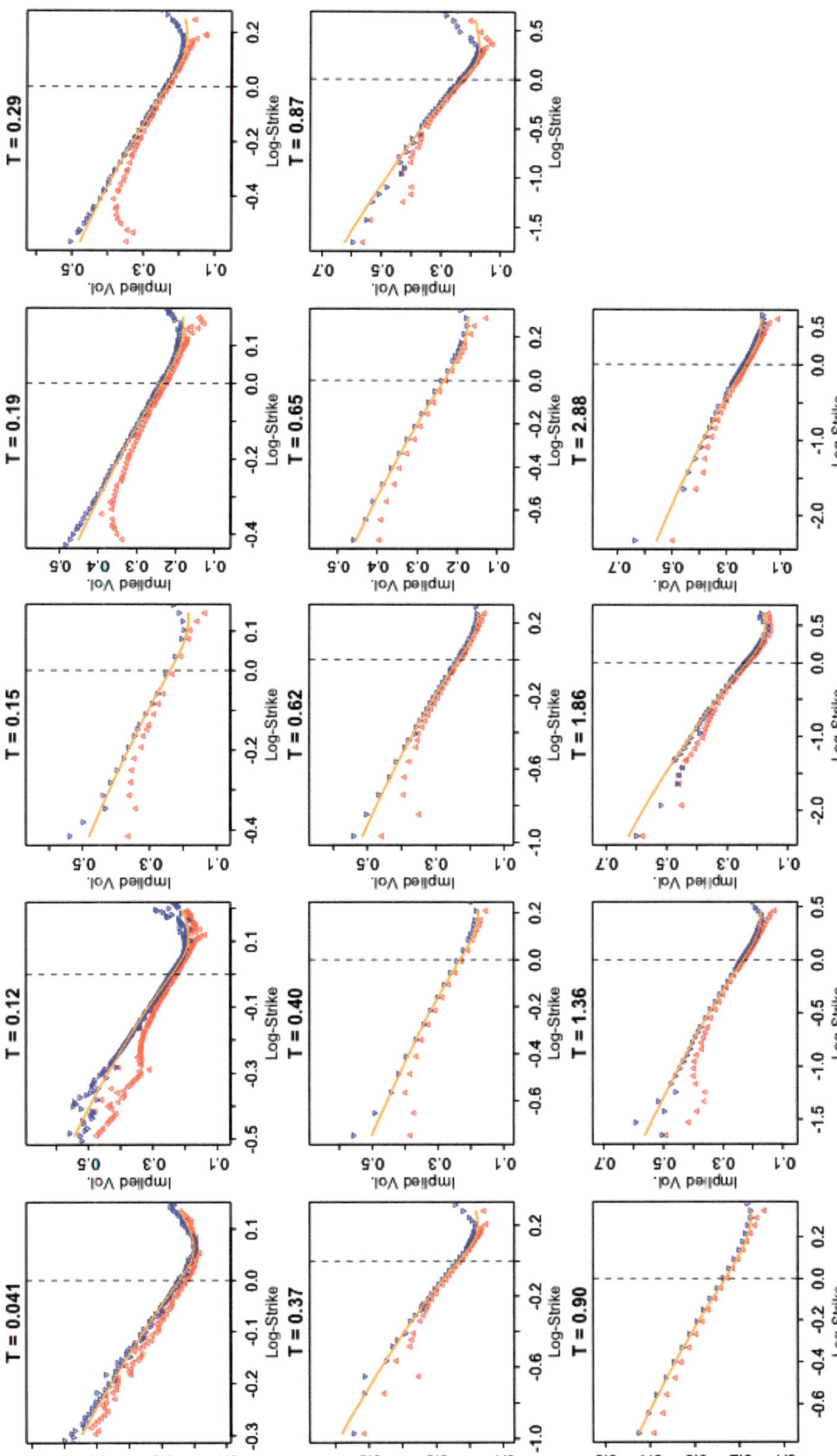

Figure 2.4: SPX smiles as of February 4, 2010: Red and blue points represent bid and offer SPX implied volatilities; orange smiles are from the rBergomi simulation.

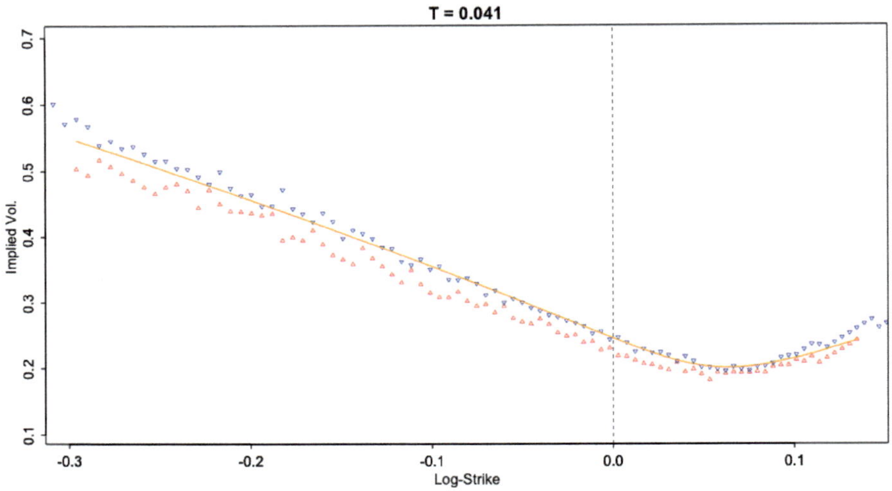

Figure 2.5: Shortest dated smile as of February 4, 2010: Red and blue points represent bid and offer SPX implied volatilities; orange smile is from the rBergomi simulation.

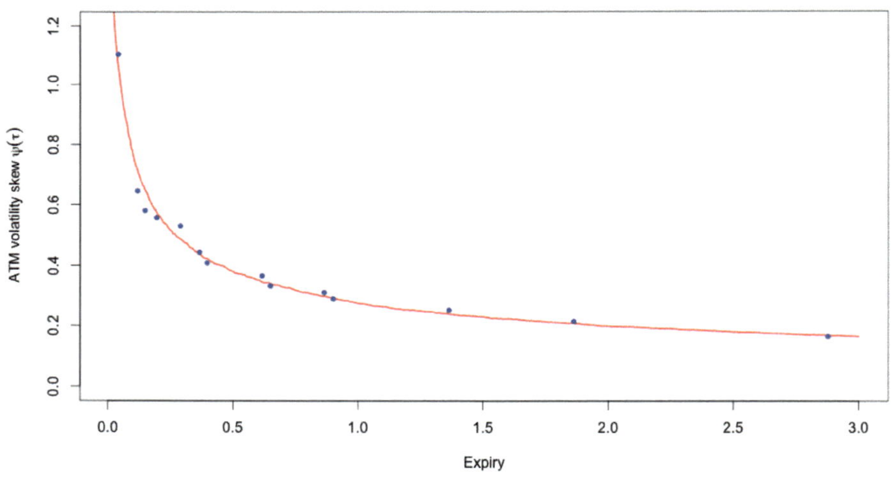

Figure 2.6: As of February 4, 2010: Blue points are empirical skews; the red line is from the rBergomi simulation. Time measured in years.

impute its parameters H and η by examining the term structure of VIX variance swaps;[31] if VIX

[31]The VIX log-strip forms the basis for the VVIX (VIX of VIX) index computation. Indeed, following CBOE (www.cboe.com), the VVIX term structure is computed every day (t) for various maturities (T) of VIX options using the usual log-strip formula that is used for the construction of VIX. More specifically, given $T > t$ and assuming that VIX options with expiry T are traded, the $\mathrm{VVIX}_{t,T}$ is given by

$$\mathrm{VVIX}^2_{t,T}(T-t) = -2\mathbb{E}_t\left[\log\sqrt{\zeta(T)} - \log\sqrt{\zeta(t)}\right],$$

where $\zeta(s)$ denotes the square of VIX at s, and $\mathbb{E}_t[\log\sqrt{\zeta(T)}]$ can be expressed in terms of put and call prices on VIX with expiry T. The usual VVIX index (at a given t) then corresponds to $\mathrm{VVIX}_{t,t+\Delta}$ for Δ equal to one month.

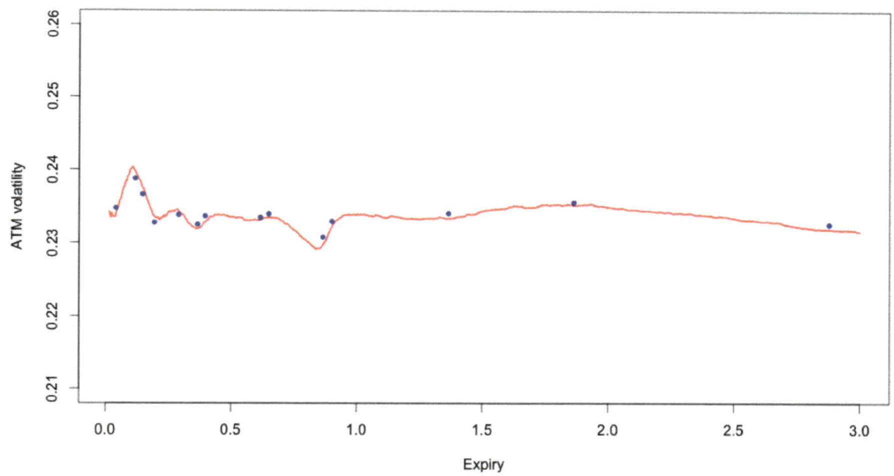

Figure 2.7: As of February 4, 2010: Blue points are empirical ATM volatilities; the red line is from the rBergomi simulation. Time measured in years.

corresponds to volatility, then VIX of VIX should correspond to "volatility of volatility." If it were possible to impute H and η in this way, model calibration would obviously be considerably easier. Denote the terminal value of the VIX futures by $\sqrt{\zeta(T)}$. Then, by definition,[32]

$$\zeta(T) = \frac{1}{\Delta}\int_T^{T+\Delta} \mathbb{E}[V_u|\mathcal{F}_T]\mathrm{d}u,$$

where Δ is one month. In the rBergomi model,

$$V_u = \xi_t(u)\,\mathcal{E}\left(\eta\,\sqrt{2H}\int_t^u \frac{\mathrm{d}W_s}{(u-s)^\gamma}\right)$$

with $\gamma = \frac{1}{2} - H$. Instantaneous variances V_u are thus lognormally distributed. It should therefore be a good approximation (and so it turns out) to assume that the VIX payoff and its square $\zeta(T)$ are also lognormally distributed. In that case, the terminal distribution of $\zeta(T)$ is completely determined by $\mathbb{E}\left[\zeta(T)|\mathcal{F}_t\right]$ and $\mathbb{V}[\log\zeta(T)|\mathcal{F}_t]$. It is immediate that

$$\mathbb{E}\left[\zeta(T)|\mathcal{F}_t\right] = \frac{1}{\Delta}\int_T^{T+\Delta} \xi_t(u)\mathrm{d}u.$$

To estimate the conditional variance of $\zeta(T)$, we approximate the arithmetic mean by the geometric mean as follows:

$$\zeta(T) \approx \exp\left\{\frac{1}{\Delta}\int_T^{T+\Delta} \mathbb{E}[\log V_u|\mathcal{F}_T]\mathrm{d}u\right\}.$$

After some computation detailed in Appendix 2.10.3, we obtain

$$\mathbb{V}[\log\zeta(T)|\mathcal{F}_t] \approx \eta^2\,(T-t)^{2H} f^H\left(\frac{\Delta}{T-t}\right)$$

[32]See Chapter 11 of [163] for more details.

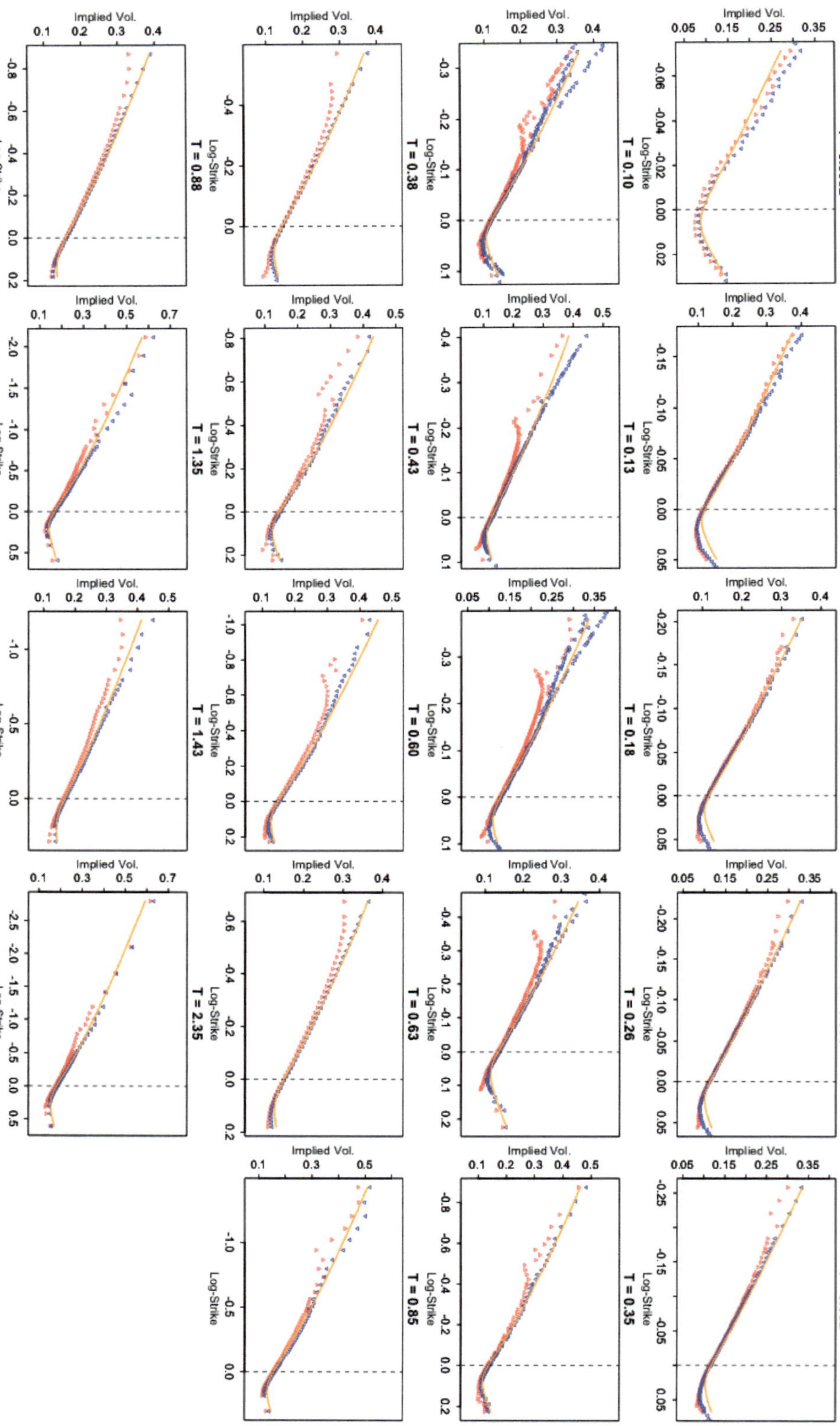

Figure 2.8: SPX smiles as of August 14, 2013: Red and blue points represent bid and offer SPX implied volatilities; orange smiles are from the rBergomi simulation.

where

$$f^H(\theta) = \frac{D_H^2}{\theta^2} \int_0^1 \left[(1+\theta-x)^{H+\frac{1}{2}} - (1-x)^{H+\frac{1}{2}} \right]^2 dx. \tag{2.17}$$

It is straightforward to show that $f^H(\theta) \to 1$ as $\theta \to 0$ which is the limit in which $\zeta(T) \to V_T$. In Appendix 2.10.3 we show further how to express $f^H(\theta)$ explicitly in terms of the hypergeometric function. However, the above form (2.17) is more convenient for computation.

The VIX variance swaps (VVIX^2) are then given by

$$\text{VVIX}_{t,T}^2(T-t) \approx \mathbb{V}\left[\log \sqrt{\zeta(T)}\middle|\mathcal{F}_t\right] \approx \frac{1}{4}\eta^2 (T-t)^{2H} f^H\left(\frac{\Delta}{T-t}\right). \tag{2.18}$$

VIX variance swaps may also be estimated directly from market prices of options on VIX using the log-strip in the usual way as

$$\text{VVIX}_{t,T}^2(T-t) = -2\mathbb{E}\left[\log \sqrt{\zeta(T)} - \log \sqrt{\zeta(t)}\middle|\mathcal{F}_t\right]. \tag{2.19}$$

By comparing the model VVIX term structure (2.18) with the market VVIX term structure (2.19), we can in principle fix the model parameters H and η.

2.6.1 ▪ The VVIX term structure in practice

Recall that the parameters we used to obtain the reasonably impressive SPX fits of Section 2.5.2 were just guessed; specifically, these parameters were as follows:

	H	η
February 4, 2010	0.07	1.9
August 14, 2013	0.05	2.3

In Figure 2.9, we show plots of equation (2.18) with the above parameters, superimposed on the empirically estimated variances of VIX. At least for February 4, 2010, given the qualitative

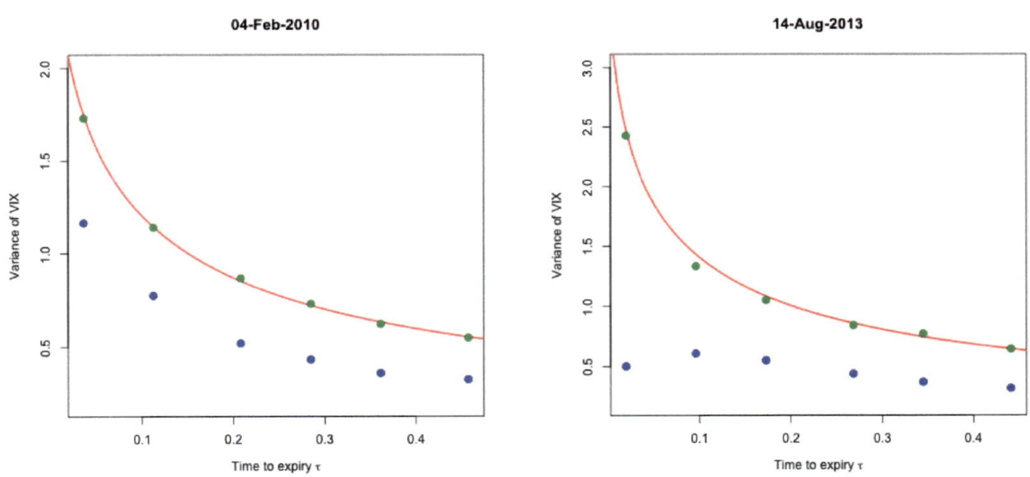

Figure 2.9: Empirical VVIX term structure data (blue points) and rBergomi (using (2.18), red line, and from simulation, green points) estimates of $\mathbb{V}[\text{VIX}]$ as of February 4, 2010 (left) and August 14, 2013 (right). Time measured in years.

agreement between the shape of the curve (2.18) and empirical estimates, it might be possible to argue some consistency of the rBergomi model with observation; for August 14, 2013, however, there is not even qualitative agreement. Whether this disagreement between model and market should be ascribed to a misspecified model, wrong market prices, or indeed both of these, is left for future research.

More specifically, assuming the RFSV model is a good description of the instantaneous variance process under $\mathbb{P}$, the aim of the present chapter is to explore the consequences for option pricing under $\mathbb{Q}$. The rBergomi model results from assuming a deterministic change of measure. Figure 2.9 shows not only that the rBergomi model cannot generate the observed upward-sloping VIX option smile, as explained in Section 2.2.1, but also that it cannot explain the level of VIX volatility either. This could be either because the RFSV model is misspecified, failing to capture important elements of the dynamics of instantaneous variance, or because the market prices of options on VIX are inconsistent with the market prices of options on SPX.

2.7 ▪ Volatility of volatility expansion

As simulation of the rBergomi model is so slow, one potential alternative is to estimate rBergomi parameters using an asymptotic expansion. Given a model expressed in forward variance curve form, Bergomi and Guyon [58] derive the following expression for the Black–Scholes implied volatility smile to second order in volatility of volatility:

$$\sigma_{\text{imp}}(k,t) = \widehat{\sigma}_T + \mathcal{S}_T\, k + C_T\, k^2 + O(\eta^3) \tag{2.20}$$

where

$$\begin{aligned}
\widehat{\sigma}_T &= \sqrt{\frac{w}{T}}\left\{1 + \frac{1}{4w}C^{x\xi} + \frac{1}{32w^3}\left(12(C^{x\xi})^2 - w(w+4)C^{\xi\xi} + 4w(w-4)C^{\mu}\right)\right\},\\
\mathcal{S}_T &= \sqrt{\frac{w}{T}}\left\{\frac{1}{2w^2}C^{x\xi} + \frac{1}{8w^3}\left(4wC^{\mu} - 3(C^{x\xi})^2\right)\right\},\\
C_T &= \sqrt{\frac{w}{T}}\frac{1}{8w^4}\left(4\,w\,C^{\mu} + wC^{\xi\xi} - 6(C^{x\xi})^2\right),
\end{aligned} \tag{2.21}$$

and

$$w = \int_0^T \xi_0(s)\mathrm{d}s$$

is total variance to expiration T. The autocorrelation functionals $C^{x\xi}$, $C^{\xi\xi}$, and C^{μ} have the following explicit expressions:

$$\begin{aligned}
C^{x\xi} &= \int_0^T \mathrm{d}t \int_t^T \mathrm{d}u \frac{\mathbb{E}\left[\mathrm{d}x_t \mathrm{d}\xi_t(u)\right]}{\mathrm{d}t},\\
C^{\xi\xi} &= \int_0^T \mathrm{d}t \int_t^T \mathrm{d}s \int_t^T \mathrm{d}u \frac{\mathbb{E}\left[\mathrm{d}\xi_t(s)\mathrm{d}\xi_t(u)\right]}{ds},\\
C^{\mu} &= \int_0^T \mathrm{d}t \int_t^T \mathrm{d}u \frac{\mathbb{E}\left[\mathrm{d}x_t \mathrm{d}\xi_t(u)\right]}{\mathrm{d}t}\,\frac{\delta C_t^{x\xi}}{\delta \xi_t(u)}
\end{aligned} \tag{2.22}$$

where the notation $\delta/\delta\xi_t(u)$ denotes a functional derivative.

In the case of the rBergomi model (2.15), we have

$$\begin{aligned}
\frac{\mathrm{d}S_t}{S_t} &= \sqrt{\xi_t(t)}\,\mathrm{d}Z_t,\\
\frac{\mathrm{d}\xi_t(u)}{\xi_t(u)} &= \eta\sqrt{2H}\frac{\mathrm{d}W_t}{(u-t)^{\gamma}}
\end{aligned}$$

with $\mathbb{E}[\mathrm{d}Z_t \mathrm{d}W_t] = \varrho \mathrm{d}t$ so that

$$\frac{\mathbb{E}\left[dx_t\, d\xi_t(u)\right]}{\mathrm{d}t} = \varrho\, \eta\, \sqrt{2H}\, \sqrt{\xi_t(t)} \frac{\xi_t(u)}{(u-t)^\gamma}.$$

The various autocorrelation functionals (2.22) may then be computed; explicit computations are presented in Appendix 2.10.2.

2.7.1 ▪ Special case: Flat variance curve

The special case $\xi_0(u) = \overline{\sigma}^2$, $u \geq 0$, where the initial forward variance curve is flat, is particularly instructive. First, from (2.28), we have

$$\begin{aligned} C^{x\xi} &= \varrho\eta\sqrt{2H}\int_0^T \sqrt{\xi_t(s)}\mathrm{d}s \int_s^T \xi_t(u)\frac{\mathrm{d}u}{(u-s)^\gamma} \\ &= \varrho\eta\overline{\sigma}^3 E_H T^{H+\frac{3}{2}}, \end{aligned} \tag{2.23}$$

where we have further defined

$$E_H = \frac{D_H}{H+\frac{3}{2}}.$$

Also, $w = \overline{\sigma}^2 T$. Substituting back into (2.21) and then (2.20) gives, to first order in η,

$$\sigma_{\mathrm{imp}}(k,t) = \overline{\sigma} + \frac{\varrho\eta E_H}{2} T^{H-\frac{1}{2}}\left(k + \frac{w}{2}\right) + O(\eta^2) \tag{2.24}$$

In particular, we see that to first order in η, the term structure of at-the-money volatility skew is given by

$$\psi(\tau) = \left.\frac{\partial\sigma^2_{\mathrm{imp}}(k,\tau)}{\partial k}\right|_{k=0} = \frac{\varrho\eta E_H}{2}\tau^{-\gamma}$$

with $\gamma = \frac{1}{2} - H$. Similarly, substituting $\xi_0(u) = \overline{\sigma}^2$ into (2.29) and (2.30), respectively, gives the terms required for computation of the second-order contribution:

$$\begin{aligned} C^{\xi\xi} &= \eta^2\overline{\sigma}^4 D_H^2 \frac{T^{2+2H}}{2+2H}, \\ C^{\mu} &= \frac{1}{2}\varrho^2\eta^2\overline{\sigma}^4 D_H^2 \left\{1 + \frac{\Gamma(H+\frac{3}{2})^2}{\Gamma(2H+3)}\right\}\frac{T^{2+2H}}{2+2H}. \end{aligned}$$

It follows that to second order in η, the term structure of at-the-money volatility skew is given by

$$\psi(\tau) = \frac{\varrho\eta}{2}E_H\frac{1}{\tau^\gamma} + \frac{1}{4}\varrho^2\eta^2\overline{\sigma}\tau^{2H}\left[\frac{D_H^2}{1+H}\left\{1 + \frac{\Gamma(H+\frac{3}{2})^2}{\Gamma(2H+3)}\right\} - \frac{3}{2}E_H^2\right]. \tag{2.25}$$

Numerical test

The dimensionless Bergomi–Guyon expansion parameter is $\lambda = \eta T^H$. When H is very small, $\lambda \sim \eta$ for all reasonable expirations; with $H < 0.1$ as in Section 2.5.2, $\lambda \sim 1.9$, which is not small enough for the asymptotic expansion to converge, even at-the-money. With the much smaller value $\eta = 0.4$, we see in Figure 2.10 very good agreement between the Bergomi–Guyon asymptotic skew formula (2.25) and the simulation.

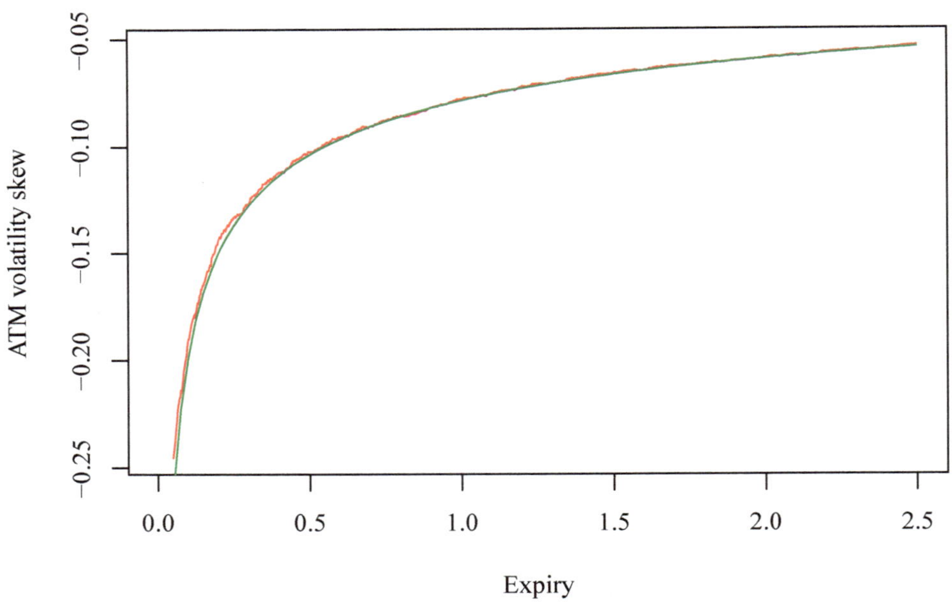

Figure 2.10: The Bergomi–Guyon second-order ATM skew approximation is in green; ATM skews from Monte Carlo simulation are in red. Parameters used were $H = 0.1, \eta = 0.4, \varrho = -0.85, \overline{\sigma} = 0.235$.

We thus conclude that both our Bergomi–Guyon computations and the simulation are likely to be correct. Sadly, however, the Bergomi–Guyon expansion does not converge with values of η consistent with the SPX volatility surface, so the Bergomi–Guyon expansion is not useful in practice for calibration of the rBergomi model.

2.8 ▪ Forecasting the variance swap curve

Recall that in the RFSV model (2.1), $\log V_t \approx 2\,\nu\, W_t^H + C$ for some constant C. In [254], it is shown, assuming $H \in (0, \frac{1}{2})$, $\Delta > 0$, that $W_{t+\Delta}^H$ is conditionally Gaussian with conditional expectation[33]

$$\mathbb{E}[W_{t+\Delta}^H|\mathcal{F}_t] = \frac{\cos(H\pi)}{\pi}\Delta^{H+\frac{1}{2}} \int_{-\infty}^{t} \frac{W_s^H}{(t-s+\Delta)(t-s)^{H+\frac{1}{2}}} \mathrm{d}s$$

and conditional variance

$$\mathbb{V}[W_{t+\Delta}^H|\mathcal{F}_t] = c\,\Delta^{2H}$$

where

$$c = \frac{\Gamma(\frac{3}{2}-H)}{\Gamma(H+\frac{1}{2})\,\Gamma(2-2H)}.$$

Thus, we obtain the following natural form for the RFSV predictor of the variance:

$$\mathbb{E}^{\mathbb{P}}[\,V_{t+\Delta}|\mathcal{F}_t] = \exp\left\{\mathbb{E}^{\mathbb{P}}\left[\log(V_{t+\Delta})\middle|\mathcal{F}_t\right] + 2\,c\,\nu^2\Delta^{2H}\right\} \tag{2.26}$$

[33]Trivially, $\mathbb{E}[W_{t+\Delta}^H|\mathcal{F}_t] = W_t^H$ when either $\Delta = 0$ or $H = \frac{1}{2}$. This corresponds to the singular behavior of the integrand, as either $\Delta \to 0$ or $H \to \frac{1}{2}$, with the necessary compensation given by $\Delta^{H+\frac{1}{2}}\,\cos(H\pi) \sim 0$ in these regimes.

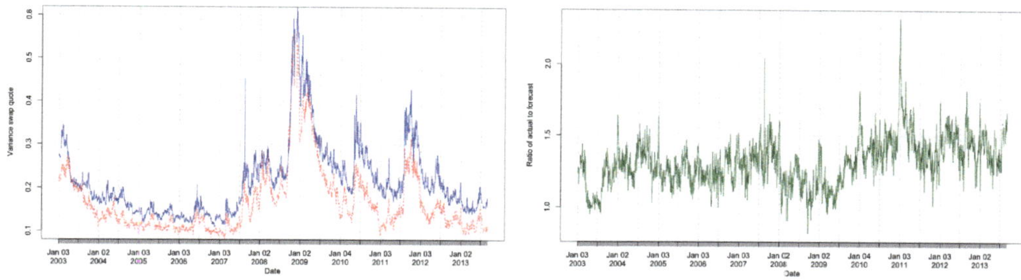

(a) 3-month variance swaps (b) Ratio of actual to forecast

Figure 2.11: Plot (a) shows actual (proxy) 3-month variance swap quotes in blue vs. forecast in red. Plot (b) shows the ratio between 3-month actual variance swap quotes and 3-month forecasts.

where

$$\mathbb{E}^{\mathbb{P}}\left[\log V_{t+\Delta} \middle| \mathcal{F}_t\right] = \frac{\cos(H\pi)}{\pi} \Delta^{H+\frac{1}{2}} \int_{-\infty}^{t} \frac{\log V_s}{(t-s+\Delta)(t-s)^{H+\frac{1}{2}}} \mathrm{d}s. \tag{2.27}$$

The fair value of a τ-maturity variance swap is given (approximately) by

$$\mathcal{V}_t(\tau) = \frac{1}{\tau} \int_t^{t+\tau} \mathbb{E}^{\mathbb{Q}}[V_s|\mathcal{F}_t]\mathrm{d}s$$

where $\mathbb{Q}$ is the risk-neutral measure. If it were possible to ignore the change of measure so that

$$\mathbb{E}^{\mathbb{Q}}[V_s|\mathcal{F}_t] = \mathbb{E}^{\mathbb{P}}[V_s|\mathcal{F}_t],$$

it would be possible to forecast variance swap curves using (2.26). In fact, we will see that from the data, $\mathbb{Q}$ is close to $\mathbb{P}$ in this sense. We now proceed to compare forecast and actual variance swap curves.

SPX variance curve forecasts are formed using the predictor (2.26) from the time series of daily realized variance estimates from the same Oxford-Man dataset that was used in [166].

As for market variance swap curves, although there is an active market, it is not easy to obtain high quality variance swap quote data, and, in any case, the bid/ask spread is wide. We thus choose to proxy the value of a τ-maturity variance swap by the value of a τ-expiration log contract as explained, for example, in Chapter 11 of [163]. Pricing the log contract for a particular expiration τ requires us to know the prices of τ-expiration options for all strikes; of course, prices are only quoted for a finite number of strikes. We therefore choose to interpolate and extrapolate observed implied volatilities using the arbitrage-free SVI parameterization of the volatility surface as explained in [165]. With closing prices of SPX options for all available strikes and expirations sourced from OptionMetrics (www.optionmetrics.com) via Wharton Research Data Services (WRDS), we follow the procedure just described to compute proxy variance swap curves each day from January 4, 1996 to August 30, 2013, a total of 4,443 days. We also need a sufficient history to be able to compute a forecast. We end up with 2,681 days of forecast and actual variance swap curves from Jan 3, 2003 to August 31, 2013.

Plotting actual vs. forecast 3-month variance swap curves in Figure 2.11, we immediately see that the actual variance swap curve is a factor (of roughly 1.4) higher than the forecast one, which

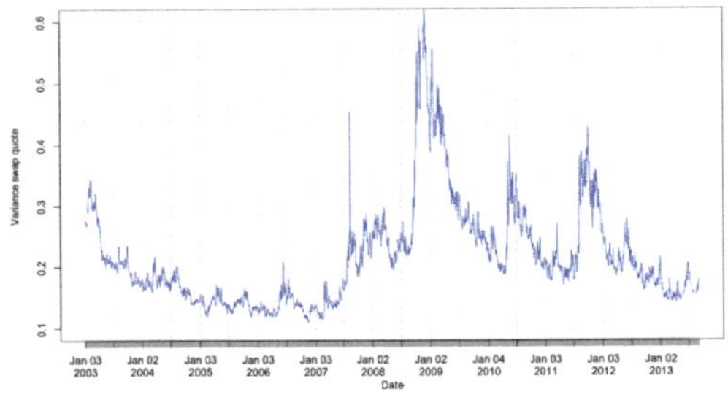

(a) Actual 3-month variance swaps

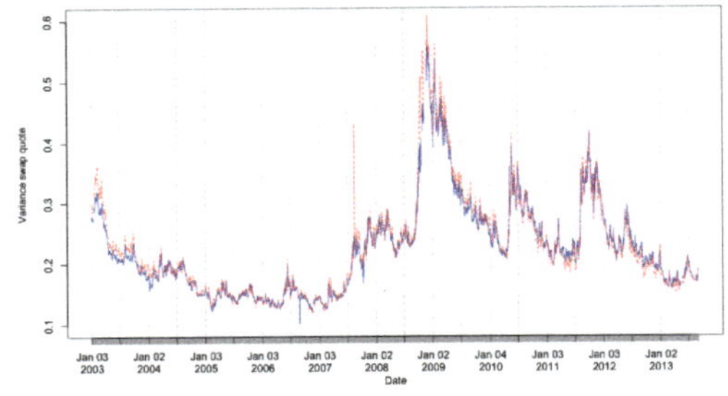

(b) Actual vs. rescaled forecast 6-month variance swaps

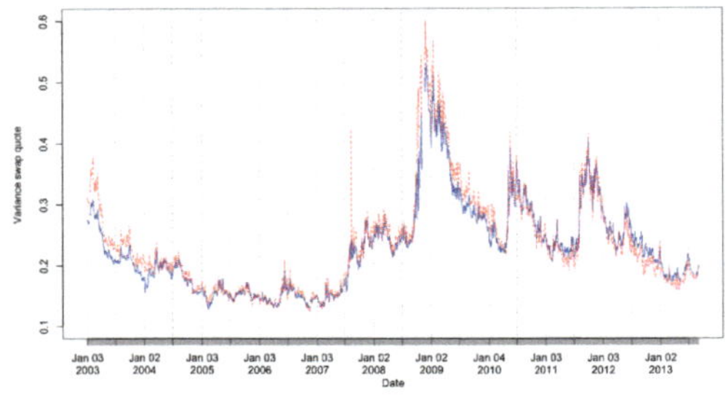

(c) Actual vs. rescaled forecast 9-month variance swaps

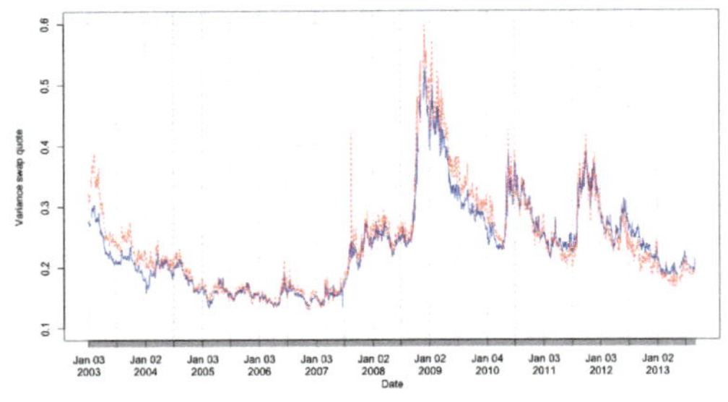

(d) Actual vs. rescaled forecast 12-month variance swaps

Figure 2.12: Plot (a) shows actual (proxy) 3-month variance swap quotes. Plots (b)–(d) show actual variance swap quotes for 6, 9, and 12 months, respectively, in blue, with forecast variance swap quotes multiplied by the 3-month actual to forecast ratio in red.

we may attribute to overnight movements of the index. Recall that realized variance estimates are intraday from open to close. Integrated variance forecasts must therefore be rescaled to obtain close-to-close integrated variance forecasts, as explained, for example, in [105], or alternatively using an econometric model such as the HEAVY model of [249].

In Figure 2.12, we see that the 6-month, 9-month, and 12-month forecasts rescaled by the 3-month ratio of actual to forecast seem to be very consistent with actual variance swap quotes. This implies that although we can only forecast variance swap curves up to a factor, we can accurately forecast their shapes. We now demonstrate this further with two dramatic examples where the variance swap curve moved significantly from one day to the next. We will see that in both these cases, the evolution of the variance swap curve seems to be consistent with our model paradigm.

2.8.1 ▪ The collapse of Lehman Brothers

As an interesting experiment,[34] consider the evolution of the S&P variance swap curve over the weekend of the collapse of Lehman Brothers. As of market close on Friday, September 12, 2008, it was generally expected that Lehman Brothers would be rescued over the weekend. As of market close on Monday, September 15, however, there had been no rescue, and the market was

[34] Suggested by Peter Leoni of KU Leuven.

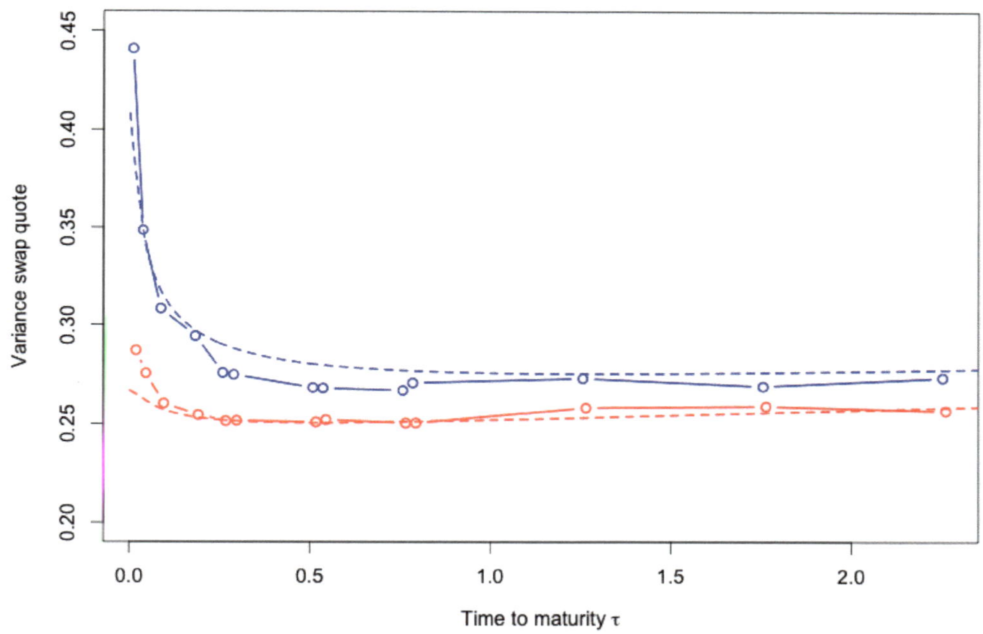

Figure 2.13: S&P variance swap curves as of September 12, 2008 (red) and September 15, 2008 (blue). The dashed curves are RFSV model forecasts rescaled by the 3-month ratio (1.29) as of the Friday close. Time measured in years.

in crisis. In Figure 2.13, we plot the actual variance swap curves as of the Friday and Monday market closes together with forecast curves rescaled by the 3-month ratio as of the close on Friday, September 12, 2008 (which was 1.29). Perhaps surprisingly, it appears that most of the evolution of the variance swap curve may be explained by a single extra data point—intraday realized variance from open to close of trading on Monday, September 15, 2008.

2.8.2 ▪ The Flash Crash

In the so-called Flash Crash of Thursday, May 6, 2010, major U.S. equity indices suddenly dropped by about 10% intraday only to recover within 30 minutes or so. Consequently, intraday realized variance for May 6 was much higher than normal. In Figure 2.14, using the same methodology as in Section 2.8.1, we plot the actual variance swap curves as of the Wednesday and Friday market closes together with forecast curves rescaled by the 3-month ratio as of the close on Wednesday, May 5 (which was 2.52). We see that the actual variance curve as of the close on Friday is consistent with a forecast from the time series of realized variance that *includes* the anomalous price action of Thursday, May 6. In Figure 2.15 we see that actual variance swap curve as of the following Monday close is no longer consistent with the forecast. However, if we drop the May 6 data point, we get a forecast that is much closer to the actual variance swap curve. The obvious explanation is that volatility traders realized over the weekend that the anomalous intraday price action of the Flash Crash should not influence future realized variance projections, adjusting index option quotes accordingly.

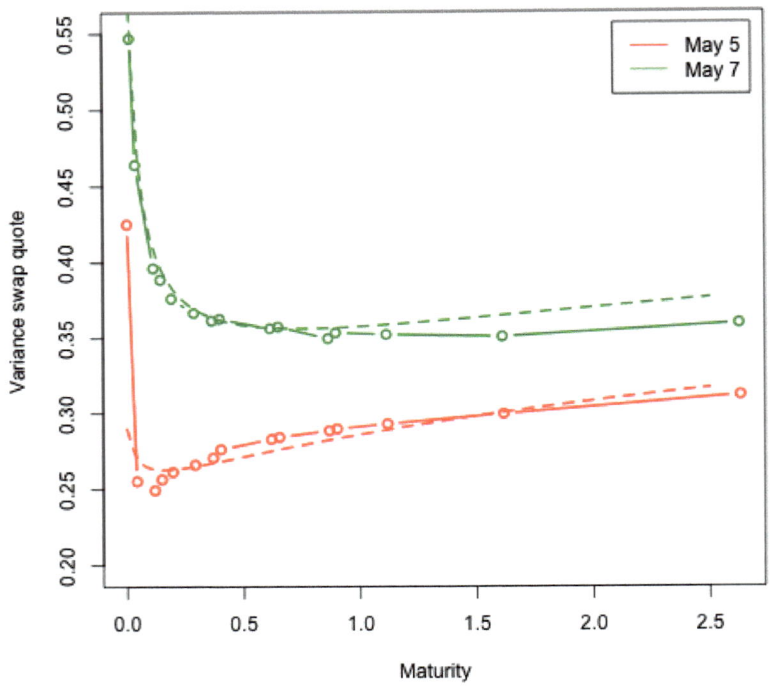

Figure 2.14: S&P variance swap curves as of May 5, 2010 (red) and May 7, 2010 (green). The dashed curves are RFSV model forecasts rescaled by the 3-month ratio (2.52) as of the close on Wednesday, May 5. Time measured in years.

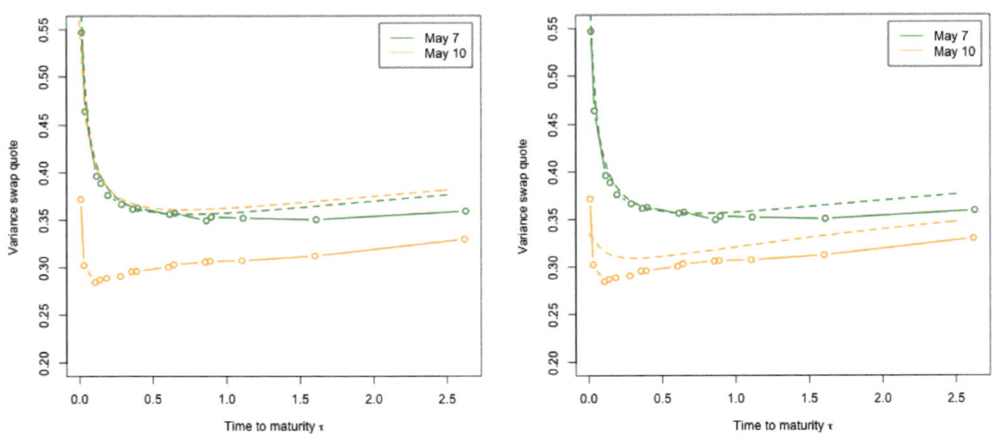

(a) Including Flash Crash

(b) Excluding Flash Crash

Figure 2.15: May 7, 2010 variance curves are in green; May 10 variance curves are in orange. Solid lines are actual curves, and dashed lines are forecast curves. In plot (a), we see that the May 10 actual curve is completely inconsistent with a forecast based on a realized variance dataset that includes the Flash Crash. In contrast, in plot (b), we see that the Monday, May 10 actual curve is consistent with a forecast using a dataset that excludes the Flash Crash.

2.9 ▪ Summary and conclusions

The rough fractional stochastic volatility (RFSV) model of [166] is remarkably consistent with the time series of realized variance for a wide range of different underlying assets. In this chapter, we have shown that this model written under the physical measure $\mathbb{P}$ leads naturally to an options pricing model under $\mathbb{Q}$ that is remarkably consistent with the observed shape of the implied volatility surface in the particular case of SPX. A special case of this model, where we assume a deterministic change of measure between $\mathbb{P}$ and $\mathbb{Q}$, turns out to be a non-Markovian extension of the well-known Bergomi model, which we consequently dub the rough Bergomi (or rBergomi) model. The rBergomi model is particularly tractable and seems to fit the SPX volatility surface very well, despite our lack at this stage of an efficient computational algorithm. We computed terms of Bergomi–Guyon expansion up to second order in volatility of volatility, but the expansion parameter $\lambda = \eta\tau^H \approx 2$ required to fit SPX option prices is too big for this asymptotic expansion to be valid. However, we do not see agreement between the rBergomi model and the term structure of VIX volatility (VVIX).

Taken together, the present work and the econometric analysis of [166] offer a (perhaps even the first) promising paradigm for the understanding of asset price formation all the way from a basic microstructure description at the order book level to option pricing. Last but not least, our framework allows for accurate prediction of the volatility surface from high frequency price data. More efficient computational methods and a more thorough investigation of the market implied change of measure $d\mathbb{Q}/d\mathbb{P}$ are left for future research.

2.10 ▪ Appendix

2.10.1 ▪ Volatility, variance, and VIX

Here we briefly recall some different notions of variance and volatility that are used throughout the chapter. For the sake of concreteness, we assume a general stochastic volatility model

$$\frac{\mathrm{d}S_t}{S_t} = \mu_t \mathrm{d}t + \sqrt{V_t}\, \mathrm{d}Z_t$$

for some continuous stochastic *instantaneous variance* process V_t and a Brownian motion Z_t—all understood under the physical measure $\mathbb{P}$. *Integrated variance* corresponds to the quadratic variation of $\log(S_t)$ and thus is given by $\int_t^T V_s ds$ in the above model. *Realized variance* is the most familiar and widely used of many possible statistical estimators of integrated variance. (See, for example, McAleer and Medeiros [238] and Gatheral and Oomen [169] for more details.) In particular, the Oxford-Man Institute provides a large database of daily estimates of integrated variance for a number of indices, including realized variance and realized kernel estimates.

Forward variance is the expectation under the pricing measure of future instantaneous variance. That is,

$$\xi_t(u) = \mathbb{E}^{\mathbb{Q}}[V_u|\mathcal{F}_t], \; u \geq t.$$

The *variance swap* is a forward contract on realized variance, the fair strike of which is given by

$$\mathcal{V}_t(T) = \int_t^T \xi_t(u)\mathrm{d}u.$$

Volatility is the square-root of variance so, for example, instantaneous volatility is the square-root of instantaneous variance. The CBOE Volatility Index (VIX) is given by a computation that approximates the square-root of the fair strike of a one-month variance swap on the S&P index (for more details see www.cboe.com and, for example, Chapter 11 of [163]). Analogously, the CBOE VVIX index approximates the square-root of the fair strike of a one-month variance swap on the VIX index. Finally, *implied volatility* is the value of the volatility parameter in the Black-Scholes formula required to match the market price of a European option.

2.10.2 ▪ Computation of Bergomi–Guyon autocorrelation functionals in the rBergomi model

Computation of $C^{x\xi}$

$$\begin{aligned} C_t^{x\xi} &= \int_t^T \mathrm{d}s \int_s^T \mathrm{d}u \frac{\mathbb{E}\left[\mathrm{d}x_s \mathrm{d}\xi_s(u)\right]}{\mathrm{d}s} \\ &= \varrho\,\eta\,\sqrt{2H} \int_t^T \mathrm{d}s \int_s^T \mathbb{E}\left[\left.\sqrt{\xi_s(s)}\xi_s(u)\right|\mathcal{F}_t\right] \frac{\mathrm{d}u}{(u-s)^\gamma} \\ &= \varrho\eta\,\sqrt{2H} \int_t^T \sqrt{\xi_t(s)}\mathrm{d}s \int_s^T \xi_t(u) \frac{\mathrm{d}u}{(u-s)^\gamma} + O(\eta^3). \end{aligned} \tag{2.28}$$

Computation of $C^{\xi\xi}$

By definition,

$$\begin{aligned} C^{\xi\xi} &= \int_0^T \mathrm{d}t \int_t^T \mathrm{d}u \int_t^T \mathrm{d}s \frac{\mathbb{E}\left[\mathrm{d}\xi_t(s)\mathrm{d}\xi_t(u)\right]}{\mathrm{d}t} \\ &= \int_0^T \mathrm{d}t \int_t^T \mathrm{d}u \int_t^T \mathrm{d}s \frac{2H\eta^2}{(u-t)^\gamma (s-t)^\gamma} \xi_t(s)\xi_t(u) \\ &= 2H\eta^2 \int_0^T \mathrm{d}t \left(\int_t^T \frac{\xi_t(u)}{(u-t)^\gamma} \mathrm{d}u\right)^2 \\ &= 2H\eta^2 \int_0^T \mathrm{d}t \left(\int_t^T \frac{\xi_0(u)}{(u-t)^\gamma} \mathrm{d}u\right)^2 + O(\eta^4). \end{aligned} \tag{2.29}$$

Computation of C^{μ}

By definition,

$$C^{\mu} = \int_0^T \mathrm{d}t \int_t^T \mathrm{d}u \frac{\mathbb{E}\left[\mathrm{d}x_t \mathrm{d}\xi_t(u)\right]}{\mathrm{d}t} \frac{\delta C_t^{x\xi}}{\delta \xi_t(u)},$$

and from (2.28),

$$\begin{aligned} \frac{\delta C_t^{x\xi}}{\delta \xi_t(v)} &= \varrho\eta\,\sqrt{2H}\left\{\int_t^T \mathrm{d}s\,\sqrt{\xi_t(s)}\frac{1}{(v-s)^\gamma}\mathbf{1}_{v>s} + \frac{1}{2\sqrt{\xi_t(v)}}\int_v^T \xi_t(u)\frac{\mathrm{d}u}{(u-v)^\gamma}\right\} \\ &= \varrho\eta\,\sqrt{2H}\left\{\int_t^v \mathrm{d}s\,\sqrt{\xi_t(s)}\frac{1}{(v-s)^\gamma} + \frac{1}{2\sqrt{\xi_t(v)}}\int_v^T \xi_t(u)\frac{\mathrm{d}u}{(u-v)^\gamma}\right\}. \end{aligned}$$

Thus,

$$
\begin{aligned}
C^{\mu} &= 2H\varrho^2\eta^2 \int_0^T \sqrt{\xi_t(t)}\mathrm{d}t \int_t^T \mathrm{d}u\, \frac{\xi_t(u)}{(u-t)^{\gamma}} \\
&\quad \times \left\{ \int_t^u \mathrm{d}s\, \sqrt{\xi_t(s)} \frac{1}{(u-s)^{\gamma}} + \frac{1}{2\sqrt{\xi_t(u)}} \int_u^T \xi_t(s) \frac{\mathrm{d}s}{(s-u)^{\gamma}} \right\} \\
&= 2H\varrho^2\eta^2 \int_0^T \sqrt{\xi_t(t)}\mathrm{d}t \int_t^T \frac{du}{(u-t)^{\gamma}} \\
&\quad \times \left\{ \int_t^u \sqrt{\xi_t(s)} \frac{\xi_t(u)}{(u-s)^{\gamma}} \mathrm{d}s + \frac{1}{2} \sqrt{\xi_t(u)} \int_u^T \frac{\xi_t(s)}{(s-u)^{\gamma}} \mathrm{d}s \right\} \\
&= 2H\varrho^2\eta^2 \int_0^T \sqrt{\xi_0(t)}\mathrm{d}t \int_t^T \frac{\mathrm{d}u}{(u-t)^{\gamma}} \\
&\quad \times \left\{ \int_t^u \sqrt{\xi_0(s)} \frac{\xi_0(u)}{(u-s)^{\gamma}} \mathrm{d}s + \frac{1}{2} \sqrt{\xi_0(u)} \int_u^T \frac{\xi_0(s)}{(s-u)^{\gamma}} \mathrm{d}s \right\} + O(\eta^4).
\end{aligned}
\tag{2.30}
$$

For any given initial forward variance curve $\xi_0(u)$, the above expressions for $C^{x\xi}$, $C^{\xi\xi}$, and C^{μ} may be easily computed numerically.

2.10.3 ▪ Approximate variance of VIX

Let $y_u = \log V_u$, and consider the following approximation of the arithmetic mean by the geometric mean:

$$
\psi(T) = \frac{1}{\Delta} \int_T^{T+\Delta} \mathbb{E}[V_u | \mathcal{F}_T] \mathrm{d}u \approx \exp\left\{ \frac{1}{\Delta} \int_T^{T+\Delta} \mathbb{E}[y_u | \mathcal{F}_T] \mathrm{d}u \right\}.
$$

Apart from $\mathcal{F}_t$-measurable terms (abbreviated as "drift"), we have

$$
\begin{aligned}
\int_T^{T+\Delta} \mathbb{E}[y_u | \mathcal{F}_T] \mathrm{d}u &= \eta\sqrt{2H} \int_t^T \frac{\mathrm{d}W_s}{(u-s)^{\gamma}} \mathrm{d}u + \text{drift} \\
&= \eta\sqrt{2H} \int_t^T \int_T^{T+\Delta} \frac{\mathrm{d}u}{(u-s)^{\gamma}} \mathrm{d}W_s + \text{drift} \\
&= \eta \frac{\sqrt{2H}}{1-\gamma} \int_t^T \left[(T+\Delta-s)^{1-\gamma} - (T-s)^{1-\gamma} \right] \mathrm{d}W_s + \text{drift}.
\end{aligned}
$$

This gives

$$
\begin{aligned}
\mathbb{V}[\log\psi(T) | \mathcal{F}_t] &\approx \frac{\eta^2 D_H^2}{\Delta^2} \int_t^T \left[(T+\Delta-s)^{H+\frac{1}{2}} - (T-s)^{H+\frac{1}{2}} \right]^2 \mathrm{d}s \\
&= \eta^2 (T-t)^{2H} f^H\!\left(\frac{\Delta}{T-t} \right)
\end{aligned}
$$

where $D_H := \frac{\sqrt{2H}}{H+\frac{1}{2}}$ and

$$
f^H(\theta) = \frac{D_H^2}{\theta^2} \int_0^1 \left[(1+\theta-x)^{H+\frac{1}{2}} - (1-x)^{H+\frac{1}{2}} \right]^2 \mathrm{d}x.
$$

To compute this integral explicitly, we use the fact that for $\kappa = H + \frac{1}{2}$,

$$\int_t^T (T + \Delta - s)^\kappa (T - s)^\kappa \mathrm{d}s = \int_0^\tau (s + \Delta)^\kappa s^\kappa \mathrm{d}s$$
$$= \frac{\tau(\Delta\tau)^\kappa}{1+\kappa} {}_2F_1\left(-H - \frac{1}{2}, H + \frac{3}{2}, H + \frac{5}{2}, -\frac{\tau}{\Delta}\right).$$

Thus, we get

$$\mathbb{V}[\log \psi(T)|\mathcal{F}_t] \approx \frac{2H\eta^2}{\Delta^2}\left[\frac{(\tau+\Delta)^{2(H+1)} - \Delta^{2(H+1)} + \tau^{2(H+1)}}{2(H+1)} - 2\frac{\tau(\tau\Delta)^{H+\frac{1}{2}}}{H+\frac{3}{2}} {}_2F_1\left(-\left(H + \frac{1}{2}\right), H + \frac{3}{2}, H + \frac{5}{2}, -\frac{\tau}{\Delta}\right)\right]$$

with $\tau := T - t$.

Acknowledgments

C.B. and P.F. gratefully acknowledge financial support by Deutsche Forschungsgemeinschaft through Research Grant BA 5484/1-1. We are also very grateful to the anonymous referees for their many detailed comments and suggestions which helped improve our paper.

Chapter 3

No-arbitrage implies power-law market impact and rough volatility[35]

Paul Jusselin, Mathieu Rosenbaum

Introduction

It is now well admitted that volatility is rough. This stylized fact, first established in [166] and confirmed in [51, 232], means that the (log-)volatility process of an asset essentially behaves as a fractional Brownian motion (fBm for short) with Hurst parameter of order 0.1. Recall that an fBm $(W_t^H)_{t\geq 0}$ with Hurst parameter $H \in (0,1)$ is a Gaussian process that can be written under the Mandelbrot–Van Ness representation as

$$W_t^H = \int_{-\infty}^{0} \left((t-s)^{H-\frac{1}{2}} - (-s)^{\frac{1}{2}}\right) \mathrm{d}B_s + \int_0^t (t-s)^{H-\frac{1}{2}} \mathrm{d}B_s,$$

with $(B_t)_{t\geq 0}$ a classical Brownian motion. For any $\varepsilon > 0$, the sample paths of $(W_t^H)_{t\geq 0}$ are almost surely $(H-\varepsilon)$-Hölder (and not H-Hölder). Therefore, the trajectories are very rough when H is small.

Various rough volatility models have been recently introduced in the literature, notably for the purpose of risk management of derivatives, such as the rough Heston model of [129] where the asset price $(P_t)_{t\geq 0}$ satisfies

$$\frac{\mathrm{d}P_t}{P_t} = \sqrt{V_t}\left(\varrho \mathrm{d}B_t^2 + \sqrt{1-\varrho^2}\mathrm{d}B_t^1\right)$$

with

$$V_t = V_0 + \frac{\lambda}{\Gamma(H+\frac{1}{2})} \int_0^t (t-s)^{H-\frac{1}{2}} \left(\theta(t) - V_s\right) \mathrm{d}s + \frac{\nu}{\Gamma(H+\frac{1}{2})} \int_0^t (t-s)^{H-\frac{1}{2}} \sqrt{V_s} \mathrm{d}B_s^1, \qquad (3.1)$$

where B_1 and B_2 are independent Brownian motions, λ and ν two positive constants, θ a deterministic nonnegative function, and $\varrho \in (-1,1)$ a correlation factor. The particular interest of this

[35]This chapter is derived from "No-arbitrage implies power-law market impact and rough volatility," *Mathematical Finance*, 30 (2020), pp. 1309–1336, https://doi.org/10.1111/mafi.12254.

model is that, as for the classical Heston model, semi-explicit pricing and hedging formulas can be obtained; see [129, 131].

A puzzling question is the origin of the universal rough volatility property of financial assets. A first explanation is proposed in [126]. In this work, the authors place themselves in a highly endogenous market, meaning that most orders are sent in reaction to other orders and without economic motivation. They show that in this context, the widely used trading practice of metaorders splitting (see below for the definition of a metaorder) leads to the rough Heston dynamic (3.1) for the macroscopic price. However, this result is found using a quite specific parametric model for the high frequency price.

In this paper, we wish to obtain a fundamental explanation underlying the rough volatility property. In fact, we prove that in a quite general framework, rough volatility is simply a consequence of the no-arbitrage principle together with the existence of market impact.

Market impact is the fact that, on average, a buy order moves the price up and a sell order moves the price down. There are two main viewpoints on market impact. The first is to consider that agents receive information and reveal their information through transactions, hence impacting the market. The other is purely mechanical, not referring to any notion of information, and considering that prices move up and down only through volume pressures, following supply and demand. We adopt the mechanical paradigm in this work. Because the impact of a single order is very difficult to assess, one usually considers large sets of orders split by brokers, so-called metaorders. Empirical studies of market impact have shown that for a buy metaorder (and symmetrically for a sell metaorder) market impact can be decomposed in two phases: a transient phase with a concave rise of the price during the metaorder execution, and a decay phase, where the price decreases towards a long-term level after the execution is completed; see [24, 272, 120, 230].

Let us consider a buy (say) metaorder and let $(q_t)_{t\geq 0}$ be the cumulative volume of this metaorder executed between the initial time 0 and time t. The market impact function of this metaorder is defined as

$$\mathrm{MI}(t) = \mathbb{E}\left[P_t^{(q_s)_{s\leq t}} - P_0\right],$$

where we put the superscript $(q_s)_{s\leq t}$ on P to insist on the fact that the price dynamic depends on the execution process of the metaorder. Of course, the above formula only makes sense in a model where $P_t^{(q_s)_{s\leq t}}$ is a well-defined stochastic process, as will be the case in the next sections.

The permanent market impact (PMI for short) of this metaorder is given by the quantity

$$\mathrm{PMI} = \lim_{t\uparrow\infty} \mathrm{MI}(t).$$

Intuitively, it is quite clear that in the long run, the permanent impact of a metaorder with volume Q should be the same as that of two consecutive orders of volume $Q/2$. This is formalized in [205, Thm. 1] and further developed in [164] where it is shown that under mild modeling assumptions, the absence of price manipulation on a market[36] implies that the PMI is proportional to the total volume of the metaorder. In particular, it does not depend on the metaorder execution strategy. From now on, we take this linear PMI property for granted. This has consequences for the price dynamics. In particular, we now assume that the price P is a martingale. We take this martingale hypothesis as a simplifying and convenient version of the classical mathematical finance condition of no almost sure arbitrage which states that price should be a semimartingale. In this setting, it is shown in Theorem 2.1 in [213] that under the purely mechanical view for market impact, up to a normalizing constant,

$$P_t = \lim_{s\uparrow\infty} \mathbb{E}\left[V_s^a - V_s^b|\mathcal{F}_t\right], \tag{3.2}$$

[36]A price manipulation is a round-trip (strategy starting and finishing with null inventory) whose expected cost is negative.

where V^a (resp., V^b) is the cumulated volume of buy (resp., sell) market orders since the initial time 0, and $(\mathcal{F}_t)_{t\geq 0}$ corresponds to the filtration generated by the order flow process. Hence, the price moves when orders arrive on the market because market participants revise their anticipation about the long-term cumulative imbalance of the order flow. Note that to derive (3.2), we only use the set of assumptions that we consider here as our no-arbitrage conditions: martingale price and linear permanent market impact.

As for the transient part of the market impact, empirical measurements show that, provided the execution rate of the metaorder is relatively constant, the function *MI* is close to a power law with respect to time, that is, $\mathrm{MI}(t) \sim t^{1-\alpha}$ with $\alpha \in (0, 1)$; see [24, 65, 230, 272]. More precisely, the coefficient α is found to be about $\frac{1}{2}$ so that the so-called square-root law is approximately satisfied. Actually, it is proved in [257] that under some leverage neutrality assumption, the square-root law can be simply derived from dimensional analysis.

We show in this work that under the no-arbitrage assumption (represented by the linear permanent impact, the martingale price, and thus (3.2)), the market impact function must indeed be a power law of the form $\mathrm{MI}(t) \sim t^{1-\alpha}$. Then we prove that for any $\alpha \in (0, 1)$, the scaling limit of the price (3.2) exists and satisfies

$$\widehat{P}_t = B_{X_t}$$

with

$$X_t = \frac{2}{\delta}\int_0^t F^{\alpha,\lambda}(s)\mathrm{d}s + \frac{1}{\delta\sqrt{\lambda}}\int_0^t F^{\alpha,\lambda}(t-s)\mathrm{d}W_{X_s}, \tag{3.3}$$

where W and B are two Brownian motions, δ and λ are two positive constants and $F^{\alpha,\lambda}$ is the Mittag–Leffler cumulative distribution function; see Appendix 3.4.1 for definition. The correlation between the Brownian motions B and W is stochastic and related to the order flow imbalance. The above equation is a generalization of the rough Heston model (3.1). Indeed, we can show that when $\alpha > \frac{1}{2}$, after differentiation, equation (3.3) can be rewritten under the form of (3.1) (up to a stochastic correlation factor) with associated Hurst parameter $H = \alpha - \frac{1}{2}$. For $\alpha \leq \frac{1}{2}$, we prove that X is not continuously differentiable but has Hölder regularity $2\alpha - \varepsilon$ for any $\varepsilon > 0$. Therefore, we call (3.3) the *hyper-rough Heston model* when $\alpha \leq \frac{1}{2}$. Hence, we are able to define rough Heston models for Hurst parameter in $(-\frac{1}{2}, \frac{1}{2}]$.

To obtain our results, we only need to specify a model for the order flow dynamics. We indeed see from equation (3.2) that we do not need to model metaorders individually. Only the aggregated order flow matters in order to derive the price dynamic. More precisely, we consider for buy and sell market order arrivals two independent Hawkes processes N^a and N^b and assume that each order is of unit size; see [126, 215]. Recall that a Hawkes process N is a self-exciting point process whose intensity $(\lambda_t)_{t\geq 0}$ is defined by

$$\lambda_t = \mu + \int_0^t \varphi(t-s)\mathrm{d}N_s,$$

with μ a positive constant and φ a nonnegative locally integrable function. Such a dynamic is a generalization of the Poisson process which is usually considered when modeling order flows; see, among others, [101, 102, 270]. It is nonparametric and very flexible so that it is really reasonable to assume that the actual order flow can be well approximated by a Hawkes-based model; see [26, 25]. Note that we will not put any restriction on the Hawkes parameters μ and φ, except the nonnegativity of μ, the local integrability of φ, and the fact that they are the same for the buy and sell flows. In this case, it is shown in [213] that the price process (3.2) satisfies

$$P_t = P_0 + \int_0^t \xi(t-s)\mathrm{d}(N^a - N^b)_s, \tag{3.4}$$

with

$$\xi(t) = 1 + \left(1 + \int_0^{+\infty} \psi(u)\mathrm{d}u\right) \int_t^{+\infty} \varphi(u)\mathrm{d}u \tag{3.5}$$

and

$$\psi = \sum_{i \geq 1} (\varphi)^{*i},$$

where $(\varphi)^{*1} = \varphi$, and, for $k \geq 2$, $(\varphi)^{*k}$ denotes the convolution product of $(\varphi)^{*(k-1)}$ with φ.

Using a rescaling procedure to describe the macroscopic behavior of (3.4), we show that only one very subtle specification of the Hawkes processes can lead to a non-trivial market impact, which has to be power law. Furthermore, it implies that the market is highly endogenous. In addition, depending on the market impact shape, the scaling limit of the price is a rough or hyper-rough Heston model (3.3), with a one-to-one correspondence between the exponent of the impact function and the Hurst parameter of the volatility.

The chapter is organized as follows. In Section 3.1, we show that under the assumption that the market impact function is not degenerate, it can only be a power law with parameter $1-\alpha$ for some $\alpha \in (0,1)$. Then in Section 3.2, we explain that the macroscopic limit of (3.4) is a rough or hyper-rough Heston model with Hurst parameter $H = \alpha - \frac{1}{2}$.

3.1 ▪ Market impact is power law

In this section, we show that if there exists a nondegenerate market impact function, it has to be a power law. Moreover, we will see that it implies a highly endogenous market. By nondegenerate we essentially mean a market impact function which is ultimately decreasing for buy metaorders (and conversely for sell metaorders); see Assumption 3.1. This is the formalization of the two-phase behavior of market impact discussed in the introduction.

3.1.1 ▪ Asymptotic framework and metaorders modeling

Let T be our final horizon time for the metaorders we will define in the sequel. Recall that the market order flow on $[0,T]$ (and after T) is given by two Hawkes processes with the same parameters, N^a for the buy market orders and N^b for the sell orders, with respective intensities λ^a and λ^b. As the time-length of a metaorder is typically large compared to the interarrivals of individual market orders, it is natural to consider that T goes to infinity.

We want to work in a general setting which enables us to be compatible with empirical studies showing that markets are highly endogenous. In the Hawkes process context, the degree of endogeneity of the market is measured by the L^1 norm of φ, denoted by $\|\varphi\|_1$; see [132, 197, 215, 214]. Therefore, a highly endogenous market corresponds to the case where $\|\varphi\|_1$ is close but smaller than unity. So we allow the model parameters to possibly depend on T. Thus, from now on, we use the superscript T for all quantities that could depend on T. In particular, $\|\varphi^T\|_1$ may go to one as T tends to infinity. We also write $N^{a,T}$, $N^{b,T}$, μ^T, φ^T to describe the market order flow and model parameters corresponding to the time horizon T, and we set $\varphi^T = a^T\varphi$ for φ a nonnegative function such that $\|\varphi\|_1 = 1$ and $(a^T)_{T\geq 0}$ is a real sequence in $(0,1)$. Note that we do not impose that a^T goes to one. In fact, we will show that "a^T tends to one" is not needed for the existence of a nondegenerate market impact function.

We finally need to define a formalism for a sequence of buy (say) metaorders which will be added to the global order flow. We assume that a metaorder is split through market orders of size one over $[0,T]$. In the spirit of [213], we consider that the arrival times of the market orders are given by a nonhomogeneous Poisson process with intensity

$$\nu^T(t) = I^T f\left(\frac{t}{T}\right),$$

where f is a nonnegative continuous function on $[0,1]$ with integral one, and I^T is a sequence of nonnegative real numbers such that the expected total volume of the metaorder is $I^T T$. The order of magnitude of its duration is T. Note that this is slightly smaller than T as the metaorder will end after the last jump time of the Poisson process before T. We allow f to be different from a constant to get more realistic splitting schemes than those given by constant rate Poisson processes. Suitable choices for f may be exponentially decaying (arrival price benchmark) or linearly decaying (VWAP benchmark); see, for example, [6, 204].

To compute the market impact function in practice, one typically considers the empirical mean of the price movements over many metaorders with similar durations and volumes counted in proportion of the total traded volume. So in our setting, it is natural to take $I^T \times T$ (the order of magnitude of the total volume of our metaorder) to be essentially proportional to the total number of other orders executed over $[0,T]$. To do so we take the intensities proportional

$$I^T = \gamma\beta^T, \text{ with } \beta^T = \mu^T(1-a^T)^{-1},$$

where $\gamma < 1$, and β^T is the long-term average intensity of the Hawkes process $N^{a,T}$ ($\beta^T = \lim_{t\uparrow\infty}(1/t)\int_0^t \lambda_s^a \mathrm{d}s$; see [23]). Thus, the proportion of the order flow which is due to the considered metaorder is essentially $\gamma/(1+\gamma)$, and γ will be considered reasonably small.

3.1.2 ▪ Market impact in the Hawkes setting

In this section the parameter T is fixed. Assuming that the volume of our metaorder is small enough, the total order flow is not deeply modified by it. Hence, other agents do not observe significant changes in the order flow dynamics. So the way the market reacts to the incoming orders remains unchanged. Recall that in our model, the market reaction to the order flow (without our metaorder) is given by (3.4).

We work under the setting of the previous section assuming that the number of shares bought through our metaorder is a nonhomogeneous Poisson process $(n_t^T)_{t\geq 0}$. Therefore we obtain

$$P_t^T = P_0 + \int_0^t \xi^T(t-s)\mathrm{d}(N^{a,T} - N^{b,T} + n^T)_s,$$

where $(N_t^{a,T}, N_t^{b,T})_{t\geq 0}$ corresponds to the aggregated order flows of all other agents. Indeed, all orders being anonymous in the market, our metaorder cannot be distinguished from the global order flow. Thus, the market digests the order flow through the kernel ξ, as if it is a bivariate Hawkes process with parameters μ and φ (γ being small).

Now we are in position to properly compute the market impact function of our metaorder. We have

$$\mathrm{MI}^T(t) = \mathbb{E}[P_t^T - P_0] = \int_0^t \xi^T(t-s)\mathbb{E}[\mathrm{d}n_s^T].$$

This equation, together with (3.5), shows that for any $t \geq 0$, the market impact function can be decomposed into two terms as follows:

$$\mathrm{MI}^T(t) = \mathrm{PMI}^T(t) + \mathrm{TMI}^T(t),$$

where

$$\mathrm{PMI}^T(t) = \mathbb{E}[n_t^T]$$

and

$$\mathrm{TMI}^T(t) = \int_0^t \Gamma^T(t-s)\mathbb{E}[\mathrm{d}n_s^T],$$

with

$$\Gamma^T(s) = (1 - a^T)^{-1} \int_s^{+\infty} \varphi^T(u)\mathrm{d}u,$$

where we have used the fact that

$$\int_0^{+\infty} \psi^T(u)\mathrm{d}u = \sum_{k\geq 1}^{+\infty} \left(\int_0^{+\infty} \varphi^T(u)\mathrm{d}u \right)^k = \frac{a^T}{1 - a^T}.$$

Note that the definition of $\mathrm{PMI}^T(t)$ is compatible with that of the PMI given in the introduction. Indeed, with the order intensity from our metaorder being eventually null, and because $\Gamma^T(t)$ tends to zero as t goes to infinity, we get

$$\lim_{t\uparrow\infty} \mathrm{TMI}^T(t) = 0.$$

The effect of the term TMI^T is thus only temporary. That is why this term is called the transient part of the market impact.

3.1.3 ▪ Scaling limit of the market impact

We now rescale the market impact function as the horizon time T goes to infinity. If the sequence of rescaled market impact functions converges, we call its limit the macroscopic market impact function.

First, we reparameterize in time and consider $(\mathrm{MI}^T(f, tT))_{t\in\mathbb{R}_+}$ (we set the function f as a parameter of MI^T to insist on the fact that the market impact function depends on the metaorder strategy). Thus, $t = 1$ corresponds essentially to the end of the metaorder. Regarding the scaling in space, because in our framework the size of a metaorder is measured relative to the total volume, which is of order $T\beta^T$ on $[0, T]$, we finally define our rescaled market impact function $\overline{\mathrm{MI}}^T$ on $\mathbb{R}_+$ by

$$\overline{\mathrm{MI}}^T(f, t) = \frac{\mathrm{MI}^T(f, tT)}{T\beta^T} = \overline{\mathrm{PMI}}^T(f, t) + \overline{\mathrm{TMI}}^T(f, t),$$

with

$$\overline{\mathrm{PMI}}^T(f, t) = \gamma \int_0^t f(x)\mathrm{d}x$$

as the rescaled permanent impact and

$$\overline{\mathrm{TMI}}^T(f, t) = \gamma \frac{a^T(1 - a^T)^{-1}}{T} \int_0^{Tt} f\left(t - \frac{x}{T}\right) \int_x^{+\infty} \varphi(u)\mathrm{d}u\mathrm{d}x$$

as the rescaled transient impact. Note that the permanent impact term does not depend on T. Thus, there always exists a macroscopic PMI function, and the convergence of the sequence $(\overline{\mathrm{MI}}^T(f, \cdot))_{T\geq 0}$ is equivalent to that of $(\overline{\mathrm{TMI}}^T(f, \cdot))_{T\geq 0}$. Motivated by the empirical results on market impact [24, 65, 175, 230, 66] discussed in the introduction, we make the following natural assumption.

Assumption 3.1. *For constant execution rate, that is, $f = \mathbf{1}_{[0,s]}$ for some $s \in (0, 1]$, the scaling limit of the market impact function exists pointwise and is nonincreasing after time s. Furthermore, there exists $t > s$ such that the value of this limiting function at time t is smaller than that at time s.*

There do exist some model parameters such that Assumption 3.1 is satisfied, for example, any kernel φ such that $\varphi(t) \sim_{+\infty} ct^{-\alpha-1}$ with $c > 0$. Assumption 3.1 implies that for $f = \mathbf{1}_{[0,s]}$ with $s \in (0, 1]$, we can define the macroscopic market impact function $\widehat{\mathrm{MI}}(f, t)$ and its transient and permanent components $\widehat{\mathrm{TMI}}(f, t)$ and $\widehat{\mathrm{PMI}}(f, t)$ as

$$\widehat{\mathrm{MI}}(f,t) = \lim_{T\uparrow\infty} \overline{\mathrm{MI}}^T(f,t),$$
$$\widehat{\mathrm{TMI}}(f,t) = \lim_{T\uparrow\infty} \overline{\mathrm{TMI}}^T(f,t),$$
$$\widehat{\mathrm{PMI}}(f,t) = \lim_{T\uparrow\infty} \overline{\mathrm{PMI}}^T(f,t).$$

Using Tauberian theorems (see Appendix 3.4.2), we obtain the following result.

Theorem 3.2. *Under Assumption* 3.1, *for any nonnegative function f defined on $\mathbb{R}_+$, continuous on* $[0, 1]$, *and supported on* $[0, 1]$, *the macroscopic market impact function and its transient part exist. More precisely, there exists a parameter $\alpha \in (0, 1]$ such that for any $t > 0$, when $\alpha < 1$,*

$$\lim_{T\uparrow\infty} \overline{\mathrm{TMI}}^T(f,t) = \gamma K(1-\alpha)\int_0^t f(t-u)u^{-\alpha}\mathrm{d}u, \tag{3.6}$$

for some $K > 0$, and when $\alpha = 1$,

$$\lim_{T\uparrow\infty} \overline{\mathrm{TMI}}^T(f,t) = \gamma K f(t). \tag{3.7}$$

Furthermore, the Hawkes kernel φ necessarily satisfies

$$\int_0^t \int_s^{+\infty} \varphi(u)\mathrm{d}u\mathrm{d}s = t^{1-\alpha} L(t),$$

where L is a slowly varying function (see definition in Appendix 3.4.2*). Finally, we necessarily have*

$$(1-a^T)^{-1}T^{-\alpha}L(T) \to K,$$

and consequently $a^T \to 1$ (see Proposition 3.17 *in Appendix* 3.4.2*).*

Considering, for example, $f = \mathbf{1}_{[0,1]}$, Theorem 3.2 shows that under no-arbitrage, together with the assumption of the existence of the macroscopic market impact function, the transient part of the market impact is power law, while the permanent part is linear. Moreover, equation (3.6) shows that the decay of the market impact is essentially a power law with exponent $-\alpha$; see Figure 3.1 for an illustration.

We see that the connection between a_T and T is completely specified in Theorem 3.2. For given α, there is only one asymptotic regime leading to a nontrivial limit. Note that the fact that a_T goes to one implies that the nonlinear transient part of the market impact (case $\alpha < 1$) can arise only in a highly endogenous market. This is actually very natural for the following reason. A nonlinear transient impact means that the market reacts differently to a child order in the end of the metaorder compared to a child order in the beginning of the metaorder. For this to be possible, correlations in the order flow must survive all over the time-length of our metaorder. In our probabilistic setting, using the population approach to Hawkes processes (see [28, 132, 214, 215]), it is easily seen that such a property can only hold provided a_T goes to one.

In this regard the case $\alpha = 1$ is quite degenerate because the market somehow has no memory and reacts the same way to market orders, independently of their positions within the metaorder. Even more, the price instantaneously decreases to its permanent level when the metaorder is completed. This means that the market is able to detect instantaneously the end of a metaorder, which seems unrealistic and incompatible with empirical measurements.

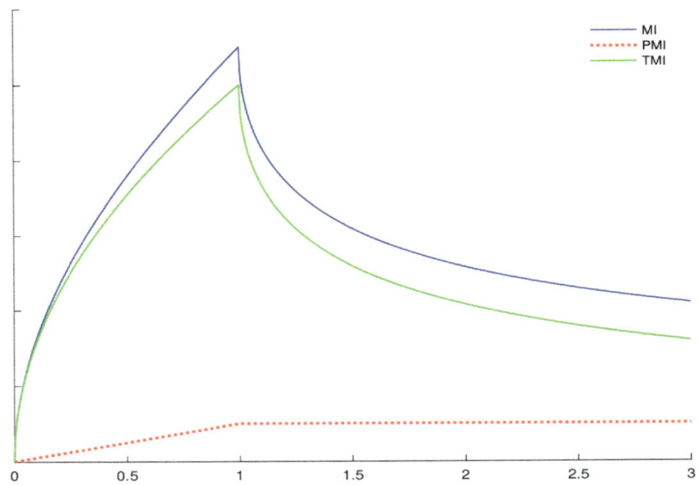

Figure 3.1: Illustration of the decomposition of the macroscopic market impact function for a metaorder executed uniformly over $[0, 1]$, with $\alpha = 0.5$. Time is on the x-axis.

3.2 ▪ Macroscopic limit of the price

We finally show in this section that under Assumption 3.1, the macroscopic price, that is, the limit as time goes to infinity of the properly rescaled microscopic price (3.4), is diffusive with rough or hyper-rough volatility. Moreover, we make explicit the link between the market impact shape exponent and the Hurst parameter of the volatility.

3.2.1 ▪ Scaling limit of the price process

We start with an assumption which is necessary to get a nontrivial long-term limit for the price (3.4).

Assumption 3.3. *For some $\delta > 0$, we have*

$$(1 - a^T)\mu^T T \underset{T\uparrow\infty}{\to} \delta.$$

Assumption 3.3 is classical in the context of Hawkes processes with kernel whose L^1 norm tends to one; see [215]. Indeed, it ensures that the number of events does not explode asymptotically.

According to equation (3.2), price and volume are homogeneous. Therefore we rescale the price the same way as the metaorders. Taking for simplicity and without loss of generality $P_0 = 0$, for $t \in [0, 1]$ we define

$$\overline{P}_t^T = \frac{1}{T\beta^T} P_{tT}^T = \frac{1 - a^T}{T\mu_T} \int_0^t \xi^T\big(T(t-s)\big)\mathrm{d}\big(N_{tT}^{a,T} - N_{tT}^{b,T}\big),$$

where

$$\xi^T(t) = \left(1 + \int_0^{+\infty} \psi^T(u)\mathrm{d}u\right)\left(1 - \int_0^t \varphi^T(u)\mathrm{d}u\right).$$

Let α be the parameter of the market impact function in Theorem 3.2, K the constant introduced in equations (3.6) and (3.7), and $\lambda = (K\Gamma(2-\alpha))^{-1}$. We have the following result for the macroscopic limit of the price process, whose proof is given in Section 3.3.2.

Theorem 3.4. *Under Assumptions* 3.1 *and* 3.3, *the sequence of rescaled price processes* $(\overline{P}^T)_{T\geq 0}$ *converges in law for the Skorokhod topology towards a process* $\widehat{P}$ *such that for* $t \in [0,1]$,

$$\widehat{P}_t = \frac{1}{\sqrt{\delta}}\left(B^a_{X^a_t} - B^b_{X^b_t}\right),$$

where B^a *and* B^b *are two independent Brownian motions such that* $B^a_{X^a_t}$ *and* $B^b_{X^b_t}$ *are two martingales,* X^a *is increasing and satisfies*

$$X^a_t = \int_0^t F^{\alpha,\lambda}(s)\mathrm{d}s + \frac{1}{\sqrt{\delta\lambda}}\int_0^t F^{\alpha,\lambda}(t-s)\mathrm{d}B^a_{X^a_s},$$

and X^b *is increasing and a solution of the same equation as above, replacing the superscript* a *by* b. *In particular, there exists a Brownian motion* W *such that the integrated variance* $X = (X^a + X^b)/\delta$ *of* $\widehat{P}$ *is a solution of the stochastic rough Volterra equation*

$$X_t = \frac{2}{\delta}\int_0^t F^{\alpha,\lambda}(s)\mathrm{d}s + \frac{1}{\delta\sqrt{\lambda}}\int_0^t F^{\alpha,\lambda}(t-s)\mathrm{d}W_{X_s}. \tag{3.8}$$

Moreover, for any $\varepsilon > 0$, *the process* X *has Hölder regularity* $1 \wedge (2\alpha - \varepsilon)$. *It is continuously differentiable for* $\alpha > \frac{1}{2}$ *and not continuously differentiable for* $\alpha \leq \frac{1}{2}$.

Theorem 3.4 shows that the no-arbitrage principle, together with the existence of market impact, implies that the macroscopic price[37] is a diffusive process whose cumulative variance is a solution of a stochastic rough Volterra equation (except when $\alpha = 1$ which corresponds to the classical Heston model; see Corollary 3.5). Note that X plays the role of an integrated variance and that when $\alpha \leq \frac{1}{2}$, it is not continuously differentiable. Thus, in that case, the spot variance is not well-defined, and only its integrated version makes sense. This is why for $\alpha \leq \frac{1}{2}$, we call this model the hyper-rough volatility model (more precisely, hyper-rough Heston model; see below).

From Theorem 3.2 in [215], we have that for $\alpha > \frac{1}{2}$, the process X^a is almost surely differentiable, and its derivative Y^a is the unique solution of

$$Y^a_t = \frac{\lambda}{\Gamma(\alpha)}\left(\int_0^t (t-s)^{\alpha-1}(1-Y^a_s)\mathrm{d}s + \frac{1}{\sqrt{\delta\lambda}}\int_0^t (t-s)^{\alpha-1}\sqrt{Y^a_s}\mathrm{d}B^a_s\right).$$

The same result holds for Y^b, replacing the superscript a by b. We deduce that when $\alpha > \frac{1}{2}$, the integrated volatility admits a derivative, and the macroscopic limit of the price follows a rough Heston model. More precisely, we have the following corollary.

Corollary 3.5. *When* $\alpha > \frac{1}{2}$, *the process* X *is differentiable almost surely, and its derivative* Y *is the unique solution of the stochastic rough Volterra equation*

$$Y_t = \frac{Y^a_t + Y^b_t}{\delta} = \frac{\lambda}{\Gamma(\alpha)}\left(\int_0^t (t-s)^{\alpha-1}\left(\frac{2}{\delta} - Y_s\right)\mathrm{d}s + \frac{1}{\delta\sqrt{\lambda}}\int_0^t (t-s)^{\alpha-1}\sqrt{Y_s}\mathrm{d}W_s\right),$$

[37] Note that under our completely symmetric setting, price can become negative. This is, however, obviously not very important for our purposes here.

with W a Brownian motion. Furthermore, the dynamic of the price $\widehat{P}$ is

$$\mathrm{d}\widehat{P}_t = \frac{1}{\sqrt{\delta}}\left(\sqrt{Y_t^a}\mathrm{d}B_t^a - \sqrt{Y_t^b}\mathrm{d}B_t^b\right).$$

This result highlights the fact that at the macroscopic limit, the correlation ϱ_t between the two Brownian motions driving price and volatility is stochastic. More precisely, we have

$$\varrho_t = \frac{Y_t^a - Y_t^b}{Y_t^a + Y_t^b}.$$

Hence, the correlation sign depends on that of $Y_t^a - Y_t^b$. The process Y^a (resp., Y^b) corresponding to the volatility of the ask (resp. bid) side of the market (see Step 4 in Section 3.3.2), this can be interpreted in terms of order flow dynamics. Indeed, suppose that $Y^a \gg Y^b$ and that price is increasing. Then the instantaneous imbalance has the same sign as price returns. Thus, the volatility increases as the order flow excites the price dynamic. Conversely, if the price increases and $Y^a \ll Y^b$, the volatility decreases as the order flow tends to compensate the upward price variation.

To prove the convergence in law in Theorem 3.4, we show that $(\overline{P}^T)_{T\geq 0}$ is tight and that all limit points have the same law. This is done using the characteristic function of Hawkes processes in the spirit of [129]. A direct proof would consist of obtaining uniqueness in law for solutions of equation (3.8) as done in [4] for $\alpha > \frac{1}{2}$. However, such an approach seems quite intricate to adapt for $\alpha \leq \frac{1}{2}$. We have the following result, whose proof is given in Section 3.3.3.

Theorem 3.6. *Let X be the cumulated variance process given in Theorem 3.4 and h a continuously differentiable function from $\mathbb{R}_+$ to $\mathbb{R}$ such that $h(0) = 0$. The function*

$$\psi(h,t) = \mathbb{E}\left[\exp\left\{\mathrm{i}\int_0^t h(t-s)\mathrm{d}X_s\right\}\right]$$

satisfies

$$\psi(h,t) = \exp\left\{\int_0^t g(s)\mathrm{d}s\right\},$$

with g the unique continuous solution of the Volterra–Riccati equation

$$g(t) = \int_0^t f^{\alpha,\lambda}(t-s)\left(\delta^{-1}\frac{1}{4}g(s)^2 + \delta^{-1}2\mathrm{i}h(s)\right)\mathrm{d}s, \tag{3.9}$$

where $f^{\alpha,\lambda}$ is the Mittag–Leffler density function; see Appendix 3.4.1.

Theorem 3.6 extends some already known results about characteristic functions related to rough Heston models for $\alpha > \frac{1}{2}$; see [128, 129, 4]. Note that the characteristic function of the macroscopic price process $\widehat{P}_t$ can also be obtained using the same type of proof as that for Theorem 3.6.

3.2.2 ▪ Conclusion

We have considered three main assumptions in this work:

- No-arbitrage, in the sense that the price process is a martingale and PMI is linear.

- Existence of market impact with a transient component.
- The order flow can be fitted by a Hawkes process (with no restriction on the Hawkes parameters).

We have shown that in this quite general framework, the market impact function can only be a power law with exponent $1-\alpha$ for some $\alpha \in (0,1)$ (here we drop the case $\alpha = 1$ which leads to a somehow degenerate market impact function). The parameter α also appears necessarily in the tail of the kernel of the Hawkes process driving the order flow: $\varphi(x) \sim x^{-(1+\alpha)}$ as x goes to infinity. Furthermore, this also implies that the market is highly endogenous. Even more interestingly, we obtain that the macroscopic behavior of the price is that of a rough or hyper-rough Heston model with Hurst parameter $H = \alpha - \frac{1}{2}$.

The relationship between market impact, tail of Hawkes kernel, and volatility Hurst parameter enables us to confront our results to empirical measurements. In [24] it is found that the market impact function fits a power law with exponent 0.45. In [166] it is shown that volatility is rough with a Hurst parameter of order 0.1. Finally, in [197] the authors calibrate a Hawkes process on market orders arrival and obtain that the kernel decays as a power-law function with exponent around -1.45. All these measurements are compatible with our results (and suggest that market impact is close to square-root).

3.3 ▪ Proofs

3.3.1 ▪ Proof of Theorem 3.2

Let $f = \mathbf{1}_{[0,s]}$, $s \in (0,1]$. From Assumption 3.1, we have the pointwise convergence of $(\overline{\mathrm{MI}}^T(f,\cdot))_{T\geq 0}$. As previously explained, this is equivalent to the convergence of $(\overline{\mathrm{TMI}}^T(f,\cdot))_{T\geq 0}$. Moreover, with $(\overline{\mathrm{PMI}}^T(f,\cdot))_{T\geq 0}$ being independent of T, Assumption 3.1 implies that the sequence of functions

$$\overline{\mathrm{TMI}}^T(f,t) = \gamma \int_0^t a^T(1-a^T)^{-1} \int_{yT}^{+\infty} \varphi(u)\mathrm{d}u f(t-y)\mathrm{d}y \tag{3.10}$$

converges pointwise. With the function φ being nonnegative and integrable, $\overline{\mathrm{TMI}}^T(f,\cdot)$ is nonnegative, nondecreasing, and concave on $[0,s]$ and then nonincreasing. Hence, $\overline{\mathrm{TMI}}^T(f,\cdot)$ reaches its maximum in s. By pointwise convergence, $\widehat{\mathrm{TMI}}(f,\cdot)$ has the same properties. Because we have assumed that $\widehat{\mathrm{MI}}(f,t) < \widehat{\mathrm{MI}}(f,s)$ for some $t > s$ and $\widehat{\mathrm{PMI}}(f,\cdot)$ is nondecreasing, we deduce that $\widehat{\mathrm{TMI}}(f,s) > 0$.

Let $g(t) = \gamma^{-1}\widehat{\mathrm{TMI}}(\mathbf{1}_{[0,t]},t)$ for $t \in (0,1]$, and consider

$$R(t) = \int_0^t \int_y^{+\infty} \varphi(u)\mathrm{d}u\mathrm{d}y > 0.$$

According to equation (3.10), for $t \in (0,1]$ we have

$$\frac{R(Tt)}{R(T)} \underset{T\uparrow\infty}{\longrightarrow} \frac{g(t)}{g(1)} > 0. \tag{3.11}$$

By the characterization theorem (see Theorem 3.18 in Appendix 3.4.2), we deduce that the previous limit holds for all $t > 0$ with some suitable extension of the function g. Moreover, there exist some $\beta \in \mathbb{R}$, $K > 0$, and L a slowly varying function such that for $t > 0$,

$$g(t) = Kt^{\beta}, \ R(t) = L(t)t^{\beta}.$$

Note that for $t \in (0,1]$, we have $g(t) = \widehat{\mathrm{TMI}}(\mathbf{1}_{[0,1]}, t)$, which is concave. Thus $\beta \in [0,1]$. Taking $s = t = 1$ in the pointwise convergence (3.10), we get

$$a^T \frac{(1-a^T)^{-1}}{T} \int_0^T \int_y^{+\infty} \varphi(u)\mathrm{d}u\mathrm{d}y = a^T(1-a^T)^{-1}T^{\beta-1}L(T) \underset{T\uparrow\infty}{\to} g(1) = K > 0. \tag{3.12}$$

Consider now the sequence of functions

$$\overline{\Gamma}^T(y) = a^T(1-a^T)^{-1} \int_{Ty}^{+\infty} \varphi(u)\mathrm{d}u.$$

We get from (3.11), (3.12), and the property of slowly varying function that for any $t > 0$,

$$\lim_{T\uparrow\infty} \int_0^t \overline{\Gamma}^T(y)\mathrm{d}y = Kt^\beta.$$

Suppose that $\beta \neq 0$. Let $0 \leq a < b$ and $y \in [a,b]$. We have

$$\lim_{T\uparrow\infty} \int_a^y \frac{\overline{\Gamma}^T(u)}{\int_a^b \overline{\Gamma}^T(v)\mathrm{d}v}\mathrm{d}u = \frac{y^\beta - a^\beta}{b^\beta - a^\beta}.$$

The right-hand side is the cumulative distribution function of a random variable with support on $[a,b]$ whose law is denoted by $m^\beta_{a,b}$. Hence, we have the convergence in law

$$\mathbf{1}_{[a,b]} \frac{\overline{\Gamma}^T(u)\mathrm{d}u}{\int_a^b \overline{\Gamma}^T(v)\mathrm{d}v} \underset{T\uparrow\infty}{\to} m^\beta_{a,b}(\mathrm{d}u).$$

So for any bounded continuous function g on $[a,b]$, we get

$$\lim_{T\uparrow\infty} \int_a^b \frac{\overline{\Gamma}^T(u)}{\int_a^b \overline{\Gamma}^T(v)\mathrm{d}v} g(u)\mathrm{d}u = \int_a^b g(u)m^\beta_{a,b}(\mathrm{d}u).$$

Consequently,

$$\lim_{T\uparrow\infty} \int_a^b \overline{\Gamma}^T(u)g(u)\mathrm{d}u = K\int_a^b g(u)m^\beta_{a,b}(\mathrm{d}u)\big(b^\beta - a^\beta\big) = K\beta \int_a^b g(u)u^{\beta-1}\mathrm{d}u.$$

Now let h be a nonnegative measurable function defined on $\mathbb{R}_+$, continuous on $[0,1]$, and supported on $[0,1]$. For $t \leq 1$ we have

$$\int_0^t h(t-u)\overline{\Gamma}^T(u)\mathrm{d}u \underset{T\uparrow\infty}{\to} K\beta \int_0^t h(t-u)u^{\beta-1}\mathrm{d}u$$

and for $t > 1$,

$$\int_0^t h(t-u)\overline{\Gamma}^T(u)\mathrm{d}u = \int_{t-1}^t h(t-u)\overline{\Gamma}^T(u)\mathrm{d}u \underset{T\uparrow\infty}{\to} K\beta \int_{t-1}^t h(t-u)u^{\beta-1}\mathrm{d}u.$$

Finally, for any $t \geq 0$

$$\widehat{\mathrm{TMI}}(h,t) = \lim_{T\uparrow\infty} \overline{\mathrm{TMI}}^T(h,t) = \gamma K\beta \int_0^t h(t-u)u^{\beta-1}\mathrm{d}u.$$

Thus when $\beta \in (0, 1]$, we have the existence of a macroscopic limit for the transient part of the market impact function (and therefore for the market impact function). Note that for $\beta = 1$

$$\widehat{\mathrm{TMI}}(f, t) = \gamma K \int_0^t f(u)\mathrm{d}u.$$

Consequently, in that case, $\widehat{\mathrm{TMI}}(\mathbf{1}_{[0,1]}, \cdot)$ is a nondecreasing function. This is in contradiction with Assumption 3.1; hence β cannot be equal to 1.

Suppose that $\beta = 0$. For any $t > 0$ we have

$$\mathbf{1}_{[0,t]} \frac{\overline{\Gamma}^T(u)\mathrm{d}u}{\int_0^t \overline{\Gamma}^T(v)\mathrm{d}v} \underset{T\uparrow\infty}{\rightarrow} \delta_0(\mathrm{d}u),$$

where δ_0 is the Dirac measure in 0. Then for any bounded continuous function g,

$$\lim_{T\uparrow\infty} \int_0^t \frac{\overline{\Gamma}^T(u)}{\int_0^t \overline{\Gamma}^T(v)\mathrm{d}v} g(u)\mathrm{d}u = g(0).$$

Now let f be a nonnegative measurable function defined on $\mathbb{R}_+$, continuous on $[0, 1]$, and supported on $[0, 1]$. For $t \leq 1$ we have

$$\int_0^t f(t - s)\overline{\Gamma}^T(s)\mathrm{d}s \underset{T\uparrow\infty}{\rightarrow} Kf(t),$$

and for $t > 1$

$$0 \leq \int_0^t f(t - s)\overline{\Gamma}^T(s)\mathrm{d}s \leq \int_0^t \tilde{f}(t - s)\overline{\Gamma}^T(s)\mathrm{d}s \underset{T\uparrow\infty}{\rightarrow} 0,$$

with $\tilde{f}$ a nonnegative continuous extension of $f\mathbf{1}_{[0,1]}$ on $\mathbb{R}_+$ supported on $[0, 1 + \frac{t-1}{2}]$. Finally, for any $t \geq 0$

$$\widehat{\mathrm{TMI}}(f, t) = \gamma K f(t).$$

Consequently for $\beta = 0$, we also have the existence of a macroscopic limit for the transient part of the market impact function (and therefore of the market impact function). We obtain the result, letting $\alpha = 1 - \beta$.

3.3.2 ▪ Proof of Theorem 3.4

We proceed in five steps.

Step 1: We first prove a preliminary result on the characteristic function of Hawkes processes that we use later in Step 3.

Step 2: We rewrite the sequence $(\overline{P}^T)_{T\geq 0}$ in a convenient way.

Step 3: We adapt results from [129] and [215] on scaling limits of nearly unstable heavy-tailed Hawkes processes to our more general framework.

Step 4: We deduce from the previous steps the convergence in law for the Skorokhod topology of the sequence $(\overline{P}^T)_{T\geq 0}$ and make explicit the equation satisfied by the limit.

Step 5: We prove the results on the regularity of solutions of equation (3.8).

For simplicity and without loss of generality we take $P_0 = 0$.

Step 1

We derive a result on the characteristic function of Hawkes processes using arguments similar to those introduced in [129]. Recall that the notation $*$ stands for the convolution product on $\mathbb{R}_+$. More precisely, for f and g suitable measurable functions and m a measure,

$$(f * g)(t) = \int_0^t f(t-s)g(s)\mathrm{d}s$$

and

$$(f * \mathrm{d}m)(t) = \int_0^t f(t-s)m(\mathrm{d}s).$$

We have the following proposition.

Proposition 3.7. *Let N be a Hawkes process with parameters (ν, φ), with ν a locally integrable nonnegative function and φ a nonnegative measurable function such that $\|\varphi\|_1 < 1$. For any continuous function h from $\mathbb{R}_+$ into $\mathbb{R}$,*

$$L(h,t) = \mathbb{E}[\exp((\mathrm{i}h * \mathrm{d}N)(t))]$$

satisfies

$$L(h,t) = \exp\left\{\int_0^t (C(h,s)-1)\nu(t-s)\mathrm{d}s\right\},$$

where C is a solution of the equation

$$C(h,\cdot) = \exp\big(\mathrm{i}h + (C(h,\cdot)-1) * \varphi\big).$$

Proof. Let $\tilde{N}$ be a Hawkes process with parameters (φ, φ) and N^0 be a Poisson process with intensity ν. Let $(\tilde{N}^j)_{j\in\mathbb{N}^\star}$ be independent copies of $\tilde{N}$, also independent of N^0. Using the population interpretation of Hawkes processes (see Appendix C.1 in [131]), we deduce the following equality in law:

$$N_t \overset{\mathcal{L}}{=} N_t^0 + \sum_{j=1}^{N_t^0} \tilde{N}^j_{t-T_j},$$

where $(T_j)_{j\in\mathbb{N}^\star}$ are the jump times of the process N^0. Consequently,

$$(\mathrm{i}h * \mathrm{d}N)(t) \overset{\mathcal{L}}{=} (\mathrm{i}h * \mathrm{d}N^0)(t) + \sum_{j=1}^{N_t^0} (\mathrm{i}h * \mathrm{d}\tilde{N}^j)(t-T_j).$$

Then taking the exponential and conditional expectation with respect to N^0 we get

$$\begin{aligned}\mathbb{E}\left[\mathrm{e}^{(\mathrm{i}h*\mathrm{d}N)(t)}|N^0\right] &= \mathrm{e}^{(\mathrm{i}h*\mathrm{d}N^0)_t}\prod_{j=1}^{N_t^0}\tilde{L}(h,t-T_j)\\ &= \exp\left\{\big((\mathrm{i}h + \log(\tilde{L}(h,\cdot))) * \mathrm{d}N^0\big)(t)\right\},\end{aligned}$$

where $\tilde{L}$ is defined as L with $\tilde{N}$ instead of N. Note that

$$\big((\mathrm{i}h + \log(\tilde{L}(h,\cdot))) * \mathrm{d}N^0\big)(t) = \sum_{j=1}^{N_t^0} \mathrm{i}h(t-T_j) + \log(\tilde{L}(h,t-T_j))$$

and that $Re\big(\log\big(\tilde{L}(h,\cdot)\big)\big) \le 0$ as $|\tilde{L}(h,\cdot)| \le 1$. Thus using Proposition 3.20 in Appendix 3.4.4, we get

$$L(h,t) = \exp\left\{\int_0^t \big(e^{ih(t-s)}\tilde{L}(h,t-s) - 1\big)\nu(s)ds\right\}.$$

In the same way, we have

$$\tilde{L}(h,t) = \exp\left\{\int_0^t \big(e^{ih(t-s)}\tilde{L}(h,t-s) - 1\big)\varphi(s)ds\right\}.$$

Thus, setting

$$C(h,t) = e^{ih(t)}\tilde{L}(h,t),$$

we obtain

$$L(h,t) = \exp\left\{\int_0^t \big(C(h,s) - 1\big)\nu(t-s)ds\right\}$$

and

$$C(h,\cdot) = \exp\big\{ih + \big(C(h,\cdot) - 1\big) * \varphi\big\}. \qquad \square$$

Step 2

We consider the price model (3.4). Let $M^{a,T}$ be defined by

$$M_t^{a,T} = N_t^{a,T} - \int_0^t \lambda_s^{a,T} ds.$$

We define $M^{b,T}$ the same way, replacing the superscript a by b in the above equation. We have the following result.

Lemma 3.8. *The price process* (3.4) *can be written as*

$$P_t^T = \left(1 + \int_0^{+\infty} \psi^T(v)dv\right)(M_t^{a,T} - M_t^{b,T}).$$

Proof. We have

$$P_t^T = \int_0^t \left(1 + \int_0^{+\infty} \psi^T(v)dv\right)\left(1 - \int_0^{t-u} \varphi^T(v)dv\right)d(N^{a,T} - N^{b,T})_u.$$

We first deal with the term T_1 defined by

$$\begin{aligned} T_1 &= \int_0^t \left(1 + \int_0^{+\infty} \psi^T(v)dv\right)\int_0^{t-u} \varphi^T(v)dv d(N^{a,T} - N^{b,T})_u \\ &= \left(1 + \int_0^{+\infty} \psi^T(v)dv\right)\int_0^t \int_u^t \varphi^T(v-u)dv d(N^{a,T} - N^{b,T})_u. \end{aligned}$$

Using the Fubini–Tonelli theorem we get

$$T_1 = \left(1 + \int_0^{+\infty} \psi^T(v)dv\right)\int_0^t \int_0^v \varphi^T(v-u)d(N^{a,T} - N^{b,T})_u dv.$$

Thus we deduce

$$T_1 = \left(1 + \int_0^{+\infty} \psi^T(v)dv\right)\int_0^t \big(\lambda_v^{a,T} - \mu - \lambda_v^{b,T} + \mu\big)dv.$$

Finally,

$$P_t^T = \left(1 + \int_0^{+\infty} \psi^T(v)\mathrm{d}v\right) \int_0^t \left(\mathrm{d}N_v^{a,T} - \lambda_v^{a,T}\mathrm{d}v - \mathrm{d}N_v^{b,T} + \lambda_v^{b,T}\mathrm{d}v\right)$$
$$= \left(1 + \int_0^{+\infty} \psi^T(v)\mathrm{d}v\right)(M_t^{a,T} - M_t^{b,T}). \qquad \square$$

Lemma 3.8 leads to

$$\overline{P}_t^T = \frac{1-a^T}{T\mu^T}\left(1 + \int_0^{+\infty} \psi^T(v)\mathrm{d}v\right)(M_{tT}^{a,T} - M_{tT}^{b,T}).$$

Step 3

We temporarily drop the superscripts a and b. Indeed, the results are valid for both buy and sell order flows. Consider the sequences

$$X_t^T = \frac{1-a^T}{T\mu^T} N_{tT}^T, \quad \Lambda_t^T = \frac{1-a^T}{T\mu^T}\int_0^{tT} \lambda_s^T \mathrm{d}s, \quad Z_t^T = \sqrt{\frac{T\mu^T}{1-a^T}}(X_t^T - \Lambda_t^T) \tag{3.13}$$

defined for $t \in [0, 1]$. The following result is borrowed from [215].

Proposition 3.9. *The sequence (Λ^T, X^T, Z^T) is tight. Furthermore, for any limit point (Λ, X, Z) of (Λ^T, X^T, Z^T), Z is a continuous martingale, $[Z, Z] = X$, and $\Lambda = X$.*

In addition, we have the following proposition which extends Theorem 3.1 in [215].

Proposition 3.10. *Under Assumptions* 3.1 *and* 3.3, *for any limit point (X, Z) of (X^T, Z^T), there exists a Brownian motion B on $(\Omega, \mathcal{A}, \mathbb{P})$ (up to extension of the space) such that*

$$Z_t = B_{X_t},$$

and X is a solution of the stochastic rough Volterra equation

$$X_t = \int_0^t F^{\alpha,\lambda}(t-s)\mathrm{d}s + \frac{1}{\sqrt{\delta\lambda}}\int_0^t F^{\alpha,\lambda}(t-s)\mathrm{d}B_{X_s}. \tag{3.14}$$

Moreover, for any $\varepsilon > 0$, the process X has Hölder regularity $1 \wedge (2\alpha - \varepsilon)$.

Note that here we are under more general assumptions than in Theorem 3.1 in [215]. Indeed, in [215] we have

$$\int_t^{+\infty} \varphi(s)\mathrm{d}s = Kt^{-\alpha},$$

while we only know that

$$\int_0^t \int_s^{+\infty} \varphi(u)\mathrm{d}u\mathrm{d}s = L(t)t^{1-\alpha},$$

with L a slowly varying function. To prove Proposition 3.10, it is enough to get the following lemma. The rest of the proof is similar to that in [215].

Lemma 3.11. *The sequence of functions $\varrho^T(t) = \frac{1-a^T}{a^T}\psi^T(Tt)T$ converges weakly towards $f^{\alpha,\lambda}$. Furthermore $\int_0^t \varrho^T(s)\mathrm{d}s$ converges uniformly towards $F^{\alpha,\lambda}$.*

Proof. The function ϱ^T is nonnegative with integral equal to one. So, it can be interpreted as the density of a random variable. Hence, it is enough to show that its Laplace transform converges pointwise to get weak convergence. For $z > 0$ we have

$$\hat{\varrho}^T(z) = \frac{1-a^T}{a^T}\hat{\psi}^T\left(\frac{z}{T}\right) = \frac{\hat{\varphi}(\frac{z}{T})}{1-a^T(1-a^T)^{-1}(\hat{\varphi}(\frac{z}{T})-1)}.$$

Let

$$R(t) = \int_0^t \int_s^{+\infty} \varphi(u)\mathrm{d}u\mathrm{d}s.$$

Recall that from Theorem 3.2, $R(t) = t^{1-\alpha}L(t)$. By Karamata's Tauberian theorem (see Theorem 3.19 in Appendix 3.4.2), we have

$$\hat{R}(z) \sim_{0^+} z^{\alpha-2}L\left(\frac{1}{z}\right)\Gamma(2-\alpha).$$

Integrating by parts twice we obtain

$$\hat{R}(z) = \int_0^{+\infty} \mathrm{e}^{-zs}R(s)\mathrm{d}s = \frac{1}{z^2}\big(1-\hat{\varphi}(z)\big).$$

So, we get

$$a^T(1-a^T)^{-1}(1-\hat{\varphi}(\frac{z}{T})) \underset{T\uparrow\infty}{\sim} a^T(1-a^T)^{-1}T^{-\alpha}L(T)\frac{L(\frac{T}{z})}{L(T)}z^\alpha\Gamma(2-\alpha).$$

We have shown in Theorem 3.2 that

$$a^T(1-a^T)^{-1}T^{-\alpha}L(T) \underset{T\uparrow\infty}{\to} K.$$

With the function L being slowly varying (see Appendix 3.4.2), we deduce

$$\lim_{T\uparrow\infty} a^T(1-a^T)^{-1}\left(1-\hat{\varphi}\left(\frac{z}{T}\right)\right) = z^\alpha\Gamma(2-\alpha)K,$$

and finally,

$$\lim_{T\uparrow\infty}\hat{\varrho}^T(z) = \frac{1}{1+K\Gamma(2-\alpha)z^\alpha} = \frac{\lambda}{\lambda+z^\alpha} = \hat{f}^{\alpha,\lambda},$$

with $\lambda = (K\Gamma(2-\alpha))^{-1}$. The uniform convergence in Lemma 3.11 is obviously deduced from Dini's theorem. □

We finally show that the sequence $(X^T, Z^T)_{T\geq 0}$ converges in law for the Skorokhod topology. We already know that it is tight, so it is enough to prove that all the limit points have the same law.

Let (X, Z) be a limit point of $(X^T, Z^T)_{T\geq 0}$. Using Proposition 3.10, together with the stochastic Fubini theorem (see [274]), we have

$$X_t = \int_0^t f^{\alpha,\lambda}(t-s)\left(s+\frac{1}{\sqrt{\delta\lambda}}Z_s\right)\mathrm{d}s.$$

From Example 42.2 in [267], this leads to

$$D^\alpha X_t + \lambda X_t - \lambda t = \sqrt{\frac{\lambda}{\delta}}Z_t,$$

where D^α is the fractional derivative operator defined in Appendix 3.4.3. Thus, the law of (X, Z) is uniquely determined by the law of X. Consequently, it is enough to prove uniqueness in law for limit points of $(X^T)_{T\geq 0}$ to get convergence in law of $(X^T, Z^T)_{T\geq 0}$. For this we prove that the characteristic function of any limit point X of the sequence $(X^T)_{T\geq 0}$ is a functional of the solution of a fractional Riccati equation. Uniqueness in law is then a consequence from the uniqueness of the solution of this equation.

Proposition 3.12. *Let X be a limit point of $(X^T)_{T\geq 0}$ and h be a continuously differentiable function from $\mathbb{R}_+$ to $\mathbb{R}$ such that $h(0) = 0$. The function*

$$\psi(h,t) = \mathbb{E}[\exp(ih * \mathrm{d}X)_t]$$

satisfies

$$\psi(h,t) = \exp\left\{\int_0^t g(s)\mathrm{d}s\right\},$$

with g the unique continuous solution of the rough Volterra Riccati equation

$$g = f^{\alpha,\lambda} * \left(\delta^{-1}\frac{1}{2}g^2 + ih\right). \tag{3.15}$$

To show this result, we are inspired by the methodology of [129]. However, note again that we are in a more general setting.

Proof. Recall that

$$X_t^T = \frac{1-a^T}{T\mu^T} N_{tT}^T.$$

We introduce the quantities

$$\begin{aligned} h^T(t) &= \frac{1-a^T}{T\mu^T} h\left(\frac{t}{T}\right), \\ L^T(h^T,t) &= \mathbb{E}\left[\exp(ih^T * \mathrm{d}N^T)(t)\right], \\ \psi^T &= L^T(h^T, tT). \end{aligned}$$

For every T, according to Proposition 3.7, there exists a function C^T solution of

$$C^T = \exp\left\{ih^T + \left(C^T - 1\right) * \varphi^T\right\}$$

such that

$$L^T(h^T,t) = \exp\left(\int_0^t \left(C^T(s) - 1\right)\mu^T \mathrm{d}s\right).$$

Now define the sequence g^T as

$$g^T(s) = C^T(sT) - 1.$$

We have

$$\psi^T = \exp\left\{\frac{g^T}{1-a^T} * (T(1-a^T)\mu^T \mathbf{1}_{\mathbb{R}_+})\right\}, \tag{3.16}$$

$$g^T + 1 = \exp\left\{\frac{1-a^T}{T\mu^T} ih + g^T * (T\varphi^T(\cdot T))\right\}. \tag{3.17}$$

An immediate adaptation of Proposition 6.4 in [129] gives that for any $s \in [0,t]$,

$$|g^T(s)| \leq c(h)(1-a^T), \tag{3.18}$$

with $c(h)$ a positive constant depending only on h. Hence, for T large enough we have

$$\log(1+g^T) = g^T - \frac{1}{2}(g^T)^2 - \epsilon^T(h,\cdot), \tag{3.19}$$

with $|\epsilon^T(h,\cdot)| \leq c(h)(1-a^T)^3$. According to equations (3.16) and (3.19), for every $s \in [0,t]$ we get

$$g^T(s) = \frac{1}{2}g^T(s)^2 + \epsilon^T(h,s) + \frac{1-a^T}{T\mu^T}\mathrm{i}h(s) + g^T * \big(T\varphi^T(\cdot T)\big)(s).$$

Using that

$$\sum_{i\geq 1}(T\varphi^T(\cdot T))^{*i} = T\psi^T(\cdot T),$$

we deduce from Lemma 4.1 in [214] that

$$g^T(s) = \big(T\psi^T(\cdot T)\big) * \left(\frac{1}{2}(g^T)^2 + \epsilon^T(h,\cdot) + \frac{1-a^T}{T\mu^T}\mathrm{i}h\right)(s) + \frac{1}{2}g^T(s)^2 + \epsilon^T(h,s) + \frac{1-a^T}{T\mu^T}\mathrm{i}h(s).$$

Consequently, letting $\theta_T = (1-a^T)^{-1}g^T$ we get

$$\theta_T(s) = \big(T(1-a^T)\psi^T(\cdot T)\big) * \left(\frac{1}{2}\theta_T^2 + \frac{1}{\delta}\mathrm{i}h\right)(s) + r_1^T(s),$$

with

$$\begin{aligned} r_1^T(s) &= \big(T(1-a^T)\psi^T(\cdot T)\big) * \left(\epsilon^T(h,\cdot)(1-a^T)^{-2} + \left(\frac{1}{T(1-a^T)\mu^T} - \delta^{-1}\right)\mathrm{i}h\right)(s) \\ &\quad + (1-a^T)^{-1}\frac{1}{2}(g^T(s))^2 + (1-a^T)^{-1}\epsilon^T(h,s) + \frac{1}{T\mu^T}\mathrm{i}h(s). \end{aligned}$$

Because a^T goes to 1, we know from Lemma 3.11 that in the sense of weak convergence,

$$T(1-a^T)\psi^T(\cdot T) \underset{T\uparrow\infty}{\to} f^{\alpha,\lambda}.$$

Finally, we have

$$\theta_T = f^{\alpha,\lambda} * \left(\frac{1}{2}\theta_T^2 + \frac{1}{\delta}\mathrm{i}h\right) + r_1^T + r_2^T,$$

where

$$r_2^T = \big(T(1-a^T)\psi^T(\cdot T) - f^{\alpha,\lambda}\big) * \left(\frac{1}{2}\theta_T^2 + \frac{1}{\delta}\mathrm{i}h\right).$$

We now prove that $(r_1^T)_{T\geq 0}$ and $(r_2^T)_{T\geq 0}$ go to 0 in $C^0([0,t],\mathbb{R})$ for the sup-norm.

Using Assumption 3.3, the second part of Lemma 3.11, and equation (3.18), we get that $(r_1^T)_{T\geq 0}$ goes to zero in $C^0([0,t],\mathbb{R})$. The sequence $(\theta_T)_{T\geq 0}$ is bounded for the sup-norm according to equation (3.18). Moreover, according to Lemma 3.13 (see after the proof) θ_T is differentiable, and $(\theta_T')_{T\geq 0}$ is bounded for the sup-norm. By integration by parts we have

$$r_2^T(t) = \left(\int_0^{\cdot} T(1-a^T)\psi^T(sT)\mathrm{d}s - F^{\alpha,\lambda}\right) * \left(\theta_T'\theta_T + \frac{1}{\delta}\mathrm{i}h'\right)(t),$$

where we have used the fact that $\theta_T(0) = 0$ and $h(0) = 0$. We then conclude that $(r_2^T)_{T\geq 0}$ converges towards 0 in $C^0([0,t],\mathbb{R})$ using dominated convergence. Lemma 3.13, together with the Ascoli theorem, gives that the sequence $(\theta_T)_{T\geq 0}$ is relatively compact in $\left(C^0([0,t],\mathbb{R}), \|\,\|_\infty\right)$. Moreover, for any limit point θ of the sequence $(\theta_T)_{T\geq 0}$, we have that θ is a solution of

$$\theta = f^{\alpha,\lambda} * \left(\frac{1}{2}\theta^2 + \frac{1}{\delta}ih\right).$$

The above equation has a unique continuous solution in $C^0([0,t],\mathbb{R})$; see Section 6.2.4 in [129]. Thus the sequence $(\theta_T)_{T\geq 0}$ converges towards this solution.

Finally, note that

$$\psi^T(t) = \mathbb{E}\left[\exp\left\{ih * \mathrm{d}X^T\right\}(t)\right].$$

Thus, convergence in law of $(X^T)_{T\geq 0}$ towards X implies that $(\psi^T)_{T\geq 0}$ converges pointwise towards the function ψ. Passing to the limit in (3.16) we get

$$\psi(t) = \exp\left((\theta * (\delta \mathbf{1}_{\mathbb{R}_+}))_t\right) = \exp\left(\delta \int_0^t \theta(s)\mathrm{d}s\right).$$

Letting $g = \delta\theta$, we have the result. □

It is enough to characterize the law of X to know $\psi(h,t)$ for any $t \in \mathbb{R}_+$ and $h \in C_0^1([0,t],\mathbb{R})$. Therefore, uniqueness in law for the limit points of $(X^T)_{T\geq 0}$ is a corollary from the uniqueness of continuous solution for the Volterra–Riccati equation (3.15); see Section 6.2.4 in [129].

We now give the lemma we used in the proof of Proposition 3.12.

Lemma 3.13. *The functions $(\theta_T)_{T\geq 0}$ are continuously differentiable, and $(\theta_T')_{T\geq 0}$ is bounded in $C^0([0,t],\mathbb{R})$.*

Proof. Using the proof of Proposition 3.7 we have

$$\theta_T = (1-a^T)^{-1}\left(\mathbb{E}\left[\exp\left\{(ih + ih * \mathrm{d}\tilde{N}_{\cdot T}^T)\frac{1-a^T}{T\mu^T}\right\}\right] - 1\right),$$

with $\tilde{N}$ a Hawkes process with parameters (φ^T, φ^T) where $\varphi^T = a^T\varphi$. Because $h(0) = 0$, $h * \mathrm{d}\tilde{N}_{\cdot T}^T$ admits a derivative and for any $s \in [0,t]$,

$$(h * \mathrm{d}\tilde{N}_{\cdot T}^T)'(s) = (h' * \mathrm{d}\tilde{N}_{\cdot T}^T)(s).$$

Furthermore we have

$$|(h' * \mathrm{d}\tilde{N}_{\cdot T}^T)(s)| \leq \|h'\|_\infty \tilde{N}_{tT}^T.$$

Using that

$$\tilde{\lambda}_s^T = \psi^T(s) + \int_0^t \psi^T(t-s)\mathrm{d}\tilde{M}_s^T,$$

we get

$$(1-a^T)\mathbb{E}[\tilde{N}_{tT}^T] \leq (1-a^T)\mathbb{E}\left[\int_0^{tT} \tilde{\lambda}_s^T \mathrm{d}s\right] \leq \int_0^{tT} (1-a^T)\psi^T(s)\mathrm{d}s \leq 1.$$

Consequently, using derivation for integral with parameters, θ^T is differentiable, and

$$\theta_T' = (1-a^T)^{-1}\mathbb{E}\left[\left(ih' + ih' * \mathrm{d}\tilde{N}_{\cdot T}^T\right)\frac{1-a^T}{T\mu^T}\exp\left((ih + ih * \mathrm{d}\tilde{N}_{\cdot T}^T)\frac{1-a^T}{T\mu^T}\right)\right].$$

Thus we have for all $s \in [0,t]$,

$$|\theta_T'(s)| \leq \frac{1}{T\mu^T(1-a^T)}(1-a^T)\mathbb{E}[||h'||_\infty + ||h'||_\infty \tilde{N}_{tT}^T].$$

The right-hand side is finite and independent of s; consequently the sequence $(\theta_T')_{T\geq 0}$ is bounded in $C^0([0,t],\mathbb{R})$. □

Finally, we have proved that the sequence $(X^T, Z^T)_{T\geq 0}$ converges in law for the Skorokhod topology.

Step 4

Consider the sequence $(X^{a,T}, Z^{a,T})_{T\geq 0}$ (resp., $(X^{b,T}, Z^{b,T})_{T\geq 0}$) defined the same way as in equation (3.13) with $(N^{a,T})_{T\geq 0}$ (resp., $(N^{b,T})_{T\geq 0}$) instead of $(N^T)_{T\geq 0}$. According to Lemma 3.8 we have

$$\overline{P}_t^T = \frac{1-a^T}{T\mu^T}\left(1+\int_0^{+\infty}\psi^T(u)\mathrm{d}u\right)\left(M_{tT}^{a,T} - M_{tT}^{b,T}\right) = \frac{1}{T\mu^T}\left(M_{tT}^{a,T} - M_{tT}^{b,T}\right).$$

Thus,

$$\overline{P}_t^T = \frac{1}{\sqrt{T\mu^T(1-a^T)}}\left(Z_t^{a,T} - Z_t^{b,T}\right).$$

Using Step 3, we get that $(Z^{a,T})_{T\geq 0}$, and $(Z^{b,T})_{T\geq 0}$ converge for the Skorokhod topology. With these sequences being independent, $(\overline{P}^T)_{T\geq 0}$ converges towards a process $\widehat{P}$ in the Skorokhod topology. Furthermore we deduce from Proposition 3.10 together with Assumption 3.3 that there exist two independent Brownian motions B^a and B^b such that

$$\widehat{P}_t = \frac{1}{\sqrt{\delta}}\left(B^a_{X_t^a} - B^b_{X_t^b}\right),$$

where X^a (resp. X^b) is the limit of the sequence $(X^{a,T})_{T\geq 0}$ (resp. $(X^{a,T})_{T\geq 0}$) and is solution of Equation (3.14) with Brownian motion B^a (resp. B^b). Hence $X = \frac{X^a+X^b}{\delta}$ is solution of

$$X_t = \frac{2}{\delta}\int_0^t F^{\alpha,\lambda}(t-s)\mathrm{d}s + \frac{1}{\delta\sqrt{\lambda}}\int_0^t F^{\alpha,\lambda}(t-s)\frac{1}{\sqrt{\delta}}\mathrm{d}\left(B^a_{X_s^a} + B^b_{X_s^b}\right).$$

Moreover there exists a Brownian motion W such that $W_{X_t} = \frac{1}{\sqrt{\delta}}(B^a_{X_t^a} + B^b_{X_t^b})$. Consequently

$$X_t = \frac{2}{\delta}\int_0^t F^{\alpha,\lambda}(t-s)\mathrm{d}s + \frac{1}{\delta\sqrt{\lambda}}\int_0^t F^{\alpha,\lambda}(t-s)\mathrm{d}W_{X_s}.$$

Step 5

We first recall a result from [215].

Proposition 3.14. *Let X be a solution of the stochastic Volterra equation* (3.8)*. Then for any $\varepsilon > 0$, almost surely, X has Hölder regularity $1 \wedge (2\alpha - \varepsilon)$. And if $\alpha > \frac{1}{2}$, X is almost surely differentiable.*

We now give a new result on the regularity of the solution of equation (3.8).

Proposition 3.15. *Let $\alpha \leq \frac{1}{2}$. Let X be a solution of the stochastic Volterra equation* (3.8)*. Then, almost surely, X is not continuously differentiable.*

Proof. As already seen in Step 3, X satisfies

$$D^{\alpha}X_t = -\lambda X_t + \frac{2\lambda}{\delta}t + \frac{\sqrt{\lambda}}{\delta}W_{X_t}. \tag{3.20}$$

Applying the law of iterated logarithm, for $0 \le t \le 1$ we get

$$\limsup_{s\to t^-}\frac{D^{\alpha}X_t - D^{\alpha}X_s - \frac{2\lambda}{\delta}(t-s)}{\sqrt{2(X_t - X_s)\log\log\left((X_t - X_s)^{-1}\right)}} = \frac{\sqrt{\lambda}}{\delta}.$$

Assume that X is continuously differentiable. According to Appendix 3.4.3 we have

$$D^{\alpha}X_t = \frac{1}{\Gamma(1-\alpha)}\int_0^t (t-s)^{-\alpha}X'_s \mathrm{d}s.$$

Let t be such that $X'_t \neq 0$. Such a point almost surely exists because X is not constant. Indeed, suppose it is constant, as $X_0 = 0$ it implies that $X = 0$. But obviously the null function is not a solution of equation (3.20). For such t using that

$$X_t - X_s \sim_{s\to t} (t-s)X'_t,$$

we have

$$\lim_{s\to t^-}\frac{t-s}{\sqrt{2(X_t - X_s)\log\log\left[(X_t - X_s)^{-1}\right]}} = 0.$$

Hence

$$\limsup_{s\to t^-}\frac{D^{\alpha}X_t - D^{\alpha}X_s}{\sqrt{2(X_t - X_s)\log\log\left[(X_t - X_s)^{-1}\right]}} = \frac{\sqrt{\lambda}}{\delta}. \tag{3.21}$$

We now give a bound on $|D^{\alpha}X_t - D^{\alpha}X_s|$, for $s < t$, where $||X'||_\infty$ denotes the supremum norm of X',

$$\begin{aligned}
|D^{\alpha}X_t - D^{\alpha}X_s| &= \Big|\int_0^s \left((t-u)^{-\alpha} - (s-u)^{-\alpha}\right)X'_s\mathrm{d}u + \int_s^t (t-u)^{-\alpha}X'_u\mathrm{d}u\Big| \\
&\le \int_0^s \left|(t-u)^{-\alpha} - (s-u)^{-\alpha}\right| ||X'||_\infty \mathrm{d}u + \frac{||X'||_\infty}{1-\alpha}(t-s)^{1-\alpha} \\
&\le \left|\int_0^s (t-u)^{-\alpha} - (s-u)^{-\alpha}\mathrm{d}u\right| ||X'||_\infty + \frac{||X'||_\infty}{1-\alpha}(t-s)^{1-\alpha} \\
&\le \left(\int_0^{t-s} u^{-\alpha}\mathrm{d}u + \int_s^t u^{-\alpha}\mathrm{d}u\right)||X'||_\infty + \frac{||X'||_\infty}{1-\alpha}(t-s)^{1-\alpha} \\
&\le \left(\frac{1}{1-\alpha}(t-s)^{1-\alpha} + (t-s)s^{-\alpha}\right)||X'||_\infty + \frac{||X'||_\infty}{1-\alpha}(t-s)^{1-\alpha}.
\end{aligned}$$

We get

$$\lim_{s\to t^-}\frac{D^{\alpha}X_t - D^{\alpha}X_s}{\sqrt{2(X_t - X_s)\log\log\left[(X_t - X_s)^{-1}\right]}} = 0.$$

This is in contradiction with equation (3.21), and hence X cannot be continuously differentiable. □

3.3.3 ▪ Proof of Theorem 3.6

We have seen in Section 3.3.2 that $X = (X^a + X^b)/\delta$, with X^a and X^b independent copies of the limit of the sequence $(X^T)_{T\geq 0}$. From Proposition 3.12, we immediately obtain Theorem 3.6.

3.4 ▪ Appendix

3.4.1 ▪ Mittag–Leffler functions

Let $(\alpha,\beta) \in (\mathbb{R}_+^*)^2$. The Mittag–Leffler function $E_{\alpha,\beta}$ is defined for $z \in \mathbb{C}$ by

$$E_{\alpha,\beta}(z) = \sum_{n\geq 0} \frac{z^n}{\Gamma(\alpha n + \beta)}.$$

For $(\alpha, \lambda) \in (0, 1] \times \mathbb{R}_+$, we also define

$$f^{\alpha,\lambda}(t) = \lambda t^{\alpha-1} E_{\alpha,\alpha}(-\lambda t^\alpha), \; t > 0,$$
$$F^{\alpha,\lambda}(t) = \int_0^t f^{\alpha,\lambda}(s)\mathrm{d}s, \; t \geq 0.$$

The function $f^{\alpha,\lambda}$ is a density function on $\mathbb{R}_+$ called the Mittag–Leffler density function. Its Laplace transform is

$$\hat{f}^{\alpha,\lambda}(z) = \frac{\lambda}{\lambda + z^\alpha}.$$

When $\alpha = 1$, the Mittag–Leffler density simply corresponds to the exponential law with parameter λ.

3.4.2 ▪ Tauberian theorems

The following results can be found in [63].

Definition 3.16. *A measurable function $L : \mathbb{R}_+ \to \mathbb{R}$ is slowly varying if for all $s > 0$,*

$$\frac{L(st)}{L(t)} \underset{t\uparrow\infty}{\to} 1.$$

Proposition 3.17. *Let L be a slowly varying function and $\alpha > 0$; then*

$$t^{-\alpha} L(t) \underset{t\uparrow\infty}{\to} 0.$$

Theorem 3.18 (characterization theorem). *Let U be a positive measurable function on $\mathbb{R}_+$ such that for all $s \in C$, with C a set with positive Lebesgue measure,*

$$\frac{U(ts)}{U(t)} \underset{t\uparrow\infty}{\to} g(s) > 0,$$

for some function g. Then the previous limit can be extended for all $s > 0$. Let $\tilde{g}$ be this limiting function extending g. There exist $\alpha \in \mathbb{R}$ such that $g(t) = t^\alpha$ and a slowly varying function L such that $U(t) = t^\alpha L(t)$.

Theorem 3.19 (Karamata's Tauberian theorem). *Let U be a measurable nonnegative function, $c \geq 0$, $\varrho > -1$, and assume $\hat{U}(z) = \int_0^{+\infty} \mathrm{e}^{-zs} U(s)\mathrm{d}s$ is finite for any $z > 0$. Then*

$$U(t) \sim_{+\infty} ct^\varrho \frac{L(t)}{\Gamma(1+\varrho)}$$

for L a slowly varying function implies

$$\hat{U}(z) \sim_{0^+} cz^{-\varrho-1} L\left(\frac{1}{z}\right).$$

3.4.3 ▪ Fractional derivative

For $\alpha \in [0,1)$, the fractional derivative operator D^α is defined for h λ-Hölder function (with $\lambda > \alpha$) by

$$D^\alpha f(t) = \frac{1}{\Gamma(1-\alpha)} \frac{\mathrm{d}}{\mathrm{d}t} \int_0^t (t-s)^{-\alpha} h(s) \mathrm{d}s.$$

Note that if the function h is continuously differentiable and $f(0) = 0$. The derivation for integral with parameters gives

$$D^\alpha f(t) = \frac{1}{\Gamma(1-\alpha)} \int_0^t (t-s)^{-\alpha} f'(s) \mathrm{d}s.$$

More information on fractional differential operator can be found in [267].

3.4.4 ▪ A result on inhomogeneous Poisson process

We recall the following well-known result.

Proposition 3.20 (exponential formula). *Let N be an inhomogeneous Poisson process with intensity ν and f be a complex measurable function defined on $\mathbb{R}_+$ such that $Re(f) \leq 0$. Consider the function*

$$N_f(t) = \sum_{i=1}^{N_t} f(T_i),$$

where $(T_i)_{i \in \mathbb{N}}$ are the jump times of N. For any $t \geq 0$ we have

$$\mathbb{E}\left[\mathrm{e}^{N_f(t)}\right] = \exp\left\{ \int_0^t (\mathrm{e}^{f(s)} - 1)\nu(s) \mathrm{d}s \right\}.$$

Acknowledgments

We thank Jean-Philippe Bouchaud, Omar El Euch, Masaaki Fukasawa, and Jim Gatheral for many interesting discussions. We also thank the Associate Editor and the two referees for their numerous relevant comments. The authors gratefully acknowledge the financial support of the *ERC Grant* 679836 *Staqamof* and of the chair *Analytics and Models for Regulation*.

Chapter 4

Rough Heston

Stefan Gerhold, Antoine Jacquier, Mathieu Rosenbaum

Introduction

The best-known stochastic volatility (SV) model, at least on equity markets, is the Heston model [199] introduced by Steve Heston in 1993, in which the stock price S satisfies the stochastic differential equation

$$\begin{aligned} \frac{dS_t}{S_t} &= \sqrt{V_t}dB_t, & S_0 > 0, \\ dV_t &= \lambda(\theta - V_t)dt + \nu\sqrt{V_t}\,dW_t, & V_0 > 0, \end{aligned} \tag{4.1}$$

where $(W, W^{\perp})$ is a two-dimensional Brownian motion, $B := \varrho W + \sqrt{1-\varrho^2}W^{\perp}$, $\nu > 0$ is the volatility of volatility, $\varrho \in [-1, 1]$ the correlation between spot and volatility, $\lambda \geq 0$ the mean-reversion speed, and $\theta > 0$ the long-term variance level. From the Yamada–Watanabe conditions [220, Proposition 2.13], the model admits a unique strong solution, and the variance process can be written in integral form as

$$V_u = V_t e^{-\lambda(u-t)} + \theta\big(1 - e^{-\lambda(u-t)}\big) + \nu\int_t^u e^{-\lambda(u-s)}\sqrt{V_s}\,dW_s, \qquad \text{for any } t \leq u,$$

and Itô isometry implies that the covariance structure reads, for any $t, u > 0$,

$$\mathrm{Cov}[V_u, V_t] = \frac{\nu^2\theta}{2\lambda}e^{-\lambda|t-u|} + \frac{\nu^2}{\lambda}(V_0 - \theta)e^{-\lambda(t\vee u)} - \frac{\nu^2}{2\lambda}(2V_0 - \theta)e^{-\lambda(t+u)}.$$

The Feller condition, $2\lambda\theta \geq \nu^2$, ensures that the origin is unattainable; otherwise, it is regular, hence attainable, and strongly reflecting [221].

This model is popular for several reasons. From a statistical viewpoint, it enables us to reproduce several important features of low frequency price data, such as leverage effect, time-varying volatility, or fat tails of returns. In terms of options, as a stochastic volatility model, it provides quite reasonable dynamics for the volatility surface. Actually, compared to other stochastic volatility models, its most appealing property is probably the fact that the characteristic function of the asset log-price is explicit. More precisely, it is shown in [199] that for any $u \in \mathbb{R}$,

$$\mathbb{E}\left[\left(\frac{S_t}{S_0}\right)^{\mathrm{i}u}\right] = \exp\Big(g(u,t) + V_0 h(u,t)\Big), \tag{4.2}$$

where h is solution of the Riccati equation

$$\partial_t h(u,t) = -\frac{u(u+\mathrm{i})}{2} + (\mathrm{i}u\varrho\nu - \lambda)h(u,t) + \frac{\nu^2}{2}h^2(u,t), \tag{4.3}$$

with boundary condition $h(u,0) = 0$ and

$$g(u,t) = \theta\lambda \int_0^t h(u,s)\mathrm{d}s.$$

The ordinary differential equation (ODE) (4.3) can be solved explicitly, and pricing European options is therefore amenable to using inverse Fourier techniques [85, 163].

4.1 ▪ Roughening Heston

4.1.1 ▪ The rough Heston model

In the classical Heston model (4.1), the volatility follows a Brownian diffusion, whereas, as explained in Chapters 1 and 2, the paths of the volatility process exhibit a rougher behavior. Chapter 3 further shed light on a rough version of the Heston model which naturally appears as a suitable price model based on microstructural and no-arbitrage arguments. This model, introduced in [129], takes the form[38]

$$\begin{aligned} \frac{\mathrm{d}S_t}{S_t} &= \sqrt{V_t}\mathrm{d}B_t, \\ V_u &= V_t + \frac{1}{\Gamma(H+\frac{1}{2})}\int_t^u \frac{1}{(u-s)^{\frac{1}{2}-H}}\Big\{\lambda\big(\theta^t(s) - V_s\big)\mathrm{d}s + \nu\sqrt{V_s}\mathrm{d}W_s\Big\}, \end{aligned} \tag{4.4}$$

for $u \geq t$, where $H \in (0, 1/2)$ is the Hurst exponent, Γ is the Gamma function, and the parameters λ, ν, ϱ have the same meaning as in the Heston model. The mean reversion level parameter $\theta^t(\cdot)$ is now an $\mathcal{F}_t$-measurable function which makes the model time consistent; see [131]. Setting the Hurst parameter to $1/2$, we retrieve the classical Heston model (4.1). We recognize in (4.4) the power-law kernel of the Mandelbrot–Van Ness representation of the fractional Brownian motion (fBm); hence the name "rough Heston." This model enjoys the nice statistical properties of rough volatility models. In addition, as will be seen in Section 4.2.3, it provides remarkable fits of the volatility surface while remaining almost as tractable as the classical Heston model, as explained in Section 4.2.

It is further shown in [131] that λ and $\theta^t(\cdot)$ can be directly inferred from the forward variance curve $(\xi_t(u))_{u\geq t}$ observed at time t, defined by $\xi_t(u) := \mathbb{E}[V_u|\mathcal{F}_t]$. By doing so, the model may be rewritten in the asymptotic setting as λ tends to zero in forward variance form as

$$\begin{aligned} \frac{\mathrm{d}S_t}{S_t} &= \sqrt{V_t}\mathrm{d}B_t, \\ V_u &= \xi_t(u) + \frac{\nu}{\Gamma(H+\frac{1}{2})}\int_t^u \frac{\sqrt{V_s}}{(u-s)^{\frac{1}{2}-H}}\mathrm{d}W_s, \qquad u \geq t. \end{aligned} \tag{4.5}$$

We refer the reader to Chapter 5 for more details on so-called affine forward variance models, such as the rough Heston model.

[38] Note that [129] uses a different parameterization, motivated by the microstructural foundation of the model: γ is our λ, and $\gamma\nu$ is our ν.

4.1.2 ▪ The fractional Heston model

An alternative way to include fBm in the volatility dynamic in a Heston-like environment is to consider the so-called fractional Heston model introduced in [180]. Here the log-price X_t is the unique strong solution to the stochastic differential equation

$$\begin{aligned} \mathrm{d}X_t &= -\frac{1}{2}\mathcal{V}_t\mathrm{d}t + \sqrt{\mathcal{V}_t}\mathrm{d}W_t^{\perp}, && X_0 = 0, \\ \mathrm{d}V_t &= \lambda(\theta - V_t)\mathrm{d}t + \nu\sqrt{V_t}\mathrm{d}W_t, && V_0 > 0, \\ \mathcal{V}_t &= \eta + \mathrm{I}^{H-\frac{1}{2}}[V_t], \end{aligned} \tag{4.6}$$

where the coefficients satisfy $\lambda, \theta, \nu > 0$. The operator $\mathrm{I}^{H-\frac{1}{2}}$ is the classical left fractional Riemann–Liouville integral of order $H - \frac{1}{2}$:

$$\mathrm{I}^{H-\frac{1}{2}}[\varphi](t) := \begin{cases} \displaystyle\int_0^t \frac{(t-s)^{H-\frac{3}{2}}}{\Gamma(H-1/2)}\varphi(s)\mathrm{d}s, & \text{for } H \in \left(\frac{1}{2}, 1\right), \\ \displaystyle\frac{\mathrm{d}}{\mathrm{d}t}\mathrm{I}^{H+\frac{1}{2}}[\varphi](t), & \text{for } H \in \left(0, \frac{1}{2}\right), \end{cases}$$

whenever it exists.

The additional parameter $\eta \geq 0$, as explained in [97], allows one to loosen the tight connection between the mean and the variance of V_t. Under the Feller condition for the CIR process V, since the Riemann–Liouville operator preserves positivity, $\mathcal{V}_t \geq \eta$ almost surely for all $t \geq 0$. Now, for any $t \geq 0$, we have $\mathbb{E}[\mathcal{V}_t] = \eta + \mathrm{I}^{H-\frac{1}{2}}\left(V_0\mathrm{e}^{-\lambda t} + \theta\left(1 - \mathrm{e}^{-\lambda t}\right)\right)$ and, for any $t, h \geq 0$,

$$\mathrm{Cov}\left[\mathcal{V}_{t+h}, \mathcal{V}_t\right] = \int_0^{t+h}\int_0^t \frac{(t+h-s)^{H-\frac{3}{2}}(t-u)^{H-\frac{3}{2}}}{\Gamma\left(H-\frac{1}{2}\right)^2}\mathrm{Cov}\left[V_s, V_u\right]\mathrm{d}u\,\mathrm{d}s.$$

Remark 4.1.1. *The motivation for this fractional volatility model comes from the seminal work by Comte and Renault* [98, 99] *on the Ornstein–Uhlenbeck process. Consider the unique strong solution to* $\mathrm{d}X(t) = -\lambda X(t)\mathrm{d}t + \sigma\mathrm{d}B^{\alpha}(t)$ *starting at zero, with* $\alpha \in (0, 1/2)$, $\lambda, \sigma > 0$, *and* B^{α} *a fractional Brownian motion; its fractional derivative* $X^{-\alpha}$ *defined via the identity* $X = \mathrm{I}^{1+\alpha}X^{-\alpha}$, *or*

$$X^{-\alpha}(t) = \frac{\mathrm{d}}{\mathrm{d}t}\left[\int_0^t \frac{(t-s)^{-\alpha}}{\Gamma(1-\alpha)}X_s\mathrm{d}s\right] = \int_0^t \frac{(t-s)^{-\alpha}}{\Gamma(1-\alpha)}\mathrm{d}X_s,$$

satisfies $\mathrm{d}X^{-\alpha}(t) = -\lambda X^{-\alpha}(t)\mathrm{d}t + \sigma\mathrm{d}B(t)$, *with* $X^{-\alpha}(0) = 0$, *for some standard Brownian motion* B. *Conversely, if* X *satisfies* $\mathrm{d}X(t) = -\lambda X(t)\mathrm{d}t + \sigma\mathrm{d}B(t)$, *with* $X(0) = 0$, *then*

$$\mathrm{d}\,\mathrm{I}^{1+\alpha}X(t) = -\lambda\mathrm{I}^{1+\alpha}X(t)\mathrm{d}t + \sigma\mathrm{d}B^{\alpha}(t), \quad \textit{with} \quad \mathrm{I}^{1+\alpha}X(0) = 0.$$

In the context of forward variance curve, we can further write, for $0 \leq t \leq u$,

$$\xi_t^{(H)}(u) := \mathbb{E}[\mathcal{V}_u|\mathcal{F}_t] = \mathbb{E}\left[V_t\mathrm{e}^{-\lambda(u-t)} + \theta\left(1 - \mathrm{e}^{-\lambda(u-t)}\right) + \nu\int_t^u \mathrm{e}^{-\lambda(u-s)}\sqrt{V_s}\mathrm{d}W_s\middle|\mathcal{F}_t\right].$$

As the mean reversion λ tends to zero, the forward variance reads $\xi_t(u) := \mathbb{E}[V_u|\mathcal{F}_t] = V_t$ in the standard Heston model, and hence for any $0 \leq t \leq u$,

$$V_u = \xi_t(u) + \nu\int_t^u \sqrt{V_s}\mathrm{d}W_s.$$

For the fractional Heston model, this then implies

$$\mathcal{V}_u = \eta + \mathrm{I}^{H-\frac{1}{2}}[V_u] = \eta + \mathrm{I}^{H-\frac{1}{2}}\left[V_t + \nu\int_t^u \mathrm{e}^{-\lambda(u-s)}\sqrt{V_s}\mathrm{d}W_s\right]$$
$$= \eta + \mathcal{V}_t + \mathrm{I}^{H-\frac{1}{2}}\left[\nu\int_t^u \mathrm{e}^{-\lambda(u-s)}\sqrt{V_s}\mathrm{d}W_s\right].$$

Since $\xi_t^{(H)}(u) := \mathbb{E}[\mathcal{V}_u|\mathcal{F}_t] = \eta + \mathcal{V}_t$, we can thus write the fractional Heston model in forward variance form as

$$\mathcal{V}_u = \xi_t^{(H)}(u) + \mathrm{I}^{H-\frac{1}{2}}\left[\nu\int_t^u \mathrm{e}^{-\lambda(u-s)}\sqrt{V_s}\mathrm{d}W_s\right].$$

4.2 ▪ Pricing and hedging

One of the main reasons for the popularity of the classical Heston model is the closed-form availability of its characteristic function (4.2). Indeed, once the latter is known, pricing European options is straightforward thanks to the correspondence between option prices and characteristic function, as developed by Carr and Madan [85] and Lewis [229]. It turns out that this approach is still feasible in the rough/fractional extensions of the Heston model, and we now describe their characteristic functions.

4.2.1 ▪ The characteristic function of the rough Heston model

Just as in the classical case, we can compute in quasi-closed form the characteristic function of the log-price of the stock in the rough Heston model (4.4). More precisely, using a connection between Hawkes processes and the rough Heston model, it is shown in [129] that the Heston formula (4.2) for the characteristic function of the log-price extends to the rough case as follows:

$$\mathbb{E}\left[\left(\frac{S_t}{S_0}\right)^{\mathrm{i}u}\right] = \exp\left(g(u,t) + V_0\,\mathrm{I}^{\frac{1}{2}-H}h(u,t)\right),$$

provided we replace the time derivative in the Riccati equation (4.3) by a fractional derivative $\mathrm{D}^{H+\frac{1}{2}} = (\mathrm{d}/\mathrm{d}t)\,\mathrm{I}^{\frac{1}{2}-H}$,

$$\mathrm{D}^{H+\frac{1}{2}}h(u,t) = -\frac{u(u+\mathrm{i})}{2} + (\mathrm{i}u\varrho\nu - \lambda)h(u,t) + \frac{\nu^2}{2}h^2(u,t), \tag{4.7}$$

with boundary condition $\mathrm{I}^{\frac{1}{2}-H}h(u,0) = 0$. As above,

$$g(u,t) = \lambda\theta\int_0^t h(u,s)\mathrm{d}s.$$

Even more, the conditional characteristic function, which is crucial to building hedging strategies, is also derived in [131]. In fact, it is shown that the rough Heston model is Markovian in log-price $X_t = \log(S_t)$ and the forward variance curve $(\xi_t(u))_{u\geq 0}$. In forward variance form, the characteristic function of the terminal log-spot X_T conditional on the state (X_t, ξ_t) at time t is given by

$$\mathbb{E}_{X_t,\xi_t}\left[\mathrm{e}^{\mathrm{i}uX_T}\right] = \exp\left\{\mathrm{i}uX_t + \int_t^T \mathrm{D}^{H+\frac{1}{2}}h(u,T-s)\xi_t(s)\mathrm{d}s\right\}, \tag{4.8}$$

where we set $\lambda = 0$ for simplicity. In contrast to the classical Heston case, no explicit solution to (4.7) is known. Luckily, very efficient numerical procedures have been developed to solve this issue, as detailed in Section 4.3.2. The effective domain of the characteristic function can be computed and estimated by the results of [171] and [223]. This is of interest for option pricing by Fourier inversion [171, Section 10].

With the characteristic function now in the Markovian form (4.8), hedging European options becomes obvious. Let $C_t(T) := \mathbb{E}_{X_t,\xi_t}[f(X_T)]$ denote the time t price of a European option with payoff $f(X_T)$. As of time t, the hedge portfolio has $\partial_{S_t} C_t(T)$ of stock and $\partial_{\xi_t} C_t(T)$ of the forward variance curve $\xi_t(s)$ for each $s \in (t, T]$, where ∂_{ξ_t} represents the Fréchet derivative [131]. From these expressions, it is clear that perfect replication exists only in theory. In practice, as with interest rates, one only holds a finite number of variance contracts. Note, however, that the rough Heston model has only one volatility factor. Thus, one can hedge with only one European option, as in the classical Heston case, provided the value of the option component of the hedge portfolio coincides with the theoretical value of the forward variance component. More details about the hedging aspects of rough volatility can be consulted in Chapter 6.

4.2.2 ▪ Pricing in the fractional Heston model

In the fractional Heston model (4.6), denote by $\varphi(t,u) := \mathbb{E}[S_t^u]$ the cumulant generating function of the log stock price, for any $t \geq 0$, and $\mathcal{D}_t := \{u \in \mathbb{R} : |\varphi(t,u)| < \infty\}$ its effective domain. It is interesting to note that the couple $(X, \mathcal{V})$ remains affine in the sense of [122]. However, correlation breaks the Markovian property of the process, as well as its affine property.

Theorem 4.1. *For any $t \geq 0$ and $(u, H) \in \mathcal{D}_t \times (0,1)$,*

$$\log \varphi(t,u) = \frac{u(u-1)\eta t}{2} - h(t)V_0 + g(t), \tag{4.9}$$

where the function h satisfies the ODE

$$h'(t) = -\lambda h(t) - \frac{\nu^2}{2} h(t)^2 - \frac{u(u-1)}{2\Gamma(\alpha+1)} t^{H-\frac{1}{2}}, \tag{4.10}$$

and $g(t) = -\lambda\theta \int_0^t h(s)\mathrm{d}s$, with initial conditions $g(0) = h(0) = 0$.

Remark 4.2.1. *When $\eta = 0$ and $H = \frac{1}{2}$, the Riccati equation* (4.10) *is the same as in the standard (uncorrelated) Heston model, which implies that, for zero correlation, the model* (4.6) *and the Heston model have the same marginals.*

When $\lambda = 0$, the Riccati equation (4.10) can be solved explicitly [258] as

$$\begin{aligned}\log \varphi(t,u) = \frac{u(u-1)\eta t}{2} - \frac{2V_0}{\nu^2}\partial_t \log\Bigg(\sqrt{t}\Bigg[C_1 J_{\frac{1}{H+\frac{3}{2}}}\left(\frac{\nu}{H+\frac{3}{2}}\sqrt{\frac{u(u-1)}{\Gamma\left(H+\frac{1}{2}\right)}}\, t^{\frac{1}{2}(H+\frac{3}{2})}\right) \\ + C_2 Y_{\frac{1}{H+\frac{3}{2}}}\left(\frac{\nu}{H+\frac{3}{2}}\sqrt{\frac{u(u-1)}{\Gamma\left(H+\frac{1}{2}\right)}}\, t^{\frac{1}{2}(H+\frac{3}{2})}\right)\Bigg]\Bigg),\end{aligned}$$

where J and Y are the Bessel functions of the first and second kinds, and C_1 and C_2 are constants determined by the boundary conditions. For $H = \frac{1}{2}$, $J_{\frac{1}{2}}(x) = \sqrt{\frac{2}{\pi x}}\sin(x)$, and $Y_{\frac{1}{2}}(x) = -\sqrt{\frac{2}{\pi x}}\cos(x)$, and we recover the Heston moment generating function [222] (with $\lambda = \eta = 0$), namely $\log \varphi(t,u) = \frac{V_0}{\nu}\sqrt{u(u-1)}\tan\left(\frac{\nu t}{2}\sqrt{u(u-1)}\right)$.

In the presence of correlation between the stock price process S and its instantaneous variance $\mathcal{V}$, this is no longer the case, as the affine structure of the pair $(S, \mathcal{V})$ (or at least a transformation thereof) is lost. From here on we shall concentrate our efforts on the rough Heston model, together with its pricing and hedging.

Variance swap

In the fractional Heston model (4.6), we can also derive prices for variance swaps. Variance swaps are highly liquid traded derivatives on equity markets; they are standard swaps where, over the time horizon $[0, T]$, the floating leg is equal to the (annualized) realized variance measured by the discretely monitored formula

$$\frac{252}{T}\sum_{i=1}^{n}\log\left(\frac{S_{t_i}}{S_{t_{i-1}}}\right)^2,$$

where $0 = t_0 < t_1 < \cdots < t_n = T$ is an equidistant partition with $t_i - t_{i-1} = iT/n$ corresponding to one day for some positive integer n. The early advances on the hedging and pricing of the variance swap by Neuberger [245], Dupire [123], Carr and Madan [86], and Demeterfi et al. [114] led to the instrument being used extensively by traders to express views on future realized variance and hedging volatility risk. Their usual formulation considers instead the continuously monitored version, measured by the quadratic variation $\langle \log S\rangle_T$ of the log share price process. The fair strike of a continuously monitored variance swap with maturity T then reads

$$\begin{aligned}\mathbb{E}\left[\frac{1}{T}\int_0^T \mathcal{V}_s \mathrm{d}s\right] &= \frac{1}{T}\int_0^T \left\{\eta + \mathrm{I}^{H-\frac{1}{2}}\left[V_0\mathrm{e}^{-\lambda s} + \theta(1-\mathrm{e}^{-\lambda s})\right]\right\}\mathrm{d}s\\ &= \eta + \frac{V_0 T^{H-\frac{1}{2}}}{\Gamma(H+\frac{3}{2})} + \frac{\lambda(\theta - V_0)\mathrm{e}^{-\lambda T}}{\Gamma(H+\frac{3}{2})T}\int_0^T s^{H+\frac{1}{2}}\mathrm{e}^{\lambda s}\mathrm{d}s.\end{aligned}$$

When $\eta = 0$ and $H = \frac{1}{2}$, this representation coincides with the expected variance formula in the standard Heston case [163]. Furthermore, for small $T > 0$, we have the expansion

$$\mathbb{E}\left[\frac{1}{T}\int_0^T \mathcal{V}_s \mathrm{d}s\right] = \frac{V_0 T^{H-\frac{1}{2}}}{\Gamma(H+\frac{3}{2})} + \eta + O\left(T^{H+\frac{1}{2}}\right),$$

so that the variance swap rate explodes when $H < \frac{1}{2}$ with a rate of $T^{H-\frac{1}{2}}$.

4.2.3 ▪ Real data calibration

To illustrate the quality of the rough Heston model (4.4) and to present the calibration methodology, we provide here SPX volatility surface calibration results for the rough Heston model for one given date in history chosen randomly, May 19, 2017. The only parameters that we calibrate are the Hurst parameter H, the volatility of volatility ν, and the correlation ϱ. The forward variance curve, being a state variable in the model, is fixed to match at-the-money (ATM) volatilities. From Bloomberg, on May 19, 2017, there were 27 expirations from 1 day to over 2.5 years, for a total of 2,743 options quoted. For each such strike-expiration pair, we compute bid and ask implied volatilities $\sigma_i^{\pm} := \sigma_{\mathrm{BS}}^{\pm}(k_i, T_i)$, where k_i denotes the log-strike. Given model parameters $\{H, \nu, \varrho\}$, we can obtain model implied volatilities $\sigma_i^M(H, \nu, \varrho)$. We calibrate the parameters by minimizing[39]

$$\sum_i \left[\left(\sigma_i^M(H,\nu,\varrho) - \sigma_i^+\right)^2 + \left(\sigma_i^M(H,\nu,\varrho) - \sigma_i^-\right)^2\right],$$

subject to the constraints

$$H \in \left(0, \frac{1}{2}\right], \quad \nu \geq 0, \quad \varrho \in [-1, 1],$$

[39] Alternatively, the parameters of the rough Heston model may be efficiently calibrated to the term structure of leverage swaps using a closed-form formula [9].

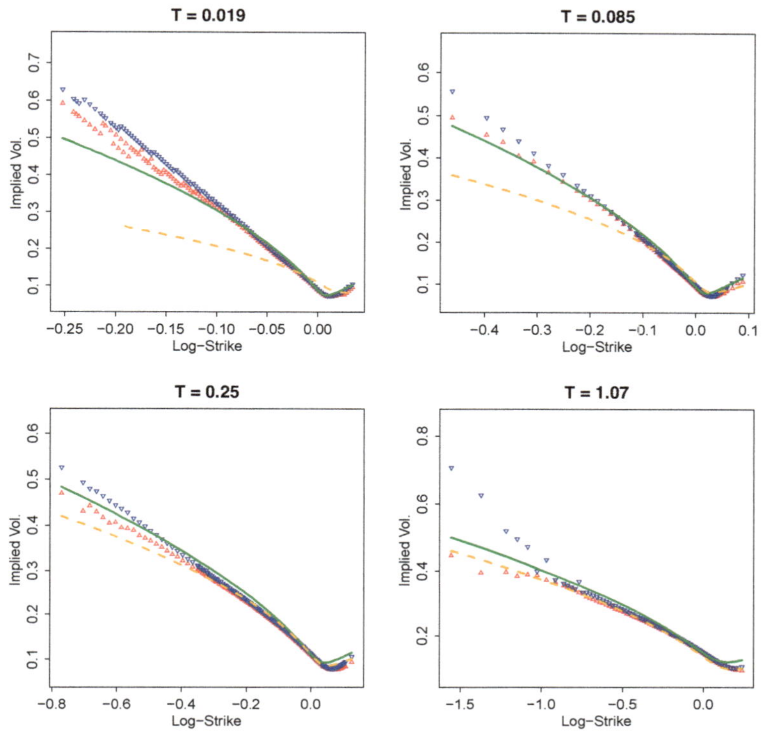

Figure 4.1: Representative SPX volatility smiles as of May 19, 2017, with maturity in years. Red and blue points represent bid and ask SPX implied volatilities; green smiles are from the rough Heston model calibrated using the Adams scheme; dashed orange lines are from the classical Heston model fitted to these four smiles.

and we obtain the optimal parameters

$$H = 0.0474; \quad \nu = 0.4061; \quad \varrho = -0.6710.$$

Figure 4.1 shows a remarkable fit to the SPX volatility surface.

4.3 ▪ Numerical approximations of the rough Heston model

We focus here on developing numerical schemes to evaluate option prices in the rough Heston model (4.4). The first approach, due to [1], is a representation of the process as an infinite superposition of standard (not rough) Heston models, while the second technique presents numerical schemes for solving the fractional Riccati equation (4.7).

4.3.1 ▪ Pathwise approximation of the rough Heston model

Because of the non-Markovian nature of the system, pathwise simulation of the rough Heston model is an intricate issue. One possibility is to use the Markovian approximation proposed in [3]. The idea behind this approximation is to remark that the fractional kernel $\frac{t^{H-1/2}}{\Gamma(H+1/2)}$ is the Laplace transform of a nonnegative measure μ:

$$K(t) = \int_0^\infty \mathrm{e}^{-\gamma t}\mu(\mathrm{d}\gamma), \qquad \text{with} \qquad \mu(\mathrm{d}\gamma) = \frac{\gamma^{-H-\frac{1}{2}}}{\Gamma\big(H+\frac{1}{2}\big)\Gamma\big(\frac{1}{2}-H\big)}\mathrm{d}\gamma.$$

Applying the stochastic Fubini theorem, we obtain that

$$V_t = g(t) + \int_0^\infty V_t^\gamma \mu(\mathrm{d}\gamma), \quad t \in [0,T],$$

with

$$\mathrm{d}V_t^\gamma = \left(-\gamma V_t^\gamma - \lambda V_t\right)\mathrm{d}t + \nu\sqrt{V_t}\mathrm{d}W_t, \quad V_0^\gamma = 0, \quad \gamma \geq 0,$$

and

$$g(t) = V_0 + \int_0^t K(t-s)\theta^0(s)\mathrm{d}s.$$

Heuristically, this exhibits an infinite-dimensional Markovian structure of the variance process where the state variables are given by the factors $(V_t^\gamma)_{\gamma\geq 0}$. In order to reduce the number of state variables, we shall approximate the measure μ by a weighted sum of Dirac measures

$$\mu^n = \sum_{i=1}^n c_i^n \delta_{\gamma_i^n}, \quad n \geq 1,$$

leading to the following approximation $V^n = (V_t^n)_{t\leq T}$ of the variance process V:

$$\begin{aligned} V_t^n &= g^n(t) + \sum_{i=1}^n c_i^n V_t^{n,i}, && t \in [0,T], \\ \mathrm{d}V_t^{n,i} &= \left(-\gamma_i^n V_t^{n,i} - \lambda V_t^n\right)\mathrm{d}t + \nu\sqrt{V_t^n}\mathrm{d}W_t, && V_0^{n,i} = 0, \end{aligned} \tag{4.11}$$

where

$$g^n(t) = V_0 + \int_0^t K^n(t-u)\theta^0(u)\mathrm{d}u, \tag{4.12}$$

and

$$K^n(t) = \sum_{i=1}^n c_i^n e^{-\gamma_i^n t}.$$

This leads us to the following definition of the multifactor approximation:

$$\mathrm{d}S_t^n = S_t^n\sqrt{V_t^n}\mathrm{d}B_t, \quad t \in [0,T],$$

with V^n given as in (4.11). This model is Markovian with $n+1$ state variables which are the spot price S^n and the factors of the variance process $V^{n,i}$ for $i \in \{1,\dots,n\}$. Then one needs to choose the positive weights $(c_i^n)_{1\leq i\leq n}$ and mean reversion coefficients $\gamma_1^n < \cdots < \gamma_n^n$. From (4.11) and (4.12), we expect that V^n converges to V if the kernel K^n is close to K for large n. Hence, inspired by [84], we naturally look for a condition on the weights and mean-reversions such that

$$\lim_{n\uparrow\infty}\int_0^T |K^n(s) - K(s)|^2\mathrm{d}s = 0.$$

This is done by taking

$$c_i^n = \int_{\eta_{i-1}^n}^{\eta_i^n}\mu(\mathrm{d}\gamma) \quad \text{and} \quad \gamma_i^n = \frac{1}{c_i^n}\int_{\eta_{i-1}^n}^{\eta_i^n}\gamma\mu(\mathrm{d}\gamma), \quad \text{for } i \in \{1,\dots,n\},$$

where the $(\eta_i^n)_{0\leq i\leq n}$ are auxiliary parameters with $\eta_0^n = 0$ and $\eta_{i-1}^n \leq \gamma_i^n \leq \eta_i^n$ for $i \in \{1,\dots,n\}$. In that case, one obtains the convergence of the approximated kernel K^n to K provided η_n^n tends to infinity and

$$\lim_{n\uparrow\infty}\sum_{i=1}^n\int_{\eta_{i-1}^n}^{\eta_i^n}(\gamma_i^n - \gamma)^2\mu(\mathrm{d}\gamma) = 0.$$

It is shown in [3] that with this choice of weights and mean reversions, the multifactor approximation is accurate and converges to the rough Heston model.

4.3.2 ▪ Solving the fractional Riccati equation

As noted previously, the Riccati equation (4.7) is of fractional type, and classical numerical schemes do not apply. We review several techniques that have been proposed to compute its solution and therefore allow for (fast) option pricing.

The Adams scheme

To use the characteristic functions presented above, one needs to solve fractional ODEs like (4.7). We consider the equation

$$\mathrm{D}^{H+\frac{1}{2}}h(u,t) = F(u,h(u,t)), \tag{4.13}$$

with boundary condition $h(u,0)=0$. The most classical scheme for solving (4.13) is the Adams method [118]. It is based on the idea that (4.13) implies the following Volterra equation:

$$h(u,t) = \frac{1}{\Gamma\left(H+\frac{1}{2}\right)}\int_0^t (t-s)^{H-\frac{1}{2}}F(u,h(u,s))\mathrm{d}s. \tag{4.14}$$

We write $g(u,t)=F(u,h(u,t))$. Over a regular discrete time grid with mesh Δ, $0\le t_0,\ldots\le t_n\le t$, we approximate

$$h(u,t_{k+1}) = \frac{1}{\Gamma\left(H+\frac{1}{2}\right)}\int_0^{t_{k+1}} (t_{k+1}-s)^{H-\frac{1}{2}}g(u,s)\mathrm{d}s$$

by

$$\frac{1}{\Gamma\left(H+\frac{1}{2}\right)}\int_0^t (t-s)^{H-\frac{1}{2}}\widehat{g}(u,s)\mathrm{d}s,$$

where, for $0\le j\le k$,

$$\widehat{g}(u,t) = \frac{t_{j+1}-t}{t_{j+1}-t_j}\widehat{g}(u,t_j) + \frac{t-t_j}{t_{j+1}-t_j}\widehat{g}(u,t_{j+1}),\quad t\in[t_j,t_{j+1}].$$

This corresponds to a trapezoidal discretization and leads to the following scheme:

$$\widehat{h}(u,t_{k+1}) = \sum_{0\le j\le k} a_{j,k+1}F(u,\widehat{h}(u,t_j)) + a_{k+1,k+1}F\left(u,\widehat{h}(u,t_{k+1})\right), \tag{4.15}$$

with

$$\begin{aligned}
a_{0,k+1} &= \frac{\Delta^{H+\frac{1}{2}}}{\Gamma\left(H+\frac{5}{2}\right)}\left[k^{H+\frac{3}{2}} - \left(k-H-\frac{1}{2}\right)(k+1)^{H+\frac{1}{2}}\right],\\
a_{j,k+1} &= \frac{\Delta^{H+\frac{1}{2}}}{\Gamma\left(H+\frac{5}{2}\right)}\left[(k-j+2)^{H+\frac{3}{2}} + (k-j)^{H+\frac{3}{2}} - 2(k-j+1)^{H+\frac{3}{2}}\right],\\
a_{k+1,k+1} &= \frac{\Delta^{H+\frac{1}{2}}}{\Gamma\left(H+\frac{5}{2}\right)},
\end{aligned} \tag{4.16}$$

for $1\le j\le k$. However, because $\widehat{h}(u,t_{k+1})$ is on both sides of (4.15), this scheme is implicit. Thus, as a first step, we compute a pre-estimation (or *predictor*) of $\widehat{h}(u,t_{k+1})$ based on a Riemann sum that we then plug into the trapezoidal quadrature. We define this predictor as

$$\widehat{h}^P(u,t_{k+1}) := \frac{1}{\Gamma\left(H+\frac{1}{2}\right)}\int_0^t (t-s)^{H-\frac{1}{2}}\widetilde{g}(u,s)\mathrm{d}s,$$

with $\widetilde{g}(u,t) = \widehat{g}(u,t_j)$ for $t \in [t_j, t_{j+1})$, for all $0 \le j \le k$. Therefore,

$$\widehat{h}^P(u,t_{k+1}) = \sum_{0\le j\le k} b_{j,k+1} F\big(u, \widehat{h}(u,t_j)\big),$$

where, for $0 \le j \le k$,

$$b_{j,k+1} = \frac{\Delta^{H+\frac{1}{2}}}{\Gamma\big(H+\frac{3}{2}\big)} \Big((k-j+1)^{H+\frac{1}{2}} - (k-j)^{H+\frac{1}{2}} \Big).$$

Thus, the final explicit numerical scheme is given by

$$\widehat{h}(u,t_{k+1}) = \sum_{0\le j\le k} a_{j,k+1} F(u, \widehat{h}(u,t_j)) + a_{k+1,k+1} F\big(u, \widehat{h}^P(u,t_j)\big),$$

where the weights $a_{j,k+1}$ are defined in (4.16).

Approximate series

Most partial differential equations (PDEs) in quantitative finance do not admit closed-form expressions, and a successful and efficient trick has been to consider instead approximate series expansions of the solution. We refer the interested reader to the monograph [140] for an overview. In the fractional case, however, the situation is more involved, and this gap has been filled in [82], at least in the fractional Riccati case (4.7). Noting that fractional power series are stable under fractional differentiation, they provide a so-called hybrid scheme for (4.7) (actually for a slightly extended version, but we only consider (4.7) here). Without loss of generality, they consider an equation of the form

$$\mathrm{D}^{H+\frac{1}{2}}\psi(t) = \lambda\psi(t)^2 + \mu\psi(t) + \nu, \qquad \text{with } \mathrm{I}^{\frac{1}{2}-H}\psi(0) = \nu, \tag{4.17}$$

for any starting point ν, where the dependence on the parameter u in (4.7) is hidden in the complex coefficients λ, μ, ν. They start by developing an ansatz of the form $\widetilde{\psi}(t) = \sum_{n\ge0} \alpha_n t^{(H+1/2)n}$, where the coefficients $(\alpha_n)_{n\ge0}$ are complex, and, by identification, show that they satisfy a sequence of convolution equations. Like every power series expansion, the latter is only valid up to its radius of convergence $R_{\lambda,\mu,\nu}$, for which the authors are able to provide explicit (in terms of the model parameters) upper and lower bounds. Under some constraints on the parameters, $R_{\lambda,\mu,\nu}$ can be computed to arbitrary precision by the algorithm in Section 7 of [171], which also uses the fractional power series.

When the maturity of the option is small (in particular, smaller than $R_{\lambda,\mu,\nu}$), the power series expansion is very accurate, with a low number of terms, and extremely fast. However, it may not hold any longer beyond $R_{\lambda,\mu,\nu}$, after which [171] switches to an Euler discretization—combined with a Richardson–Romberg extrapolation—of the Riccati equation, with the added complexity of keeping track of past values; hence the notion of hybrid scheme (not to be confused with the hybrid scheme from Chapter 7). Numerical results seem to encourage the use of this method instead of the more classical Adams scheme, originally proposed in [129]. For short maturities (up to one week), computational times are similar, but the Adams scheme is not very stable and can output arbitrageable prices. For maturities larger than a month, the Adams scheme becomes computationally prohibitive, while the hybrid scheme (where the maturity is usually larger than the radius of convergence $R_{\lambda,\mu,\nu}$) remains fast and accurate.

Padé approximations

Using the Itô decomposition derived in [8], Alòs, Gatheral, and Radoičić [9] derived the small-time expansion

$$h(u,t) = \sum_{n\geq 0} \frac{\Gamma_n^+}{\Gamma_{n+1}^+}\beta_n(u)\nu^{n+1}t^{(n+1)(H+\frac{1}{2})}, \tag{4.18}$$

for the solution to the fractional Riccati equation (4.7), where we set $\Gamma_n^\pm := \Gamma\left(1 \pm \left(H+\frac{1}{2}\right)n\right)$. The coefficients $\beta_n(u)$ are defined recursively as $\beta_0(u) = -\frac{1}{2}u(u+\mathrm{i})$, and

$$\beta_n(u) = \frac{1}{2}\sum_{i,j=0}^{n-2}\left(\mathbf{1}_{\{i+j=n-2\}}\beta_i(u)\beta_j(u)\frac{\Gamma_i^+}{\Gamma_{i+1}^+}\frac{\Gamma_j^+}{\Gamma_{j+1}^+}\right) + \mathrm{i}\varrho u\frac{\Gamma_{n-1}^+}{\Gamma_n^+}\beta_{n-1}(u).$$

In the large-time case, such an expansion is more involved, however, but based on an analogy with the standard (nonfractional) Riccati equation (4.3), Gatheral and Radoičić [170] showed that the approximation

$$h(u,t) = \underline{r}\sum_{n\geq 0}\gamma_n\frac{\nu^{-n}t^{-n(H+\frac{1}{2})}}{A_u^n\Gamma_n^-}, \tag{4.19}$$

with $A_u := \sqrt{u(u+\mathrm{i}) - \varrho^2u^2}$ and $\underline{r} := -\mathrm{i}\varrho u - A_u$, is valid for large times with an error of order $O(t^{-2H-1})$, where the coefficients γ_n satisfy the recursion $\gamma_0 = 1$, $\gamma_1 = -1$, and, for $n \geq 2$,

$$\gamma_n = -\gamma_{n-1} + \frac{\underline{r}\Gamma_n^-}{2A}\sum_{i,j=1}^{\infty}\mathbf{1}_{\{i+j=n\}}\frac{\gamma_i\gamma_j}{\Gamma_i^-\Gamma_j^-}.$$

The main result in [170] is then to find a Padé approximation (e.g., a ratio of two polynomials) for the function $h(\cdot)$ consistent with both the small-time (4.18) and the large-time (4.19) expansions. More specifically, a Padé approximation for the function $h(\cdot)$ has the form

$$h^{m,n}(u,t) = \frac{\sum_{i=1}^m p_i(u)t^{(H+\frac{1}{2})i}}{\sum_{j=1}^n q_j(u)t^{(H+\frac{1}{2})j}}.$$

Matching such an expansion to the small- and large-time expressions above yields a system of equations that can be solved in closed form, and we refer the reader to [170] for the long, albeit explicit, expressions for the coefficients $p_i(u)$ and $q_j(u)$ in the diagonal case $m = n = 3$.

4.3.3 ▪ A poor man's Heston

We end this section with some fast and almost instantaneous recipes enabling us to compute an approximation for the implied volatility in the rough Heston model for a given expiry T and parameters (H, ϱ, ν). The realized variance of the rough Heston model is given from (4.5) by

$$\int_0^T V_u \mathrm{d}u = \int_0^T \xi_0(u)\mathrm{d}u + \frac{\nu}{\Gamma(H+\frac{3}{2})}\int_0^T (T-u)^{\frac{1}{2}+H}\sqrt{V_u}\,\mathrm{d}W_u. \tag{4.20}$$

The variance of (4.20) is equal to

$$\nu^2\int_0^T \frac{(T-s)^{2H+1}}{\Gamma(H+\frac{3}{2})^2}\xi_0(s)\mathrm{d}s. \tag{4.21}$$

We now want to approximate the rough Heston smile by that produced from the classical Heston-like model (4.5) with $H = 1/2$. In that case, a natural choice is a scaled volatility of volatility parameter $\widetilde{\nu}(T)$ matching the variance (4.21), that is,

$$\nu^2 \int_0^T \frac{(T-t)^{2H+1}}{\Gamma(H+\frac{3}{2})^2}\xi_0(t)\mathrm{d}t = \widetilde{\nu}(T)^2 \int_0^T (T-t)^2\xi_0(t)\mathrm{d}t.$$

Thus,

$$\widetilde{\nu}(T) = \frac{\nu}{\Gamma(H+\frac{3}{3})}\sqrt{\frac{\int_0^T (T-t)^{2H+1}\xi_0(t)\mathrm{d}t}{\int_0^T (T-t)^2\xi_0(t)\mathrm{d}t}}. \tag{4.22}$$

For each expiry T, the characteristic function formula (4.8) is then approximated by the classical one,

$$\exp\left\{\mathrm{i}aX_0 + \int_0^T \partial_t h^{(T)}(u, T-t)\xi_0(t)\,\mathrm{d}t\right\}, \tag{4.23}$$

where $h^{(T)}(u,.)$ is now a solution of the classical Riccati equation,

$$\partial_t h^{(T)}(u,t) = -\frac{u(u+\mathrm{i})}{2} + \mathrm{i}\varrho\widetilde{\nu}(T)uh^{(T)}(u,t) + \frac{1}{2}\widetilde{\nu}(T)^2(h^{(T)})^2(u,t),$$

with boundary condition $h^{(T)}(u,0) = 0$. This equation can be solved explicitly [163, page 18] as

$$h^{(T)}(u,t) = r_-(T)\frac{1-\exp\left\{-A_u\widetilde{\nu}(T)t\right\}}{1-\frac{r_-(T)}{r_+(T)}\exp\left\{-A_u\widetilde{\nu}(T)t\right\}}, \tag{4.24}$$

with

$$A_u = \sqrt{u(u+\mathrm{i})-\varrho^2u^2} \qquad \text{and} \qquad r_\pm(T) = -\frac{\mathrm{i}\varrho u \pm A_u}{\widetilde{\nu}(T)}.$$

For a given expiry T, we can simplify even more the previous approximation by taking a flat forward variance curve $\xi_0(\cdot)$. In practice, the obvious choice is $\frac{1}{T}\int_0^T \xi_0(t)\mathrm{d}t$, the fair value of the variance swap. In that case, we obtain (4.22),

$$\widetilde{\nu}(T) = \sqrt{\frac{3}{2H+2}}\frac{\nu}{\Gamma\left(H+\frac{3}{2}\right)}\frac{1}{T^{\frac{1}{2}-H}}. \tag{4.25}$$

Under this poor man's approach, the approximate characteristic function (4.23) is identical to the characteristic function of the classical Heston model with initial variance $\frac{1}{T}\int_0^T \xi_0(t)\mathrm{d}t$, mean reversion $\lambda = 0$, correlation ϱ, and volatility of volatility $\widetilde{\nu}(T)$ given by (4.25). Option prices under rough Heston may thus be almost instantaneously approximated using any existing implementation of the classical Heston pricing model. To demonstrate the quality of approximations (4.22) and (4.25), in Figure 4.2 we have replotted the rough Heston smiles of Figure 4.1 generated using the Adams scheme together with those generated using the above approximations. We observe that both approximations are surprisingly good, even the poor man's one, although that using (4.22) is obviously closer to true rough Heston smiles.

4.4 ▪ Asymptotics for the rough Heston model

4.4.1 ▪ Tail and wing asymptotics

While path properties of rough volatility models are inherently different from those of their classical volatility counterparts, one may wonder whether, for a given maturity $T > 0$, rough smiles

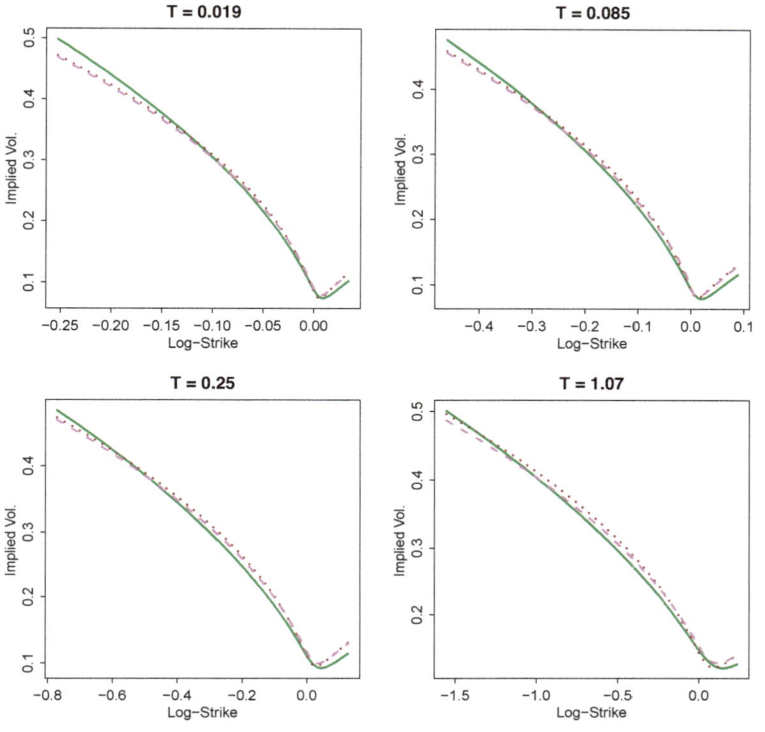

Figure 4.2: Representative SPX volatility smiles as of May 19, 2017, with time to expiration in years. Green smiles are from the rough Heston model calibrated using the Adams scheme; violet dashed smiles are generated using the approximate rough Heston characteristic function (4.23); the brown dotted smiles are generated from the poor man's existing Heston model with scaled volatility of volatility parameter.

may be similar to classical ones. General results are out of reach for now. When the volatility process possesses a specific scaling structure (essentially being of Ornstein–Uhlenbeck type), large deviations techniques can be developed for the two-dimensional process $(\varepsilon \log(S), \varepsilon^{\alpha}\sigma)$, for some $\alpha > 0$ as ε becomes small. In turn, this asymptotic tail behavior can be transformed into behavior for the large-strike implied volatility, yielding an expression of the form

$$\lim_{x\uparrow\infty} \frac{\sigma^2_{\text{imp}}(x,t)t}{x} = \frac{1}{\Lambda(1)},$$

where Λ plays the role of the large deviations rate function. We refer the interested reader to [115] for a detailed analysis in the classical diffusive case. In the fractional Ornstein–Uhlenbeck case, where a similar scaling holds, large deviations for Gaussian processes yield an analogous formula, albeit with a different function Λ [200]. In the particular case of the rough Heston model, a thorough analysis was carried out in [171, 223]. The key tool is Roger Lee's moment formula, which states the following.

Theorem 4.2 (the moment formula [228]). *Assume that the underlying stock price is a strictly positive martingale, and, for any $T > 0$, denote*

$$p_T^* := \sup\left\{p \geq 0 : \mathbb{E}\left[S_T^{1+p}\right] < \infty\right\} \qquad \textit{and} \qquad q_T^* := \sup\left\{q \geq 0 : \mathbb{E}\left[S_T^{-q}\right] < \infty\right\}.$$

Then

$$\beta_+(T) := \limsup_{x\uparrow\infty} \frac{\sigma^2_{\text{imp}}(x,T)T}{x} = \psi(p_T^*),$$

$$\beta_-(T) := \limsup_{x\downarrow-\infty} \frac{\sigma^2_{\text{imp}}(x,T)T}{|x|} = \psi(q_T^*),$$

where $\psi(u) := 2 - 4\left(\sqrt{u(1+u)} - u\right) \in [0,2]$.

The theorem thus says that the wing behavior of the smile, for any given maturity T, is fully determined by the knowledge of the moment explosions of the random variable S_T. For sufficiently "regular" moment generating functions, all limsups above are actual limits [45, 47]. In the classical Heston case, Andersen and Piterbarg [14] used the explicit representation of the Heston characteristic function to provide semi-explicit representations for p_T^* and q_T^*. In the rough Heston case, the characteristic function is available in semiclosed form, and a similar analysis can be carried out, which allows us to compare the impact of the Hurst exponent on the wing behavior of the smile. More precisely, Keller-Ressel and Majid [223] proved the following.

Proposition 4.3. *Let $\varrho < 0$, $T > 0$, and denote by $\beta^H_\pm(T)$ the limiting left/right slopes of the implied volatility smile at maturity T in the rough Heston model, and by $\beta_\pm(T)$ the corresponding slopes in the classical Heston model.*

- *For the left slope, whenever*

$$T \le \left(\frac{(H+\frac{1}{2})\Gamma\left(H+\frac{1}{2}\right)|\varrho|\pi}{\sqrt{\lambda(\lambda-\varrho\nu)}}\right)^{\frac{1}{H+\frac{1}{2}}},$$

we have

$$\beta^H_-(T) \ge \frac{T^{H-\frac{1}{2}}}{(H+\frac{1}{2})\Gamma(H+\frac{1}{2})}\beta_-\left(\frac{T^{H+\frac{1}{2}}}{(H+\frac{1}{2})\Gamma(H+\frac{1}{2})}\right).$$

- *As T tends to zero,*

$$\lim_{T\downarrow 0} \frac{\beta^H_-(T)}{T^{H+\frac{1}{2}}} \ge \frac{\widetilde{\beta}_-(0)}{(H+\frac{1}{2})\Gamma(H+\frac{1}{2})},$$

$$\lim_{T\downarrow 0} \frac{\beta^H_+(T)}{T^{H+\frac{1}{2}}} \ge \frac{C_+}{\pi - \frac{2\varrho}{\sqrt{1-\varrho^2}}}\frac{\widetilde{\beta}_+(0)}{(H+\frac{1}{2})\Gamma(H+\frac{1}{2})},$$

where

$$C_\pm := \pi - 2\arctan\left(\frac{\pm\varrho}{\sqrt{1-\varrho^2}}\right) \quad \text{and} \quad \widetilde{\beta}_\pm(0) := \lim_{T\downarrow 0}\frac{\beta_\pm(T)}{T} = \frac{\nu\sqrt{1-\varrho^2}}{2C_\pm}.$$

4.4.2 ▪ Small maturity

For fast approximate calibration and for studying the effect of model parameters, asymptotic approximations of option prices and implied volatilities under the rough Heston model (4.4) are of interest. We refer the reader to Chapter 8 for a thorough overview of asymptotics of rough volatility models, and we concentrate here on those pertaining to the rough Heston model, many of which rely of the specific affine structure of the latter.

We mainly focus on small-maturity asymptotics here, which have been the most useful ones in practice. The affine structure of the model allows for asymptotic expressions that can be evaluated fairly easily, in particular without solving a variational problem. The natural small-time regime for rough Heston arises from strikes

$$K_t = S_0 \exp\left\{xt^{\frac{1}{2}-H}\right\}, \quad \text{for } x \text{ fixed,} \tag{4.26}$$

which converge to the spot as the maturity t tends to zero. This is mathematically convenient, as it allows one to translate small-time asymptotics to small-noise asymptotics, where the volatility is driven by $\varepsilon^H B$ with $\varepsilon \downarrow 0$ instead of B. In practice, the strike parameterization (4.26) fits well to available options data, as liquid strikes tend to concentrate around the spot when maturity decreases.

Theorem 4.4 ([134]). *There is a function $I(\cdot)$ such that out-of-the-money call options in the rough Heston model satisfy, with $x > 0$ fixed,*

$$\mathbb{E}\left[\left(S_t - S_0 \exp\left\{xt^{\frac{1}{2}-H}\right\}\right)_+\right] = \exp\left\{-\frac{I(x)}{t^{2H}}(1+o(1))\right\}, \quad t \downarrow 0.$$

The implied volatility for log-strike $X_0 + xt^{\frac{1}{2}-H}$ satisfies

$$\lim_{t\downarrow 0} \sigma_{\text{imp}}\left(xt^{\frac{1}{2}-H}, t\right) = \frac{|x|}{\sqrt{2I(x)}}, \quad x \neq 0. \tag{4.27}$$

These estimates are Corollaries 3.7 and 3.8 of [134] and are a consequence of the large deviations principle proved in Theorem 3.4 of [134]. The rate function I, which governs the exponential decay of the call price, can be computed as follows. Let $\psi(p,t)$ denote the solution of the fractional ODE[40]

$$\mathrm{D}^{H+\frac{1}{2}}\psi(p,t) = \frac{1}{2}p^2 + p\varrho\nu\psi(p,t) + \frac{\nu^2}{2}\psi(p,t)^2, \quad \psi(p,0) = 0, \tag{4.28}$$

and

$$\Lambda(p,t) := \mathrm{I}^{\frac{1}{2}-H}\psi(p,t).$$

Then the rate function

$$I(x) := \sup_{p\in\mathbb{R}} \left(px - V_0\Lambda(p,1)\right) \tag{4.29}$$

is the Fenchel–Legendre transform of $V_0\Lambda(p,1)$. Computing I is greatly simplified by the scaling relation

$$\Lambda(p,1) = |p|^{2H/(H+\frac{1}{2})}\Lambda\left(\text{sgn}(p), |p|^{1/(H+\frac{1}{2})}\right),$$

which shows that it suffices to solve (4.28) for $p = \pm 1$ only and not for all p. The function $\Lambda(p,1)$ is finite on an interval (p_-, p_+) and blows up towards the boundaries; see [134] for details. Note that, in the classical Heston model, (4.26) is the fixed-strike regime, and the analogue of Λ is available in closed form [135].

[40]The initial condition in (4.7) comes from results on fractional ODEs that allow for functions that are singular at zero. For continuous functions, it is automatically satisfied. In fact, no initial condition is needed in (4.7) and (4.28), and the unique continuous solution of the fractional ODE vanishes at zero. This follows from the equivalence with Volterra integral equations [224, Chapter 3].

To avoid solving the optimization problem in (4.29), a fractional power series for $\psi(p,t)$ can be computed as explained after (4.17), leading to a power series for $\Lambda(p,1)$ and finally to a power series for the right-hand side of (4.27). The net result is the approximation

$$\sigma_{\text{imp}}\left(xt^{\frac{1}{2}-H},t\right) \approx \sqrt{V_0} + \frac{\varrho\nu}{2\Gamma(H+\frac{5}{2})\sqrt{V_0}}x + \nu^2\frac{\Gamma(2H+2)+2\varrho^2\Gamma(H+\frac{3}{2})^2\left(2-3\frac{\Gamma(2H+3)}{\Gamma(H+\frac{5}{2})^2}\right)}{8V_0^{\frac{3}{2}}\Gamma(H+\frac{3}{2})^2\Gamma(2H+3)}x^2,$$

which can be used for approximate calibration (for x and t small) to volatility level, skew, and convexity.

A slightly different rescaling and formula was obtained in [137, Section 3] based on an Edgeworth-type expansion. Assuming that $H \in (0, \frac{1}{4})$, the authors prove [137, Corollary 3.4] that

$$\sigma_{\text{imp}}\left(x\sqrt{t},t\right) = \sqrt{V_0} + \frac{\varrho\nu}{2\sqrt{V_0}\Gamma\left(H+\frac{5}{2}\right)}zt^H + O\left(t^{2H}\right),$$

for any fixed $x \neq 0$. Introduce $\eta := \nu/(\Gamma(H+\frac{1}{2})\sqrt{2H})$. This expansion then implies a finite-difference approximation of the skew as

$$\frac{\sigma_{\text{imp}}\left(x\sqrt{t},t\right)-\sigma_{\text{imp}}\left(x'\sqrt{t},t\right)}{(x-x')\sqrt{t}} \approx \frac{\varrho\nu t^{H-\frac{1}{2}}}{2\sqrt{V_0}\Gamma\left(H+\frac{5}{2}\right)} = \frac{\varrho\eta\sqrt{2H}t^{H-\frac{1}{2}}}{2\sqrt{V_0}\left(H+\frac{3}{2}\right)\left(H+\frac{1}{2}\right)},$$

for any $x \neq x'$ nonnull, which is reminiscent of the skew formulae in Chapter 8. Further results for the rough Heston model, using moderate deviations, were proved in [134, Section 3.6].

4.4.3 ▪ Other asymptotic regimes

In the *large*-time, large log-moneyness regime, the implied volatility $\sigma_{\text{imp}}(xt,t)$, for $x > 0$ fixed, converges to an explicit constant as the maturity t tends to infinity (see Section 4 of [134]). This constant is the same as for the classical Heston model, meaning that the Hurst parameter is not visible at first order.

4.5 ▪ The quadratic rough Heston model

4.5.1 ▪ Rough volatility and the Zumbach effect

Despite the success of the rough Heston model, the following subtle question raised by Jean-Philippe Bouchaud remains: Can a rough volatility model reproduce the so-called *Zumbach effect*, the observation originally due to Gilles Zumbach (see [278, 279]) that financial time series are not time-reversal invariant? To answer this question, we introduce two notions, each of which corresponds to different aspects of the Zumbach effect:

- The *weak Zumbach effect* (typically considered in the econophysics literature; see [278]: Past squared returns forecast future integrated volatilities better than past integrated volatilities forecast future squared returns. This property is not satisfied in classical stochastic volatility models. However, rough stochastic volatility models are consistent with the weak Zumbach effect; see [127] for explicit computations in the rough Heston model.

- The *strong Zumbach effect*: Conditional dynamics of volatility with respect to the past depend not only on the past volatility trajectory but also on the historical price path; specifically, price trends tend to increase volatility; see [279]. Such feedback of the historical price path on volatility also occurs on implied volatility as illustrated in [279]. Rough stochastic volatility models such as the rough Heston model are not consistent with the strong Zumbach effect; see [131].

The quest for a rough volatility model consistent with the strong Zumbach effect and the empirical success of quadratic Hawkes process-based models documented in [64] led to the development of super-Heston rough volatility models in [109]. These extensions of the rough Heston model arise as limits of quadratic Hawkes process-based microstructural models just as the rough Heston model arises as the continuous-time limit of a linear Hawkes process-based microstructural model.

4.5.2 ▪ The quadratic rough Heston process

The joint dynamics of the asset S and its spot variance V satisfy

$$dS_t = S_t \sqrt{V_t} dW_t, \; V_t = a(Z_t - b)^2 + c,$$

where W is a Brownian motion, a, b, and c are some positive constants, and Z_t is of rough Heston type, in the sense that weighted past price returns are drivers of the volatility dynamics. More precisely,

$$Z_t = \int_0^t (t-s)^{\alpha-1} \frac{\lambda}{\Gamma(\alpha)} (\theta_0(s) - Z_s) ds + \int_0^t (t-s)^{\alpha-1} \frac{\lambda}{\Gamma(\alpha)} \eta \sqrt{V_s} dW_s, \tag{4.30}$$

with $\alpha \in (1/2, 1)$, $\lambda > 0$, $\eta > 0$ and θ_0 a deterministic function. In this special case of a super-Heston rough volatility model, the asset S and its volatility depend on the history of only one Brownian motion. The model is thus a pure feedback model; volatility is driven only by the price dynamics, with no additional source of randomness. In general, of course, the volatility process does not need to depend only on the Brownian motion driving the asset price S.

As in the general case of super-Heston rough volatility models, because the effect of past returns on Z cannot be reduced to an influence of past volatility dynamics on Z, the quadratic rough Heston model also exhibits the strong Zumbach effect; see [109] for more details. In fact, the process Z_t may be understood as a weighted moving average of past price log returns. Indeed, from Lemma A.1 in [131], we have that

$$Z_t = \int_0^t f^{\alpha,\lambda}(t-s)\theta_0(s) ds + \int_0^t f^{\alpha,\lambda}(t-s)\eta \sqrt{V_s} dW_s,$$

where $f^{\alpha,\lambda}(t)$ is the Mittag–Leffler density function defined for $t \geq 0$ as

$$f^{\alpha,\lambda}(t) = \lambda t^{\alpha-1} E_{\alpha,\alpha}(-\lambda t^{\alpha}),$$

with

$$E_{\alpha,\beta}(z) = \sum_{n \geq 0} \frac{z^n}{\Gamma(\alpha n + \beta)}.$$

The variable Z_t is therefore *path-dependent*: a power-law weighted average of past returns. Therefore, long periods of trends, or sudden upward or downward moves of the price, generate large values for $|Z|$ and thus high volatility, in particular when Z is negative. Such a link is clearly observed on data. Thus, with the quadratic rough Heston dynamics, we obtain a model which exhibits

- rough volatility,
- essentially lognormal volatility (see [109] for details),
- strong Zumbach effect.

4.5.3 ▪ The quadratic rough Heston model and the joint SPX-VIX options calibration problem

The rough Heston model is known to provide excellent fits of the implied volatility surface of the SPX. However, it does not lead to reasonable calibration when we consider VIX options. The VIX index is, in fact, by definition a derivative of the SPX index S, which can be represented as

$$\mathrm{VIX}_t = \sqrt{-2\mathbb{E}[\log(S_{t+\Delta}/S_t)|\mathcal{F}_t]} \times 100\,, \tag{4.31}$$

where $\Delta = 30$ days and $\mathbb{E}$ is the risk-neutral expectation. Consequently, VIX options are also derivatives of SPX. Finding a model which jointly calibrates the prices of SPX and VIX options is known to be very challenging, especially for short maturities. As indicated by Guyon [186, Section 1, page 1]:

> The very large negative skew of short-term SPX options, which in classical continuous SV models implies a very large volatility of volatility, seems inconsistent with the comparatively low levels of VIX implied volatilities.

As explained in [167], combining rough volatility and the Zumbach effect can be key to solving the joint calibration problem with continuous-time models with continuous sample paths. As a matter of fact, it is shown in [167, 265] that the quadratic rough Heston model enables us to obtain the following with only six parameters:

- VIX smiles in the bid-ask spread,
- a global shape of the implied volatility surface of the SPX that is very well reproduced, in particular the ATM skew, which can be exactly matched,
- model SPX skews typically of orders -1.5 (shortest maturities) to -1 (longer maturities) as for market data.

An example of such joint calibration is given in Figures 4.3 and 4.4.

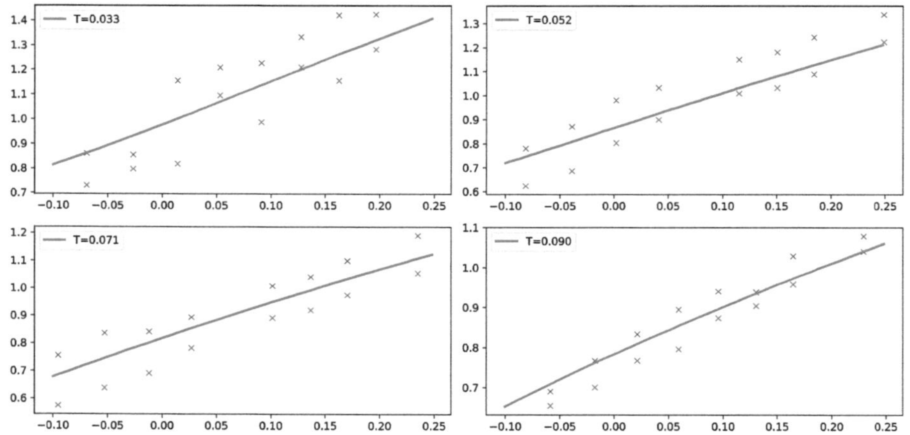

Figure 4.3: Implied volatility on VIX options for May 19, 2017. Blue and red points are bid and ask of market implied volatilities. Model implied volatility smiles from the model are in green. Strikes are in log-moneyness, expiries in years.

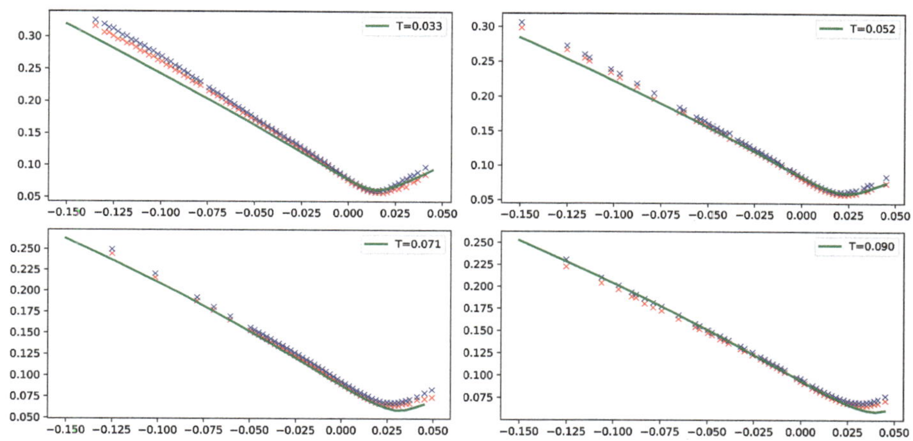

Figure 4.4: Implied volatility on SPX options for May 19, 2017. Blue and red points are bid and ask of market implied volatilities. Model implied volatility smiles from the model are in green. Strikes are in log-moneyness, expiries in years.

Chapter 5
Rough affine models

Martin Larsson, Martin Keller-Ressel, Sergio Pulido

Introduction

Affine stochastic volatility models have a long history in the quantitative finance literature; see [122, 219] and the references listed there. These models are generally of the form

$$\mathrm{d}S_t = S_t\sqrt{V_t}\mathrm{d}B_t, \tag{5.1}$$

where S is the asset price, and the (spot) variance V is modeled by an affine process. Arguably, the most prominent example is the Heston model [199], where V follows a scalar square-root diffusion. The affine property leads to tractable Fourier–Laplace transforms of various quantities of interest. For example, the log-price satisfies the exponential-affine transform formula

$$\mathbb{E}[\exp(u\log S_T)\mid \mathcal{F}_t] = \exp\left(u\log S_t + \varphi(T-t) + \psi(T-t)V_t\right), \tag{5.2}$$

where φ, ψ are the solutions to ordinary differential equations (ODEs) of Riccati type that depend on u. Similar formulas also exist for the spot variance and integrated spot variance.

Unfortunately, these models do not produce the rough trajectories of volatility that seem to occur empirically (see [166]) and have trouble capturing the term structure of implied volatilities and its skew; cf. [153]. Still, it is possible to construct stochastic volatility models with these features, and with an "affine structure" that produces formulas similar to (5.2). This has recently been done by [180, 129, 4, 168], and related ideas appear in [97]. The goal of this chapter is to explain and elucidate this "affine structure" and show how it leads to exponential-affine transform formulas.

We will give four perspectives. The point of departure, in Section 5.1, is a class of *stochastic convolution equations* for V with affine coefficients, which contains the rough Heston model of [129] as an immediate special case. This is the first perspective. The second perspective, in Section 5.2, is to view these models as *forward variance models*, which focus on the forward variance curve

$$\xi_t(T) = \mathbb{E}[V_T\mid \mathcal{F}_t]. \tag{5.3}$$

The third perspective, in Section 5.3, is to regard a modified forward variance curve as the solution of a *stochastic partial differential equation* (SPDE). Finally, the fourth perspective, in

Section 5.4, is available when the convolution kernel is the Laplace transform of a possibly signed measure. This leads to a representation as a *mixture of mean-reverting processes*. Dually, this gives multiple viewpoints on the Riccati equations that characterize the associated Fourier–Laplace functionals.

Space constraints prevent us from including rigorous proofs of all results. Still, some proofs and derivations are presented, selected because they are instructive without being too long. We occasionally use the convolution notation $(f * g)(t) = \int_0^t f(t-s)g(s)\mathrm{d}s$ for functions f and g, and similarly $(f * \mathrm{d}Z)_t = \int_0^t f(t-s)\mathrm{d}Z_s$ when Z is a semimartingale.

5.1 ▪ Stochastic convolution equations

Consider the stochastic convolution equation

$$V_t = V_0 + \int_0^t K(t-s)b(V_s)\mathrm{d}s + \int_0^t K(t-s)\sigma(V_s)\mathrm{d}W_s, \tag{5.4}$$

for some real continuous functions b and σ, kernel $K \in L^2_{\mathrm{loc}}(\mathbb{R}_+)$, initial condition $V_0 \in \mathbb{R}$, and Brownian motion W. Solutions to (5.4) are always understood to have continuous paths.

Example 5.1. Taking $b(x) = \lambda(\theta - x)$, $\sigma(x) = \zeta\sqrt{x}$, and the power-law kernel $K_{\alpha\text{-pow}}(t) = t^{\alpha-1}/\Gamma(\alpha)$ with $\alpha \in (\frac{1}{2}, 1)$, we obtain the spot variance process in the rough Heston model of [129]. With $\alpha = 1$, we recover the spot variance process in the classical Heston model.

Our focus is on the case where V is an *affine Volterra process*, which is when $b(x)$ and $\sigma(x)^2$ are affine in x. This definition naturally generalizes to higher dimension: $b(x)$ is then a vector and $\sigma(x)$ a matrix, and one requires $b(x)$ and $\sigma(x)\sigma(x)^\top$ to be affine in x. In this chapter we focus on the one-dimensional affine case, so that

$$b(x) = \beta - \lambda x \text{ and } \sigma(x)^2 = \alpha + ax \tag{5.5}$$

for some real parameters $\beta, \lambda, \alpha, a$ such that $\alpha + aV_t \geq 0$ for all $t \geq 0$. The latter condition raises delicate questions of existence of solutions to (5.4) when $a \neq 0$, which we do not address in detail. Let us, however, state the following result, whose proof can be found in [4]. Part (ii) of Theorem 5.2 requires the following assumption on the kernel:

> K is strictly positive and completely monotone. There is $\gamma \in (0, 2]$ such that $\int_0^h K(t)^2\mathrm{d}t = O(h^\gamma)$ and $\int_0^T (K(t+h) - K(t))^2\mathrm{d}t = O(h^\gamma)$ for every $T < \infty$. (5.6)

(Recall that a C^∞ function $f\colon (0,\infty) \to \mathbb{R}$ is completely monotone if $(-1)^k f^{(k)} \geq 0$, $k \geq 0$.) In particular, the power-law kernel in Example 5.1 satisfies (5.6) with $\gamma = 2\alpha - 1$.

Theorem 5.2. *Consider the equation* (5.4) *with coefficients* $b(x)$ *and* $\sigma(x)$ *as in* (5.5) *and kernel* $K \in L^2_{\mathrm{loc}}(\mathbb{R}_+)$.

(i) *Assume that* $\alpha \geq 0$ *and* $a = 0$. *Then there exists a pathwise unique strong solution* V *for any initial condition* $V_0 \in \mathbb{R}$, *the* ***Volterra Ornstein–Uhlenbeck process***.

(ii) *Assume that* $\alpha = 0$, $a > 0$, $\beta \geq 0$, *and* K *satisfies* (5.6). *Then there exists a unique in law* $\mathbb{R}_+$*-valued weak solution* V *for any initial condition* $V_0 \in \mathbb{R}_+$; *the* ***Volterra square-root process***.

In either case, the trajectories of V *are Hölder continuous of any order less than* $\gamma/2$.

Remark 5.1.1. *A solution V is called strong if it is adapted to the filtration generated by the Brownian motion W. This is not required for a weak solution, where one is free to construct the Brownian motion as needed. Pathwise uniqueness means that any two solutions V and V' to* (5.4) *driven by the same Brownian motion must have identical trajectories (outside a nullset). It is not known whether pathwise uniqueness holds in* (ii).

A Volterra–Heston model is a stochastic volatility model of the form (5.1), where the spot variance V is a Volterra square-root process. Most of this chapter is concerned with such models. The process V is generally neither a Markov process nor a semimartingale. This causes difficulties that the alternative perspectives developed in the following sections help to circumvent.

5.2 ▪ Forward variance models

A useful perspective on (5.4) is as a *forward variance model*, as noted by [39, 58, 81]. Consider a model (5.1) where the spot variance process V is given by the stochastic convolution equation (5.4) with affine drift $b(x) = \lambda(\theta - x)$, i.e.,

$$V_t = V_0 + \lambda \int_0^t K(t-s)(\theta - V_s)\mathrm{d}s + \int_0^t K(t-s)\sigma(V_s)\mathrm{d}W_s. \tag{5.7}$$

Our first goal is to derive an SDE for the forward variance $\xi_t(T)$ defined in (5.3). To this end, we remark that for any kernel $k \in L^2_{\mathrm{loc}}(\mathbb{R}_+)$ there exists a unique kernel $r \in L^2_{\mathrm{loc}}(\mathbb{R}_+)$, called the *resolvent* or *resolvent of the second kind* of k, such that

$$k(t) - r(t) = \int_0^t r(t-s)k(s)\mathrm{d}s, \quad t \geq 0.$$

Example 5.3. If $k(t) \equiv c$ is constant, then $r(t) = c\,\mathrm{e}^{-ct}$. If $k(t) = c\,t^{\alpha-1}/\Gamma(\alpha)$ is proportional to the power-law kernel, then $r(t) = ct^{\alpha-1}E_{\alpha,\alpha}(-ct^{\alpha})$ where $E_{\alpha,\alpha}$ denotes the Mittag–Leffler function.

The forward variance dynamics can now be described as follows.

Proposition 5.4. *Let R_λ be the resolvent of λK. The forward variance $\xi_t(T)$ associated to* (5.7) *satisfies*

$$\mathrm{d}\xi_t(T) = \frac{1}{\lambda} R_\lambda(T-t)\sigma(V_t)\mathrm{d}W_t$$

with initial condition

$$\xi_0(T) = V_0\left(1 - \int_0^T R_\lambda(s)\mathrm{d}s\right) + \theta \int_0^T R_\lambda(s)\mathrm{d}s.$$

If $\lambda = 0$, interpret $\lambda^{-1}R_\lambda = K$, and note that $R_\lambda = 0$ in this case.

Proof. Suppose $\lambda \neq 0$; otherwise, the proof is easier and does not use resolvents. Denote by $\mathbf{1}$ the function which takes the constant value 1. The spot variance process V is given by $V = V_0 + \lambda K * (\theta - V) + K * (\sigma(X)\mathrm{d}W)$. Therefore,

$$V - R_\lambda * V = V_0(1 - R_\lambda * \mathbf{1}) + \lambda(K - R_\lambda * K) * (\theta - V) + (K - R_\lambda * K) * (\sigma(V)\mathrm{d}W).$$

By definition of the resolvent, $K - R_\lambda * K = \frac{1}{\lambda}R_\lambda$. Plug this in and cancel the $-R_\lambda * V$ terms on both sides to get

$$V = V_0(1 - R_\lambda * \mathbf{1}) + \theta R_\lambda * \mathbf{1} + \frac{1}{\lambda} R_\lambda * (\sigma(V)\mathrm{d}W). \tag{5.8}$$

The process $M_u := \int_0^u R_\lambda(T-s)\sigma(X_s)\mathrm{d}W_s$, $u \in [0,T]$, is a martingale. Therefore, evaluating (5.8) at T and taking $\mathcal{F}_t$-conditional expectations yields

$$\xi_t(T) = \mathbb{E}[V_T \mid \mathcal{F}_t] = \xi_0(T) + \int_0^T \frac{1}{\lambda} R_\lambda(T-s)\sigma(V_s)\mathrm{d}W_s,$$

which is the claimed result. □

Remark 5.2.1. *We ignored some technical but important points in the proof. First, the associativity property* $(k_1 * k_2) * \mathrm{d}Z = k_1 * (k_2 * \mathrm{d}Z)$ *was used for certain kernels* k_1, k_2, *and* $\mathrm{d}Z = \sigma(X)\mathrm{d}W$. *This identity can be proved using the stochastic Fubini theorem. Second, we did not verify that* M *is really a martingale and not just a local martingale. For* $\sigma(x) = \alpha + ax$ *this can be done by noting that*

$$\mathbb{E}[\langle M\rangle_T] = \frac{1}{\lambda^2}\int_0^T R_\lambda(T-s)^2\, \mathbb{E}[\sigma(X_s)^2]\mathrm{d}s \le C\left(1 + \sup_{s\le T}\mathbb{E}[|X_s|]\right),$$

where one can take $C = \frac{1}{\lambda^2}(|\alpha| + |a|)\int_0^T R_\lambda(s)^2\mathrm{d}s$. *The right-hand side is finite, so* M *is actually a square-integrable martingale. Details are given by* [4].

5.2.1 ▪ Fourier–Laplace transforms and Riccati–Volterra equations

In the affine case (5.5), not only conditional expectations but also Fourier–Laplace transforms have useful representations. We now explain how such representations can be derived once the Volterra–Heston model (in log-price notation $L = \log S$) is written in forward variance form,

$$\begin{cases} \mathrm{d}L_t = -\dfrac{1}{2}V_t\mathrm{d}t + \sqrt{V_t}\mathrm{d}B_t, \\ \mathrm{d}\xi_t(T) = \dfrac{1}{\lambda}R_\lambda(T-t)\sigma\sqrt{V_t}\mathrm{d}W_t. \end{cases} \tag{5.9}$$

Here $\lambda \ge 0$, $d\langle B,W\rangle_t = \varrho dt$ for $\varrho \in [0,1]$, and the diffusion part has the affine form (5.5) with $\alpha = 0$ and $a = \sigma^2$ for some $\sigma > 0$. Define the function

$$Q(u,z) = \frac{1}{2}(u^2 - u) + \sigma\varrho uz + \frac{\sigma^2}{2}z^2, \quad u,z \in \mathbf{C}. \tag{5.10}$$

Theorem 5.5. *Consider the Volterra–Heston model* (5.9). *Fix* $T > 0$ *and* $(u,v,w) \in \mathbf{C}^3$, *and assume that the Riccati–Volterra equation*

$$\psi = vK + K * (Q(u,\psi) - \lambda\psi + w) \tag{5.11}$$

has a solution $\psi \in L^2(0,T)$. *Then the auxiliary process*

$$M_t = \exp\left(uL_t + v\xi_t(T) + w\int_0^T \xi_t(s)\mathrm{d}s + \int_t^T \xi_t(s)Q(u,\psi(T-s))\mathrm{d}s\right) \tag{5.12}$$

is a local martingale on $[0,T]$ *and satisfies*

$$\frac{\mathrm{d}M_t}{M_t} = u\sqrt{V_t}\mathrm{d}B_t + \psi(T-t)\sigma\sqrt{V_t}\mathrm{d}W_t. \tag{5.13}$$

If M *is a true martingale, the joint conditional Fourier–Laplace transform of the triplet* $(L_T, V_T, \int_0^T V_s\mathrm{d}s)$ *is* $\mathbb{E}[\exp(uL_T + vV_T + w\int_0^T V_s\mathrm{d}s) \mid \mathcal{F}_t] = M_t$.

The two crucial assumptions are, of course, that (5.11) has a solution and that the local martingale M is really a true martingale. The following theorem gives a sufficient condition that guarantees this; the proof can be found in [4].

Theorem 5.6. *Let K be a kernel satisfying (5.6), and let $(u, v, w) \in \mathbf{C}^3$ satisfy $\operatorname{Re}(u) \in [0, 1]$, $\operatorname{Re}(v) \leq 0$, and $\operatorname{Re}(w) \leq 0$. Then the Riccati–Volterra equation (5.11) has a unique global solution ψ, and the local martingale M in (5.12) is a true martingale.*

We now present the proof of Theorem 5.5 in the special case where $v = w = 0$. This simplifies the calculations, and the proof in the general case is similar.

Proof *(proof of Theorem 5.5 for $v = w = 0$).* Subtract $R_\lambda * \psi$ from both sides of (5.11), where now $v = w = 0$, and apply the resolvent equation $K - R_\lambda * K = \frac{1}{\lambda} R_\lambda$ to get an equivalent form of the Riccati–Volterra equation,

$$\psi = \frac{1}{\lambda} R_\lambda * Q(u, \psi). \tag{5.14}$$

We aim to apply Itô's formula to M_t, so we define $G_t = \int_t^T Q(u, \psi(T-s))\xi_t(s)\mathrm{d}s$. Since $\xi_t(s) = V_s$ for $s \leq t$, we can write

$$G_t = \int_0^T Q(u, \psi(T-s))\xi_t(s)\mathrm{d}s - \int_0^t Q(u, \psi(T-s))V_s\mathrm{d}s.$$

Focus on the first term. Using first Proposition 5.4, then the stochastic Fubini theorem [274, Theorem 2.2], and finally (5.14), we get

$$\begin{aligned}
&\int_0^T Q(u, \psi(T-s))\xi_t(s)\mathrm{d}s \\
&\quad = \int_0^T Q(u, \psi(T-s))\left\{\xi_0(s) + \sigma \int_0^{t\wedge s} \frac{1}{\lambda} R_\lambda(s-r)\sqrt{V_r}\mathrm{d}W_r\right\}\mathrm{d}s \\
&\quad = \int_0^T Q(u, \psi(T-s))\xi_0(s)\mathrm{d}s + \sigma \int_0^t \int_r^T Q(u, \psi(T-s))\frac{1}{\lambda} R_\lambda(s-r)\mathrm{d}s\sqrt{V_r}\mathrm{d}W_r \\
&\quad = \int_0^T Q(u, \psi(T-s))\xi_0(s)\mathrm{d}s + \sigma \int_0^t \psi(T-r)\sqrt{V_r}\mathrm{d}W_r.
\end{aligned}$$

As a result,

$$G_t = \int_0^T Q(u, \psi(T-s))\xi_0(s)\mathrm{d}s + \sigma \int_0^t \psi(T-r)\sqrt{V_r}\mathrm{d}W_r - \int_0^t Q(u, \psi(T-s))V_s\mathrm{d}s.$$

This leaves us in a position to apply Itô's formula to $M_t = \exp(uL_t + G_t)$, giving

$$\begin{aligned}
\frac{\mathrm{d}M_t}{M_t} &= u\,\mathrm{d}L_t + \mathrm{d}G_t + \frac{u^2}{2}\mathrm{d}\langle L\rangle_t + \mathrm{d}\langle L, G\rangle_t + \frac{1}{2}\mathrm{d}\langle G\rangle_t \\
&= \left\{\frac{1}{2}(u^2 - u) - Q(u, \psi(T-t)) + u\varrho\sigma\psi(T-t) + \frac{\sigma^2}{2}\psi(T-t)^2\right\}V_t\,\mathrm{d}t \\
&\quad + u\sqrt{V_t}\mathrm{d}B_t + \psi(T-t)\sigma\sqrt{V_t}\mathrm{d}W_t.
\end{aligned}$$

Comparing with (5.10) shows that the dt-term vanishes, so M is indeed the local martingale in (5.13). If M is a true martingale, we conclude that $\mathbb{E}[\exp(uL_T)|\mathcal{F}_t] = \mathbb{E}[M_T \mid \mathcal{F}_t] = M_t$, as claimed. □

Remark 5.2.2. *Setting $g = Q(u,\psi)$ and applying $Q(u,\cdot)$ to both sides of* (5.14) *yields*

$$g = Q\left(u, \frac{1}{\lambda} R_\lambda * g\right).$$

This is the "convolution Riccati equation" considered by [168], *which leads to an equivalent formulation in terms of g instead of ψ.*

5.2.2 ▪ Necessity of affine representations

We now discuss a converse to Theorem 5.5 obtained by [168]. Consider a general forward variance model of the type

$$\mathrm{d}\xi_t(T) = \eta_t(T)\mathrm{d}W_t,$$

where $\eta_t(T)$ is decreasing in T and $\xi_t(T)$ is the forward variance associated to a price process S of the form $\mathrm{d}S_t = S_t a(V_t)\mathrm{d}W_t$. Assume that the conditional cumulant generating function of the log-price $L_T = \log S_T$ is of the form

$$\mathbb{E}[\exp(uL_T) \mid \mathcal{F}_t] = \exp\left(uL_t + \int_t^T \xi_t(T-s)g(s,u)\mathrm{d}s\right)$$

for all $u \in [0,1]$ and $0 \le t \le T$, for some continuous function $g \le 0$. Under mild integrability conditions on $\eta_t(T)$, [168] then shows that, necessarily,

$$a(V_t) = a\sqrt{V_t} \quad \text{and} \quad \eta_t(T) = \kappa(T-t)\sqrt{V_t}$$

for some constant $a \ge 0$ and kernel κ. Thus the model is precisely of the form (5.9), and κ is identified with the resolvent $\frac{1}{\lambda}R_\lambda$.

5.2.3 ▪ Fractional calculus and the rough Heston model

Consider the power-law kernel $K_{\alpha\text{-pow}}(t) = t^{\alpha-1}/\Gamma(\alpha)$ used in the rough Heston model. The *Riemann–Liouville fractional integral I^α* is defined via convolution with this kernel, $I^\alpha f = K_{\alpha\text{-pow}} * f$. One then defines the Riemann–Liouville *fractional derivative D^α* as $D^\alpha f = \frac{\mathrm{d}}{\mathrm{d}t} I^{1-\alpha} f$, which provides an inverse to the fractional integral in that $D^\alpha(I^\alpha f) = I^\alpha(D^\alpha f) = f$. It follows that, in the case $v = w = 0$, (5.11) is equivalent to

$$D^\alpha \psi = Q(u,\psi) - \lambda\psi,$$

which is precisely the fractional Riccati equation derived by [129]. Using Proposition 5.4 and (5.14) we can rewrite the exponent in (5.12), for $t = 0$, as

$$\begin{aligned}
\xi_0 * Q(u,\psi) &= V_0\,\mathbf{1} * Q(u,\psi) + (\theta - V_0)(\mathbf{1} * R_\lambda * Q(u,\psi)) \\
&= V_0\,\mathbf{1} * Q(u,\psi) + \lambda(\theta - V_0)(\mathbf{1} * \psi) \\
&= V_0\, I^{1-\alpha}\psi + \lambda\theta(\mathbf{1} * \psi).
\end{aligned}$$

Thus, in the rough Heston model, the unconditional transform formula in Theorem 5.5, with $v = w = 0$, becomes

$$\mathbb{E}[\exp(uL_T)] = \exp\left\{uL_0 + \lambda\theta \int_0^T \psi(s)\mathrm{d}s + V_0\, I^{1-\alpha}\psi(T)\right\},$$

which is consistent with [129].

5.3 ▪ Modified forward process representation

Another perspective on (5.4) via a *stochastic partial differential equation* (SPDE) arises as follows. Starting with a Volterra process V of the form (5.4), define the process

$$u_t(x) = \mathbb{E}\left[V_{t+x} - \int_t^{t+x} K(t-s+x)b(V_s)\mathrm{d}s \,\middle|\, \mathcal{F}_t\right].$$

This process is considered by [2]. We call it the *modified forward process* because, had we not subtracted the time-integral, we would have obtained the so-called Musiela parameterization $\xi_t(t+x)$ of the forward process. The only term inside the conditional expectation that is not already $\mathcal{F}_t$-measurable is an integral with respect to W. This gives

$$u_t(x) = V_0 + \int_0^t K(t-s+x)b(V_s)\mathrm{d}s + \int_0^t K(t-s+x)\sigma(V_s)\,\mathrm{d}W_s, \tag{5.15}$$

which can be expressed in terms of the following SPDE.

Proposition 5.7. *The process $u_t(x)$ in* (5.15) *is a mild solution of the SPDE*

$$\mathrm{d}u_t(x) = (\partial_x u_t(x) + K(x)b(u_t(0)))\mathrm{d}t + K(x)\sigma(u_t(0))\mathrm{d}W_t \tag{5.16}$$

with initial condition $u_0(x) = V_0$ for all x.

Proof. Formally taking the differential in (5.15), using that $\partial_t K(t-s+x) = \partial_x K(t-s+x)$ and that $u_t(0) = V_t$, gives (5.16). More rigorously, note that $K(t-s+x) = T_{t-s}K(x)$, where T_{t-s} is the shift operator that maps any function f to the shifted function $f(t-s+\cdot)$. The derivative ∂_x is the infinitesimal generator of the shift semigroup $\{T_t\}_{t\geq 0}$, so, by definition, (5.15) is actually the mild formulation of the SPDE (5.16); see [108, Section 6.1]. □

5.3.1 ▪ Fourier–Laplace transforms and Riccati equations

The SPDE (5.16) suggests that the process $\{u_t(\cdot)\}_{t\geq 0}$ is an infinite-dimensional Markov process. In the affine case (5.5), we therefore expect a Fourier–Laplace transform formula such as

$$\mathbb{E}\left[\exp\left(\int_0^\infty h(x)u_T(x)\mathrm{d}x\right) \,\middle|\, \mathcal{F}_t\right] = \exp\left(\varphi(T-t) + \int_0^\infty \Psi(T-t,x)u_t(x)\mathrm{d}x\right), \tag{5.17}$$

where $\varphi(\tau)$ and $\Psi(\tau,x)$ are solutions of appropriate Riccati equations. These equations turn out to be

$$\partial_t\varphi(t) = \mathcal{R}_\varphi\big(\textstyle\int_0^\infty \Psi(t,y)K(y)\mathrm{d}y\big), \tag{5.18}$$

$$\Psi(t,x) = h(x-t)\mathbb{1}_{\{x\geq t\}} + \mathcal{R}_\Psi\big(\textstyle\int_0^\infty \Psi(t-x,y)K(y)\mathrm{d}y\big)\mathbb{1}_{\{x<t\}} \tag{5.19}$$

with $\varphi(0) = 0$ and where we define

$$\mathcal{R}_\varphi(y) = \beta y + \frac{\alpha}{2}y^2, \qquad \mathcal{R}_\Psi(y) = -\lambda y + \frac{a}{2}y^2. \tag{5.20}$$

Remark 5.3.1. *At first sight,* (5.19) *does not look like a differential equation for $\Psi(t,x)$. But, along the lines of the proof of Proposition* 5.7, (5.19) *can actually be viewed as a mild formulation of the formal PDE*

$$\partial_t\Psi(t,x) = -\partial_x\Psi(t,x) + \mathcal{R}_\Psi\left(\int_0^\infty \Psi(t,y)K(y)\,\mathrm{d}x\right)\delta_0(x)$$

with initial condition $\Psi(0,x) = h(x)$.

Let us give a derivation of the Riccati equations (5.18)–(5.19). We assume that $V_0 = 0$; this does not affect the validity of the Riccati equations but simplifies the calculations. Suppose that $\Psi(t,x)$ satisfies (5.19), and define $\mathrm{d}Z_t = b(V_t)dt+\sigma(V_t)\mathrm{d}W_t$, a semimartingale. Using (5.15), (5.19), and the stochastic Fubini theorem; then a change of variables; and finally (5.19) once again, we get

$$\begin{aligned}
&\int_0^\infty \Psi(T-t,x)u_t(x)\mathrm{d}x\\
&\quad = \int_0^t \int_{T-t}^\infty h(x-T+t)K(t-s+x)\,\mathrm{d}x\,\mathrm{d}Z_s\\
&\qquad + \int_0^t \int_0^{T-t} \mathcal{R}_\Psi\big(\textstyle\int_0^\infty \Psi(T-t-x,z)K(z)dz\big)K(t-s+x)\,\mathrm{d}x\,\mathrm{d}Z_s\\
&\quad = \int_0^t \int_{T-s}^\infty h(y-T+s)K(y)\,\mathrm{d}y\,\mathrm{d}Z_s\\
&\qquad + \int_0^t \int_{t-s}^{T-s} \mathcal{R}_\Psi\left(\int_0^\infty \Psi(T-s-y,z)K(z)\mathrm{d}z\right)K(y)\,\mathrm{d}y\,\mathrm{d}Z_s\\
&\quad = \int_0^t \int_0^\infty \Psi(T-s,x)K(x)\mathrm{d}x\,\mathrm{d}Z_s\\
&\qquad - \int_0^t \int_0^{t-s} \mathcal{R}_\Psi\big(\textstyle\int_0^\infty \Psi(T-s-y,z)K(z)dz\big)K(y)\,\mathrm{d}y\,\mathrm{d}Z_s.
\end{aligned}$$

Combining this with the stochastic Volterra equation (5.4) satisfied by V yields

$$\begin{aligned}
&\mathrm{d}\int_0^\infty \Psi(T-t,x)u_t(x)\mathrm{d}x\\
&\quad = \int_0^\infty \Psi(T-t,x)K(x)\mathrm{d}x\,\mathrm{d}Z_t - \int_0^t R_\Psi\big(\textstyle\int_0^\infty \Psi(T-t,y)K(y)\mathrm{d}y\big)K(t-s)\,\mathrm{d}Z_s\,\mathrm{d}t\\
&\quad = \int_0^\infty \Psi(T-t,x)K(x)\mathrm{d}x\,\mathrm{d}Z_t - R_\Psi\big(\textstyle\int_0^\infty \Psi(T-t,y)K(y)\mathrm{d}y\big)V_t\,\mathrm{d}t.
\end{aligned}$$

Let M_t denote the right-hand side of (5.17). Use the previous equation and (5.18) to get

$$\frac{\mathrm{d}M_t}{M_t} = \int_0^\infty \Psi(T-t,x)K(x)\mathrm{d}x\,\sigma(V_t)\,\mathrm{d}W_t. \tag{5.21}$$

Thus M is a local martingale, and $M_T = \exp(\int_0^\infty h(x)u_T(x)\mathrm{d}x)$ since $\Psi(0,x) = h(x)$. If M is a true martingale, we deduce the exponential-affine formula (5.17).

This can be used to derive the special case of Theorem 5.5 where $u = w = 0$ (note that $\alpha = 0$ and $a = \sigma^2$ in that theorem). Formally setting $h = v\delta_0$ with $v \in \mathbb{C}$ gives

$$\mathbb{E}\big[\exp(vV_T)\mid \mathcal{F}_t\big] = \exp\left(\varphi(T-t) + \int_0^\infty \Psi(T-t,x)u_t(x)\,\mathrm{d}x\right).$$

There is a connection between the Riccati equation (5.19) and the Riccati–Volterra equation (5.11) in Theorem 5.5. To wit, suppose that $\Psi(t,x)$ solves (5.19), and define

$$\psi(t) = \int_0^\infty \Psi(t,x)K(x)\mathrm{d}x. \tag{5.22}$$

Using the definition (5.20) of $\mathcal{R}_\Psi$, the definition (5.22) of ψ, (5.19), and a change of variables, we get

$$\begin{aligned} K * \left(-\lambda\psi + \frac{a}{2}\psi^2\right)(t) &= \int_0^\infty K(x)\mathcal{R}_\Psi(\psi(t-x))\mathbf{1}_{\{x<t\}}\mathrm{d}x \\ &= \int_0^\infty K(x)(\Psi(t,x) - h(x-t)\mathbf{1}_{\{x\geq t\}})\mathrm{d}x \\ &= \psi(t) - \int_0^\infty h(x)K(t+x)\mathrm{d}x. \end{aligned}$$

If $h = v\delta_0$, we deduce the Riccati–Volterra equation (5.11) for the case $u = w = 0$. Observe also that, in view of (5.22), (5.21) agrees with (5.13).

5.4 ▪ Laplace representation

Our final perspective on (5.4) is as a *mixture of mean-reverting processes.* Mathematically, this is analogous to the SPDE representation in Section 5.3. In fact, the SPDE representation and the developments here can be regarded as two instances of a single abstract *infinite-dimensional lift.* This unifying perspective is developed by [107] but goes well beyond the scope of this chapter. Still, to emphasize the analogies *we will use the notation $u_t(x)$ and $\Psi(t,x)$ also in this section, though with different meanings than in Section* 5.3. The reader will notice strong similarities to the derivations in Section 5.3. Assume that the kernel K is the Laplace transform of some measure μ, that is,

$$K(t) = \int_0^\infty \mathrm{e}^{-xt}\mu(\mathrm{d}x), \quad t > 0. \tag{5.23}$$

If μ is a positive measure, then K is completely monotone on $(0,\infty)$. Conversely, any such K is of the form (5.23), a result known as the Bernstein–Widder theorem. This clearly jibes well with Theorem 5.2(ii). On the other hand, μ could also be a signed measure, as long as K remains in $L^2_{\mathrm{loc}}(\mathbb{R}_+)$. This gives a large class of kernels, not necessarily completely monotone, that are compatible with Theorem 5.2(i).

Example 5.8. If we are in the classical case $K(t) = 1$, then $\mu = \delta_0$. In the rough Heston case $K(t) = t^{\alpha-1}/\Gamma(\alpha)$ with $\alpha \in (\frac{1}{2}, 1)$, we have $\mu(\mathrm{d}x) = \frac{x^{-\alpha}}{\Gamma(\alpha)\Gamma(1-\alpha)}\mathrm{d}x$.

To see how (5.23) leads to a (possibly infinite) mixture of mean-reverting processes, and in order to simplify the presentation, we will assume that $V_0 = 0$. The general case can be deduced by considering the process $\widetilde{V} = V - V_0$; the reader is invited to work out what happens in this general case.

Substituting (5.23) into (5.4) with $V_0 = 0$ and interchanging the time- and μ-integrals (justified by the stochastic Fubini theorem) yields the representation

$$V_t = \int_0^\infty u_t(x)\mu(\mathrm{d}x), \tag{5.24}$$

where we define, for all $t \geq 0$,

$$u_t(x) = \int_0^t \mathrm{e}^{-x(t-s)}b(V_s)\mathrm{d}s + \int_0^t \mathrm{e}^{-x(t-s)}\sigma(V_s)\mathrm{d}W_s.$$

Crucially, each process $\{u_t(x)\}_{t\geq 0}$ is a semimartingale, even if V is not. To find its dynamics move e^{-xt} outside the time-integrals and apply the product rule to get

$$\mathrm{d}u_t(x) = (-xu_t(x) + b(V_t))\mathrm{d}t + \sigma(V_t)\mathrm{d}W_t.$$

Plugging (5.5) and (5.24) into this expression gives

$$du_t(x) = \left(\beta - xu_t(x) - \lambda \int_0^\infty u_t(y)\mu(\mathrm{d}y)\right)\mathrm{d}t + \sqrt{\alpha + a\int_0^\infty u_t(y)\mu(\mathrm{d}y)}\mathrm{d}W_t. \tag{5.25}$$

As x ranges through the support of μ, (5.25) defines a (possibly infinite) coupled system of mean-reverting processes, and (5.24) expresses V as a mixture of those processes. The Gaussian case $a = 0$ is covered by results of [84, 198].

Apart from its theoretical interest, this representation can be useful for numerical purposes. The idea is to replace μ by an approximation μ_n that is supported on finitely many points $x_1, \ldots, x_n$. The system (5.25) then becomes an SDE for the n-dimensional Markov process $\{u_t(x_1), \ldots, u_t(x_n)\}_{t\geq 0}$. This can be used to approximate the affine Volterra process V. More details on this construction are given by [3, 107].

5.4.1 ▪ Fourier–Laplace transforms and Riccati equations

The drift and squared volatility in (5.25) depend on the curve $u_t(\,\cdot\,)$ in an affine way. This suggests that the process $\{u_t(\,\cdot\,)\}_{t\geq 0}$ is an affine Markov process, possibly infinite-dimensional. In particular, we expect a transform formula similar to (5.17),

$$\mathbb{E}\left[\exp\left\{\int_0^\infty h(x)u_T(x)\mu(\mathrm{d}x)\right\} \,\middle|\, \mathcal{F}_t\right] = \exp\left\{\varphi(T-t) + \int_0^\infty \Psi(T-t,x)u_t(x)\mu(\mathrm{d}x)\right\}, \tag{5.26}$$

where $\varphi(\tau)$ and $\Psi(\tau, x)$ are solutions of appropriate Riccati equations with initial conditions

$$\varphi(0) = 0, \quad \Psi(0, x) = h(x). \tag{5.27}$$

In this Markovian situation, one can apply the standard method for deriving the Riccati equations. Let M_t denote the right-hand side of (5.26). Itô's formula and (5.25) give, after some computations,

$$\begin{aligned}\frac{\mathrm{d}M_t}{M_t} = \Big[&-\partial_t\varphi(T-t) + \mathcal{R}_\varphi\big(\textstyle\int_0^\infty \Psi(T-t,y)\mu(\mathrm{d}y)\big) \\ &+ \int_0^\infty \Big(-\partial_t\Psi(T-t,x) - x\Psi(T-t,x) \\ &\qquad + \mathcal{R}_\Psi\big(\textstyle\int_0^\infty \Psi(T-t,y)\mu(\mathrm{d}y)\big)\Big) u_t(x)\mu(\mathrm{d}x)\Big]\mathrm{d}t \\ &+ \text{local martingale},\end{aligned} \tag{5.28}$$

with $\mathcal{R}_\varphi$, $\mathcal{R}_\Psi$ as in (5.20). It is remarkable that the same functions $\mathcal{R}_\varphi$ and $\mathcal{R}_\Psi$ as for the SPDE representation occur also here. This is one manifestation of the underlying abstract point of view due to [107].

Suppose that φ and Ψ solve the possibly infinite-dimensional Riccati equations

$$\begin{aligned}\partial_t\varphi(t) &= \mathcal{R}_\varphi\big(\textstyle\int_0^\infty \Psi(t,y)\mu(\mathrm{d}y)\big), \\ \partial_t\Psi(t,x) &= -x\Psi(t,x) + \mathcal{R}_\Psi\big(\textstyle\int_0^\infty \Psi(t,y)\mu(\mathrm{d}y)\big),\end{aligned} \tag{5.29}$$

with initial conditions (5.27). Then, due to (5.28), M is a local martingale with $M_T = \exp(\int_0^\infty h(x)u_T(x)\mu(\mathrm{d}x))$. If M is actually a true martingale, we obtain the transform formula (5.26), which is nothing but the martingale property $\mathbb{E}[M_T \mid \mathcal{F}_t] = M_t$. In particular, if $h(x) \equiv v$ is constant, combining (5.24) and (5.26) gives

$$\mathbb{E}\left[\exp(vV_T) \mid \mathcal{F}_t\right] = \exp\left(\varphi(T-t) + \int_0^\infty \Psi(T-t,x)u_t(x)\mu(\mathrm{d}x)\right).$$

Just as in Section 5.3, there is a connection between the solution $\Psi(t,x)$ to the Riccati equation (5.29), with $h(x) \equiv v$ constant, and the solution $\psi(t)$ to the Riccati–Volterra equation (5.11), with $u = w = 0$. The link is given by the formula

$$\psi(t) = \int_0^\infty \Psi(t,x)\mu(\mathrm{d}x),$$

which can be verified by similar calculations are in Section 5.3. This gives yet another way to derive the Fourier–Laplace transform formula.

Acknowledgments

The authors would like to thank Eduardo Abi Jaber and Christa Cuchiero for valuable discussions and suggestions. Martin Keller-Ressel gratefully acknowledges financial support from DFG grants ZUK 64 and KE 1736/1-1. Martin Larsson gratefully acknowledges financial support from SNF grant 205121_163425. Sergio Pulido gratefully acknowledges financial support from MATH AmSud project SaSMoTiDep 18-MATH-17.

Chapter 6

Hedging under rough volatility

Masaaki Fukasawa, Blanka Horvath, Peter Tankov

In this chapter we first briefly review the existing approaches to hedging in rough volatility models. Next, we present a simple but general result which shows that in a one-factor rough stochastic volatility model, any option may be perfectly hedged with a dynamic portfolio containing the underlying asset and one other, such as a variance swap. In the final section we report the results of a backtest experiment using real data, where VIX options are hedged with a forward variance swap. In this experiment, using a rough volatility model allows us to almost completely remove the bias and reduce the overall hedging error by a factor of 27% compared to traditional diffusion-based models.

6.1 ▪ Introduction

The many advantages of rough volatility models have been outlined in previous chapters. One of the only potential challenges that remain to be addressed in practice is an (apparent) difficulty in hedging derivatives in rough models. Hedging in rough volatility models can seem intricate since the dynamics of rough volatility models involve a fractional Brownian motion (fBm). In this chapter we demonstrate how this apparent challenge can be overcome in different modeling scenarios exhibiting different levels of generality, which allow us to derive (often explicit) hedging strategies. For building a hedging portfolio, one essentially needs to compute conditional expectations of the form

$$C_t = \mathbb{E}[f(S_T)|\mathcal{F}_t], \tag{6.1}$$

where f is a deterministic payoff function, and determine the associated martingale representations. Classical theory tells us that the option payoff can be replicated at time t for a price $\mathbf{P}_{T-t}f(S_t)$, where $(\mathbf{P}_t)_{t\geq 0}$ is the semigroup on $C^b(\mathbb{R}^2)$ generated by the infinitesimal generator $\mathbf{A}$ associated with the instantaneous covariance of S and with the (local) martingale problem characterizing the law of the process S. Under the assumption that an equivalent local martingale measure exists (and even beyond that; see [266]), classical theory [20, 112, 208] gives conditions under which a contingent claim can be hedged (optimally). In a classical Markovian setting, an optimal trading strategy h can be derived directly from the solution C_t to the Cauchy problem

$\partial_t C_t(s) - \mathbf{A}C_t(s) = 0$ associated with the generator $\mathbf{A}$ [266]. The non-Markovian nature (at least in the finite-dimensional sense) of rough models coming from the fractional Brownian driver leaves most of these results beyond the scope of rough volatility. In addition, this very (non-Markovian) nature of rough volatility models also prohibits the direct use of (PDE-based) efficient numerical methods for a tractable evaluation of prices (6.1) and associated hedging portfolios, which can make hedging more challenging than in classical models. In the case of affine rough models it is possible to exploit the affine structure to derive efficient pricing and hedging. For the general case, Monte Carlo methods have been derived [50, 155, 239] for pricing under rough volatility[41] which can, in some cases, be computationally slow, but by using deep neural networks it is possible to speed up these pricing methods by several orders of magnitude as demonstrated in [43, 202]. Also, deep neural networks can aid the direct computation of hedging strategies as in [80, 202]. Indeed, the deep hedging framework is applicable in great generality, including in rough volatility models as recently demonstrated in [203]. These hedging portfolios are obtained based on the idea that every investment strategy gives rise to a profit and loss, whose distribution can be optimized with respect to specific risk measures. More specifically, Horvath, Teichmann, and Zurič [203] compute hedging strategies for the rough Bergomi model, numerically building on results of Gassiat [162] and Viens and Zhang [275], and demonstrate the applicability of deep hedging for the calculation of hedging strategies in rough models.

6.1.1 ▪ The special role of the forward variance curve in rough models

As mentioned, in some cases within the rough volatility framework, it is possible to derive hedging strategies more explicitly. In the rough Heston model, for example, El Euch and Rosenbaum [131] obtain explicit hedging strategies that lead to *perfect hedging*. To obtain these results, it is essential to identify the relevant state variables: in rough Heston models these are, namely, the underlying $(S_t)_{0\leq t\leq T}$ and the so-called forward variance curve

$$(\mathbb{E}[V_{\theta+t}|\mathcal{F}_t])_{\theta\geq 0}, \tag{6.2}$$

where V_t is the instantaneous variance of the underlying price at time t. In the setting of [131], the conditional expectation (6.1) above can then be written explicitly as

$$\mathrm{C}_t = \mathrm{C}\Big(T-t, S_t, (\mathbb{E}[V_{\theta+t}|\mathcal{F}_t])_{\theta\geq 0}\Big), \tag{6.3}$$

where C is some deterministic function. Indeed, it can be observed more broadly that replicating portfolios in rough volatility models typically contain the *underlying asset* and the *forward variance curve*. In fact, not only in the case of the rough Heston model but also in all affine rough models, a close relation to the forward variance curve can be drawn from the affine structure, as highlighted in Chapter 5. This gives rise to one of the perspectives presented in Chapter 5, viewing rough affine models as forward variance models. Viens and Zhang [275] confirmed this idea for general Volterra-type stochastic differential equations. In fact, while for some practitioners the idea of using the forward variance curve for hedging (even vanilla options) may come as a surprise, the observation of the importance of the forward variance curve was already emphasized in [39] and is also very close in spirit to the approach developed by Bergomi in [57].

6.1.2 ▪ Martingale problems and Markovianity

Viens and Zhang presented in [275] a martingale approach for general[42] Volterra stochastic differential equations. While for affine rough models it has been noted that martingale problems

[41] For small values of the Hurst parameter, the computational cost of calculation of prices increases.

[42] This, in particular, includes the rough Heston model, the rough Bergomi model, the fractional Ornstein–Uhlenbeck process, as well as affine Volterra processes.

(connected to hedging problems) are more convenient thanks to the affine structure, the introduction of a martingale component is also key in the general case in [275] for recovering the flow property, which makes it possible to derive a certain "Markov" property for rough models.

Pricing and hedging *volatility options* under rough volatility models have been considered in [201], where the special role of the forward variance is reconfirmed and martingality considerations are revisited. In particular, it is shown that by focusing on the forward variance instead of the instantaneous volatility, one recovers the martingale framework and, in particular, the classical martingale representation property of option prices. This makes it possible to compute the hedge ratios and to show that options can be hedged with a finite number of liquid assets, as in the classical setting. To calibrate VIX option smiles via rough volatility, we consider extended lognormal models by adding volatility modulation through an independent stochastic factor in the Volterra integral which preserves part of the analytical tractability of the lognormal setting by extending it through an affine structure, which makes it possible to develop approximate option pricing and calibration algorithms based on Fourier transform techniques.

In this chapter, we showcase these ideas in relatively transparent and illustrative settings. We discuss the role of the forward variance curve in establishing (perfect) hedging in rough models and present a hands-on empirical study illustrating the role of the Hurst parameter (driving the roughness of the paths) in the hedging performance for hedging VIX options.

6.2 ▪ A theoretical framework

The purpose of this section is to illustrate an infinite-dimensional Markov nature of rough volatility models, which enables us to hedge options without any "memory" of the past. While fractional Brownian motions have (long or short) memory properties, we see that the memory is stored in an option market.

6.2.1 ▪ The model

Here we consider a two-factor model; on a probability space $(\Omega, \mathcal{F}, \mathbb{P})$ with a filtration $\{\mathcal{F}_t\}_{t\in\mathbb{R}}$, we consider a hypothetical option market where call and put options are traded for all strike prices $K \geq 0$ and maturities $T \geq 0$. Their prices at time $t \geq 0$ are denoted by $\mathrm{C}_t(K,T)$ and $\mathrm{P}_t(K,T)$, respectively. The underlying asset price process of the options is denoted by S, and we suppose

$$\begin{aligned}
\mathrm{C}_t(K,T) &= \mathbb{E}^{\mathbb{Q}}[(S_T - K)_+|\mathcal{F}_t], \\
\mathrm{P}_t(K,T) &= \mathbb{E}^{\mathbb{Q}}[(K - S_T)_+|\mathcal{F}_t], \\
S_t &= \mathrm{C}_t(0,T)
\end{aligned}$$

for all $K, T \geq 0$ and $t \in [0,T]$, where $\mathbb{Q}$ is an equivalent measure to $\mathbb{P}$ of which existence is assumed. Here and hereafter we assume that risk-free rates are zero for brevity.

We introduce a SABR/Bergomi-type stochastic volatility model

$$\begin{aligned}
\mathrm{d}S_t &= f(S_t)\sqrt{V_t^t}\left[\varrho \mathrm{d}W_t + \sqrt{1-\varrho^2}\mathrm{d}\overline{W}_t\right], \\
\mathrm{d}V_t^u &= V_t^u g(u-t)\mathrm{d}W_t, \quad t < u,
\end{aligned} \tag{6.4}$$

where $(W, \overline{W})$ is a 2-dimensional $\{\mathcal{F}_t\}_{t\geq 0}$-Brownian motion under $\mathbb{Q}$, f and g are deterministic Borel functions on $[0,\infty)$, and $\varrho \in (-1,1)$. We assume g is locally square integrable, so that we have the explicit expressions

$$V_t^u = V_s^u \exp\left\{\int_s^t g(u-v)\mathrm{d}W_v - \frac{1}{2}\int_s^t g(u-v)^2 \mathrm{d}v\right\} \tag{6.5}$$

for $0 \le s \le t \le u$. Note that $\mathbb{E}^{\mathbb{Q}}[V_t^t|\mathcal{F}_s] = V_s^t$ for $t \ge s$. The case $f(s) = s$, $g(u) = \eta u^{H-\frac{1}{2}}$ corresponds to the rough Bergomi model introduced in [39]. A volatility process driven by a fractional Brownian motion (fBm) can be treated in this framework. For example, if the log-volatility is a stationary fractional Ornstein–Uhlenbeck process [30]

$$\log v_t = \frac{1}{2}\int_{-\infty}^{t} g(t-s)\mathrm{d}W_s^{\mathbb{P}}, \quad \text{with} \quad g(u) = \eta u^{H-\frac{1}{2}} - \eta\lambda \mathrm{e}^{-\lambda u}\int_0^u v^{H-\frac{1}{2}}\mathrm{e}^{\lambda v}\mathrm{d}v,$$

under $\mathbb{P}$ for some $\mathbb{P}$-Brownian motion $W^{\mathbb{P}}$, and the volatility risk premium is deterministic, then we have (6.4) and (6.5) with $f(s) = s$, $V_t^u = \mathbb{E}^{\mathbb{Q}}[v_u^2|\mathcal{F}_t]$ for $u \ge t$, and a suitable family $\{V_0^t\}_{t\ge 0}$ of $\mathcal{F}_0$ measurable random variables. We call the curve

$$\widehat{V}_t : \theta \mapsto V_t^{t+\theta}$$

the forward variance curve at time t.

Proposition 6.1. *The forward variance curve $\{\widehat{V}_t\}_{t\ge 0}$ is a Markov process with state space $C[0,\infty)$.*

Proof. By (6.5), for $t \ge s$ we have

$$\widehat{V}_t(\theta) = \widehat{V}_s(\theta + t - s)\exp\left\{\int_s^t g(\theta + t - u)\mathrm{d}W_u - \frac{1}{2}\int_s^t g(\theta + t - u)^2\mathrm{d}u\right\}.$$

Since the exponential term is independent of $\mathcal{F}_s$, the result follows. □

Corollary 6.2. *$(S, \widehat{V})$ is a Markov process with state space* $[0,\infty) \times C[0,\infty)$.

Now we discuss that $\widehat{V}$ is an observable state. By Itô's formula, we have

$$\int_t^{t+\theta} V_u^u \mathrm{d}u = \int_t^{t+\theta} \frac{S_u^2}{f(S_u)^2}\mathrm{d}\langle \log S\rangle_u,$$

which is the payoff of a weighted variance swap. The fair strike of this swap is

$$\mathbb{E}^{\mathbb{Q}}\left[\int_t^{t+\theta} \frac{S_u^2}{f(S_u)^2}\mathrm{d}\langle \log S\rangle_u \middle| \mathcal{F}_t\right] = \int_t^{t+\theta} \mathbb{E}^{\mathbb{Q}}[V_u^u|\mathcal{F}_t]\mathrm{d}u = \int_t^{t+\theta} V_t^u \mathrm{d}u = \int_0^{\theta} \widehat{V}_t(s)\mathrm{d}s.$$

Therefore, the forward variance curve $\widehat{V}$ is the derivative in θ of this derivative price. It is uniquely determined by call and put option prices in a model-free manner as follows: assume $1/f$ is locally square integrable on $(0,\infty)$, and let

$$h(x) := \int_1^x \int_1^y \frac{2}{f(z)^2}\mathrm{d}z\mathrm{d}y.$$

Then, again by Itô's formula,

$$h(S_{t+\theta}) = h(S_t) + \int_t^{t+\theta} h'(S_u)\mathrm{d}S_u + \int_t^{t+\theta} \frac{S_u^2}{f(S_u)^2}\mathrm{d}\langle \log S\rangle_u,$$

and by an integration-by-parts formula,

$$\begin{aligned}h(S_{t+\theta}) &= h(S_t) + h'(S_t)(S_{t+\theta} - S_t)\\ &\quad + \int_0^{S_t} (K - S_{t+\theta})_+ h''(K)\mathrm{d}K + \int_{S_t}^{\infty} (S_{t+\theta} - K)_+ h''(K)\mathrm{d}K.\end{aligned}$$

This means a model-free replication of the weighted variance swap payoff is given as a static portfolio of call and put options with weight $h''(K) = 2/f(K)^2$. The replication price is given by

$$U_t(\theta) := 2\int_0^{S_t} \mathrm{P}_t(K, t+\theta)\frac{\mathrm{d}K}{f(K)^2} + 2\int_{S_t}^{\infty} \mathrm{C}_t(K, t+\theta)\frac{\mathrm{d}K}{f(K)^2}.$$

Finally, we get $\widehat{V}_t(\theta) = \frac{\partial}{\partial\theta}U_t(\theta)$. Consequently, for a possibly path-dependent functional $\mathrm{F} = \mathrm{F}(\{S_u\}_{u\in[t,T]})$, its conditional expectation $\mathbb{E}^{\mathbb{Q}}[\mathrm{F}|\mathcal{F}_t]$ is a function of S_t and $\{\widehat{V}_t(\theta)\}_{\theta\geq 0}$, which are observable from the option market at time t.

6.2.2 ▪ Perfect hedging

We consider an infinite-dimensional Markov model. But we have only two factors, so in light of the martingale representation theorem, every square integrable payoff is perfectly replicated with a dynamic portfolio of two traded assets. A natural choice of the two would be the underlying asset and the weighted variance swap (with a fixed maturity).

As a hedging instrument, the replication portfolio for the weighted variance swap is more convenient than the weighted variance swap itself because it is a local martingale. Let

$$U_t^T := \int_0^t (h'(S_0) - h'(S_u))\mathrm{d}S_u + 2\int_0^{S_0} \mathrm{P}_t(K,T)\frac{\mathrm{d}K}{f(K)^2} + 2\int_{S_C}^{\infty} \mathrm{C}_t(K,T)\frac{\mathrm{d}K}{f(K)^2}$$

be the time t value of the replication portfolio with maturity T initiated at time 0. We have then

$$\begin{aligned}U_t^T &= \mathbb{E}^{\mathbb{Q}}\left[\int_0^T V_u^u \mathrm{d}u \,\middle|\, \mathcal{F}_t\right]\\ &= \mathbb{E}^{\mathbb{Q}}\left[\int_0^T \left\{V_0^u + \int_0^u V_s^u g(u-s)\mathrm{d}W_s\right\}\mathrm{d}u \,\middle|\, \mathcal{F}_t\right]\\ &= \int_0^T V_0^u \mathrm{d}u + \int_0^t \mathrm{d}W_s \int_s^T V_s^u g(u-s)\mathrm{d}u.\end{aligned}$$

Therefore,

$$\mathrm{d}U_t^T = \mathcal{D}_g U_t^T \mathrm{d}W_t, \tag{6.6}$$

where

$$\mathcal{D}_g U_t^T = \int_t^T V_t^u g(u-t)\mathrm{d}u = \int_t^T \frac{\partial U_t^u}{\partial u} g(u-t)\mathrm{d}u.$$

Proposition 6.3. *For any* $\mathrm{F} \in L^2(\mathcal{F}_\tau, Q)$, $\tau \in (t,T)$, *there exists an adapted process* (H^S, H^U) *such that*

$$\mathrm{F} = \mathbb{E}^{\mathbb{Q}}[\mathrm{F}|\mathcal{F}_t] + \int_t^\tau H_v^S \mathrm{d}S_v + \int_t^\tau H_v^U \mathrm{d}U_v^T.$$

Proof. By the martingale representation theorem, there exists (H^1, H^2) such that

$$\mathrm{F} = \mathbb{E}^{\mathbb{Q}}[\mathrm{F}|\mathcal{F}_t] + \int_t^\tau H_v^1 \mathrm{d}W_v + \int_t^\tau H_v^2 \mathrm{d}\overline{W}_v.$$

Since

$$dW_t = \frac{1}{\mathcal{D}_g U_t^T} dU_t^T,$$

$$d\overline{W}_t = \frac{1}{\sqrt{1-\varrho^2}} \left[\frac{1}{f(S_t)\sqrt{V_t^t}} dS_t - \varrho dW_t \right],$$

by (6.6), we have the result. □

6.3 ▪ Hedging VIX options: Empirical analysis

In this section, we illustrate the advantages of rough volatility modeling for managing a simple VIX option. We consider three models for the VIX index: the Black–Scholes model (geometric Brownian motion), the CIR model, and the rough stochastic volatility model (where the volatility is the exponential of a fractional Brownian motion). Since we are interested in hedging short-term options, we use simplified versions of the models without drift and neglect the effect of the interest rate. As a result, all models have only one parameter to be estimated (see list below).

In each model, we perform a series of backtests of dynamic hedging of a VIX option with a forward variance swap with the same maturity as the option and with the duration corresponding to that of the VIX (1 month). In all tests, the hedging portfolio is readjusted daily using the closing prices of the hedging instruments. The test is performed 1,000 times, starting on each working day t between Jan 10, 2012 and April 29, 2016. The backtest is organized as follows:

- The parameter is estimated on the 88-day period preceding day t.
- The initial value of the hedging portfolio is initialized with the at-the-money (ATM) VIX option price with maturity 1.5 months computed within the model, and the quantity of the hedging asset in the portfolio is initialized with the corresponding model-based hedge ratio.
- For 29 working days following day t, on each day the portfolio value is readjusted following the change in the value of the hedging asset, and the hedge ratio is recomputed.
- At the end of the 29-working-days period, the P&L of the hedging portfolio is recorded, and the no-hedge P&L is recorded as the difference between the option price at the beginning and option price at the end of the period.

The backtest uses synthetic forward variance curve data computed from the historical prices of S&P index options and downloaded from the WRDS database. Detailed descriptions of models and hedging procedures are given. Table 6.1 presents the main results of the backtest. We see that while the Black–Scholes and CIR benchmarks appear to have similar performance, the rough fractional stochastic volatility (RFSV) model exhibits a much lower bias and a root mean square error (RMSE) which is 27% lower than the other two models. Figure 6.1 plots the backtest P&L evolution as a function of the starting date of the backtest. The consistently low bias of the strategy based on the RFSV model is clearly visible here.

6.3.1 ▪ Black–Scholes model

Let VIX_t be the VIX index at time t, and let F_t^T be the T-forward variance swap at time t, which refers to the same period as the VIX index, that is, $F_t^T = \mathbb{E}^{\mathbb{Q}}[\mathrm{VIX}_T^2 | \mathcal{F}_t]$. Assume that the VIX index follows the lognormal dynamics $\mathrm{VIX}_t = e^{X_t}$, where X is an Ornstein–Uhlenbeck process

Table 6.1: Empirical performance of hedging strategies based on different models.

	Black–Scholes		CIR		RFSV	
	No hedge	hedge	No hedge	hedge	No hedge	hedge
Mean	0.0144	0.00533	0.0136	0.00339	0.00991	0.00063
Std. dev.	0.0189	0.00355	0.0206	0.00650	0.0188	0.00414
RMSE	0.0238	0.00641	0.0247	0.00734	0.0212	0.00419
Reduction factor	3.7176		3.7176		5.0724	

$\mathrm{d}X_t = \kappa(\theta - X_t)\mathrm{d}t + \gamma \mathrm{d}W_t$ under the risk-neutral probability. Then, the forward variance swap has dynamics

$$\mathrm{d}F_t^T = 2\gamma F_t^T \mathrm{e}^{-\kappa(T-t)}\mathrm{d}W_t.$$

When close to term, the exponential factor can be neglected, and we obtain the simple Black–Scholes dynamics. On the other hand, γ may be estimated from the volatility of VIX:

$$\mathrm{d}\langle \mathrm{VIX}\rangle_t = \gamma^2 \mathrm{VIX}_t^2 \mathrm{d}t.$$

We are hedging a VIX option with payoff

$$(\mathrm{VIX}_T - K)_+.$$

The VIX future

$$\mathfrak{V}_t^T := \mathbb{E}^{\mathbb{Q}}[\mathrm{VIX}_T | \mathcal{F}_t]$$

similarly admits the lognormal dynamics $\mathrm{d}\mathfrak{V}_t^T = \gamma \mathfrak{V}_t^T \mathrm{e}^{-\kappa(T-t)}\mathrm{d}W_t$, so that, neglecting the interest rate and the exponential term, the option price is given by the Black–Scholes formula

$$\mathbb{E}^{\mathbb{Q}}[(\mathrm{VIX}_T - K)_+ | \mathcal{F}_t] = \mathbb{E}^{\mathbb{Q}}\left[\left(\mathfrak{V}_T^T - K\right)_+ | \mathcal{F}_t\right] = \mathfrak{V}_t^T \mathcal{N}(d_+(t)) - K\mathcal{N}(d_-(t)),$$

where

$$d_\pm(t) := \frac{\log \frac{\mathfrak{V}_t^T}{K} \pm \frac{\gamma^2(T-t)}{2}}{\gamma\sqrt{T-t}},$$

or in other words,

$$\mathbb{E}^{\mathbb{Q}}[(\mathrm{VIX}_T - K)_+ | \mathcal{F}_t] = \sqrt{F_t^T}\mathrm{e}^{-\frac{\gamma^2}{2}(T-t)}\mathcal{N}(d_+(t)) - K\mathcal{N}(d_-(t)),$$

where

$$d_\pm(t) = \frac{\log\left(\frac{\sqrt{F_t^T}\mathrm{e}^{-\frac{\gamma^2}{2}(T-t)}}{K}\right) \pm \frac{\gamma^2(T-t)}{2}}{\gamma\sqrt{T-t}},$$

and the hedge ratio is $\frac{\mathcal{N}(d_+(t))}{2\sqrt{F_t^T}}\mathrm{e}^{-\frac{\gamma^2}{2}(T-t)}$.

6.3.2 ▪ CIR model

Assume that the VIX index follows the square-root dynamics:

$$\mathrm{dVIX}_t^2 = \kappa(\theta - \mathrm{VIX}_t^2)\mathrm{d}t + \gamma \mathrm{VIX}_t \mathrm{d}W_t.$$

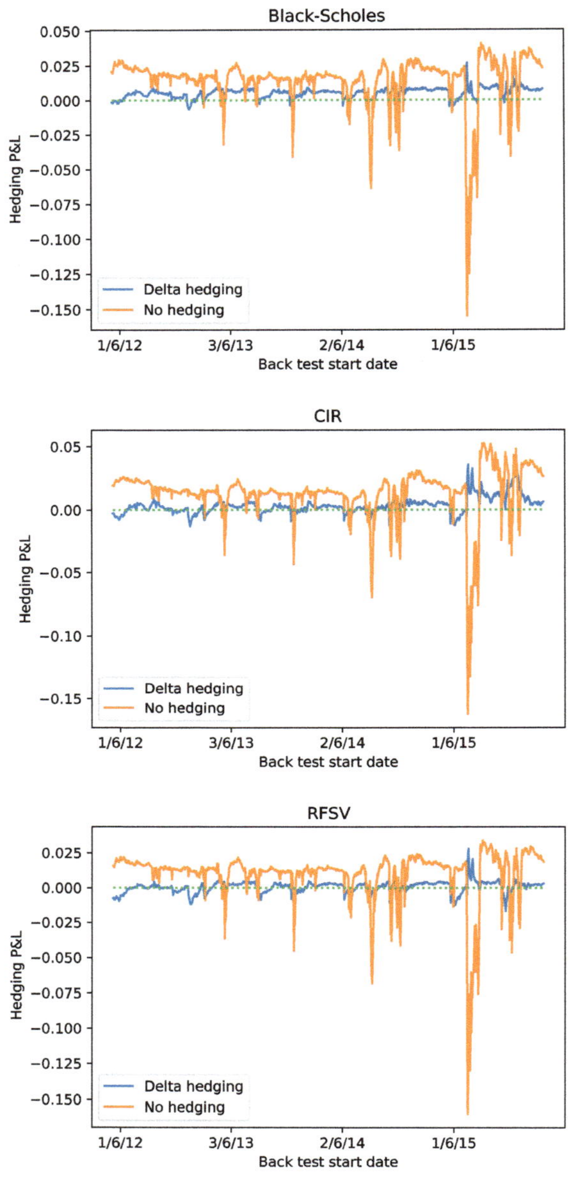

Figure 6.1: Backtesting P&L as function of time for the three models (Black–Scholes, CIR, RFSV).

Since we are hedging short-maturity options and cannot estimate κ and θ under the risk-neutral measure anyway, we assume that $\kappa = 0$ so that

$$\mathrm{dVIX}_t^2 = \gamma \mathrm{VIX}_t \mathrm{d}W_t.$$

The forward variance swap is then given by

$$F_t^T = \mathbb{E}^{\mathbb{Q}}[\mathrm{VIX}_T^2 | \mathcal{F}_t] = \mathrm{VIX}_t^2$$

and follows the dynamics

$$dF_t^T = \gamma \sqrt{F_t^T} dW_t.$$

We are hedging a VIX option with payoff

$$(\text{VIX}_T - K)_+,$$

with a forward variance swap. The price of the VIX option is given by

$$\mathbb{E}^{\mathbb{Q}}[(\text{VIX}_T - K)_+|\mathcal{F}_t] = \int_{K^2}^{\infty} (\sqrt{z} - K) p_{T-t}(F_t^T, z) dz,$$

where $p_T(x, \cdot)$ is the density of the CIR process at time T with starting value x. The parameter γ may be estimated by observing that $\langle \text{VIX} \rangle_t = \frac{\gamma^2}{4} t$.

6.3.3 ▪ Rough fractional stochastic volatility

Assume now that the VIX index is given by

$$\text{VIX}_t = C e^{X_t},$$

where $C > 0$ is a constant and X is a centered Gaussian process under the risk-neutral probability. For all $s \geq 0$, let $\mathcal{F}_s^0 := \sigma(X_r, r \leq s)$, and $\mathcal{F}_s := \cap_{s<t} \mathcal{F}_t^0$. The interest rate is taken to be zero. Fix a time horizon T, and let $Z_t(T) := \mathbb{E}^{\mathbb{Q}}[X_T|\mathcal{F}_t]$, so that $(Z_t(T))_{t\geq 0}$ is a Gaussian martingale and thus a process with independent increments, completely characterized by the function

$$c^T(t) := \mathbb{E}^{\mathbb{Q}}\left[Z_t(T)^2\right] = \mathbb{E}^{\mathbb{Q}}\left[\mathbb{E}^{\mathbb{Q}}[X_T|\mathcal{F}_t]^2\right].$$

If we assume in addition that $c^T(\cdot)$ is continuous, then $(Z_t(T))_{t\geq 0}$ is almost surely continuous. Using the total variance formula, the forward variance swap can be characterized as

$$\begin{aligned} F_t^T := \mathbb{E}^{\mathbb{Q}}\left[\text{VIX}_T^2|\mathcal{F}_t\right] = C^2 \mathbb{E}^{\mathbb{Q}}\left[e^{2X_T}|\mathcal{F}_t\right] &= C^2 \exp\left\{2\mathbb{E}^{\mathbb{Q}}[X_T|\mathcal{F}_t] + 2\mathbb{V}[X_T|\mathcal{F}_t]\right\} \\ &= C^2 \exp\left\{2(Z_t(T) + c^T(T) - c^T(t))\right\}. \end{aligned}$$

The time t price of a call on the VIX is given by $\text{C}_t := \mathbb{E}^{\mathbb{Q}}[(\text{VIX}_T - K)_+|\mathcal{F}_t]$. Note that the VIX future is a continuous lognormal martingale with $\mathbb{E}^{\mathbb{Q}}[\text{VIX}_T|\mathcal{F}_t] = \mathfrak{V}_t^T$, and, by the total variance formula,

$$\mathbb{V}[\log \text{VIX}_t|\mathcal{F}_t] = \mathbb{V}[X_T|\mathcal{F}_t] = c^T(T) - c^T(t).$$

In other words,

$$\text{C}_t = \mathfrak{V}_t^T \mathcal{N}(d_+(t)) - K\mathcal{N}(d_-(t)), \quad d_\pm(t) = \frac{\log \frac{\mathfrak{V}_t^T}{K} \pm \frac{1}{2}(c^T(T) - c^T(t))}{\sqrt{c^T(T) - c^T(t)}}.$$

Applying Itô's formula and keeping in mind the martingale property of the option price, we obtain

$$d\text{C}_t = \mathcal{N}(d_+(t)) d\mathfrak{V}_t^T.$$

In terms of forward variance swap, we then have

$$\text{C}_t = \sqrt{F_t^T} e^{-\frac{1}{2}(c^T(T) - c^T(t))} \mathcal{N}(d_+(t)) - K\mathcal{N}(d_-(t)),$$

with

$$d_\pm(t) = \frac{\log \frac{\sqrt{F_t^T}\mathrm{e}^{-\frac{1}{2}(c^T(T)-c^T(t))}}{K} \pm \frac{1}{2}(c^T(T)-c^T(t))}{\sqrt{c^T(T)-c^T(t)}},$$

and the option price dynamics take the following form

$$\mathrm{d}C_t = \frac{\mathcal{N}(d_+(t))\mathrm{e}^{-\frac{1}{2}(c^T(T)-c^T(t))}}{2\sqrt{F_t^T}}\mathrm{d}F_t^T.$$

Assuming that $X_t = \nu W_t^H$, where W is the fBm with the Hurst parameter H, we get, after some computations using the Mandelbrot–Van Ness representation,

$$\begin{aligned} c^T(t) &= \frac{\nu^2}{\Gamma^2(H+\frac{1}{2})}\left\{\int_0^\infty \left[(T+s)^{H-\frac{1}{2}} - s^{H-\frac{1}{2}}\right]^2 \mathrm{d}s + \int_0^t (T-s)^{2H-1}\mathrm{d}s\right\} \\ &= f(T) - \frac{\nu^2(T-t)^{2H}}{2H\Gamma^2(H+\frac{1}{2})} \end{aligned}$$

for some function $f(T)$, which cancels out in the difference, so that

$$c^T(T) - c^T(t) = \frac{\nu^2(T-t)^{2H}}{2H\Gamma^2(H+\frac{1}{2})}.$$

Contrary to the previous two models, this one formally has two parameters to be estimated: ν and H. To estimate the Hurst parameter, following [166], we define

$$m(q,\Delta) := \frac{1}{\lfloor T/\Delta \rfloor}\sum_{k=1}^{\lfloor T/\Delta \rfloor} \left|\log(\mathrm{VIX}_{k\Delta}) - \log(\mathrm{VIX}_{(k-1)\Delta})\right|^q$$

and estimate H from the half-slope of $m(2,\Delta)$ as a function of Δ in the log-log coordinates (see Figure 6.2, where Δ varies from 1 to 30 days). Since this procedure requires a relatively long dataset to be precise, we perform it only once, on the VIX index time series from April 17, 2001 to April 16, 2021. This gives an estimated Hurst parameter value of 0.377, and the procedure is quite stable: when estimating on the first 10 years of the dataset, one obtains 0.380, and on the last 10 years one obtains 0.379.

These estimated values of the Hurst index are much higher than the values found by [166] and many other authors using the daily time series of realized volatility (typically between 0.1 and 0.15). However, the VIX index is constructed from prices of one-month options on the S&P index, and using the implied volatility of one-month options as proxy for volatility, [232] find a value of $H = 0.32$, which is much closer to our result. The relatively high value of the Hurst index we find can thus be explained by the averaging effects associated with computing option prices.

In view of the stability of the Hurst index estimation, we fix the value $H = 0.377$ for all tests, rather than estimating it before each backtest. Note that the hedging performance of the model remains very similar for $H \in [0.37, 0.39]$. This leaves us with a single parameter, ν, to estimate before each backtest, which is estimated by

$$\widehat{\nu} = \sqrt{\frac{m(2,\Delta)}{\Delta^{2H}}}.$$

To further illustrate the dependence of the hedging performance on the value of the Hurst index and the importance of using a "rough volatility" specification, we performed the same test for H values between 0.2 and 0.5, with step of 0.01, where the value $H = 0.5$ corresponds to the Black–Scholes benchmark. Figure 6.3 shows the dependence of the hedging RMSE on the value of the Hurst parameter with the minimum attained around $H = 0.38$.

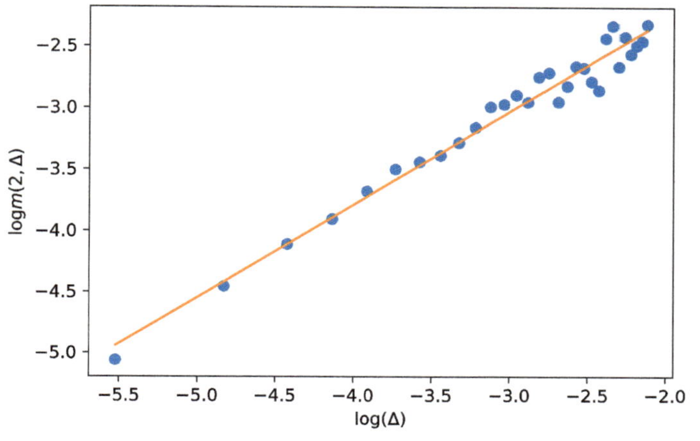

Figure 6.2: Estimation of the Hurst parameter.

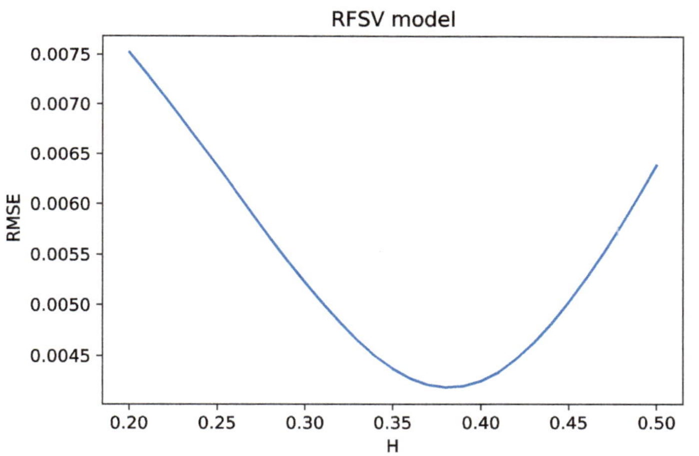

Figure 6.3: Dependence of the hedging RMSE on the Hurst parameter value.

Chapter 7

Hybrid scheme for Brownian semistationary processes[43]

Mikkel Bennedsen, Asger Lunde, Mikko Pakkanen

Introduction

We study simulation methods for *Brownian semistationary* ($\mathcal{BSS}$) processes, first introduced by Barndorff-Nielsen and Schmiegel [35, 36], which form a flexible class of stochastic processes that are able to capture some common features of empirical time series, such as stochastic volatility (intermittency), roughness, stationarity, and strong dependence. These processes have been applied in various contexts, most notably in the study of turbulence in physics [34, 103] and in finance as models of energy prices [31, 48]. A $\mathcal{BSS}$ process X is defined via the integral representation

$$X_t = \int_{-\infty}^{t} g(t-s)\sigma_s \mathrm{d}W_s, \tag{7.1}$$

where W is a two-sided Brownian motion providing the fundamental noise innovations, the amplitude of which is modulated by a stochastic volatility (intermittency) process σ that may depend on W. This driving noise is then convolved with a deterministic kernel function g that specifies the dependence structure of X. The process X can also be viewed as a *moving average* of volatility-modulated Brownian noise; setting $\sigma_s = 1$, we see that stationary *Brownian moving averages* are nested in this class of processes.

In the applications mentioned above, the case where X is not a semimartingale is particularly relevant. This situation arises when the kernel function g behaves like a power law near zero, more specifically, when for some $\alpha \in (-\frac{1}{2}, \frac{1}{2}) \setminus \{0\}$,

$$g(x) \propto x^{\alpha} \quad \text{for small } x > 0. \tag{7.2}$$

Here we write "$\propto$" to indicate proportionality in an *informal* sense, anticipating a rigorous formulation of this relationship given in Section 7.1.2 using the theory of *regular variation* [63],

[43]This chapter is derived from "Hybrid scheme for Brownian semistationary processes," *Finance and Stochastics*, 21 (2017), pp. 931–965, https://dx.doi.org/10.1007/s00780-017-0335-5.

which plays a significant role in our subsequent arguments. The case $\alpha = -\frac{1}{6}$ in (7.2) is important in statistical modeling of turbulence [103] as it gives rise to processes that are compatible with Kolmogorov's scaling law for ideal turbulence. Moreover, processes of similar type with $\alpha \approx -0.4$ have been recently used in the context of option pricing as models of *rough volatility* [11, 39, 153, 166]; see Sections 7.1.5 and 7.2.3. The case $\alpha = 0$ would (roughly speaking) lead to a process that is a semimartingale, which is thus excluded.

Under (7.2), the trajectories of X behave locally like the trajectories of a *fractional Brownian motion* (fBm) with Hurst index $H = \alpha + \frac{1}{2} \in (0,1) \setminus \{\frac{1}{2}\}$. While the *local* behavior and roughness, measured in terms of Hölder regularity, of X are determined by the parameter α, the *global* behavior of X (e.g., whether the process has long or short memory) depends on the behavior of $g(x)$ as $x \uparrow \infty$, which can be specified independently of α. This should be contrasted with fractional Brownian motion and related *self-similar* models, which necessarily must conform to a restrictive affine relationship between their Hölder regularity (local behavior and roughness) and Hurst index (global behavior), as elucidated by Gneiting and Schlather [174]. Indeed, in the realm of $\mathcal{BSS}$ processes, local and global behaviors are conveniently decoupled, which underlines the flexibility of these processes as a modeling framework.

In connection with practical applications, it is important to be able to simulate the process X. If the volatility process σ is deterministic and constant in time, then X will be strictly stationary and Gaussian. This makes X amenable to exact simulation using the Cholesky factorization or circulant embeddings; see, e.g., [21, Chapter XI]. However, it seems difficult, if not impossible, to develop an exact method that is applicable with a stochastic σ, as the process X is then neither Markovian nor Gaussian. Thus, in the general case one must resort to approximative methods. To this end, Benth, Eyjolfsson, and Veraart [53] have recently proposed a Fourier-based method of simulating $\mathcal{BSS}$ processes, and more general *Lévy semistationary* ($\mathcal{LSS}$) processes, which rely on approximating the kernel function g in the frequency domain.

In this chapter, we introduce a new discretization scheme for $\mathcal{BSS}$ processes based on approximating the kernel function g in the time domain. Our starting point is the Riemann-sum discretization of (7.1). The Riemann-sum scheme builds on an approximation of g using step functions, which has the pitfall of failing to capture appropriately the steepness of g near zero. In particular, this becomes a serious defect under (7.2) when $\alpha \in (-\frac{1}{2}, 0)$. In our new scheme, we mitigate this problem by approximating g using an appropriate power function near zero and a step function elsewhere. The resulting discretization scheme can be realized as a linear combination of Wiener integrals with respect to the driving Brownian motion W and a Riemann sum, which is why we call it a *hybrid scheme*. The hybrid scheme is only slightly more demanding to implement than the Riemann-sum scheme, and the schemes have the same computational complexity as the number of discretization cells tends to infinity.

Our main theoretical result describes the exact asymptotic behavior of the mean square error (MSE) of the hybrid scheme and, as a special case, that of the Riemann-sum scheme. We observe that switching from the Riemann-sum scheme to the hybrid scheme reduces the asymptotic root mean square error (RMSE) substantially. Using merely the simplest variant of the hybrid scheme, where a power function is used in a single discretization cell, the reduction is at least 50% for all $\alpha \in (0, \frac{1}{2})$ and at least 80% for all $\alpha \in (-\frac{1}{2}, 0)$. The reduction in RMSE is close to 100% as α approaches $-\frac{1}{2}$, which indicates that the hybrid scheme indeed resolves the problem of poor precision that affects the Riemann-sum scheme.

To assess the accuracy of the hybrid scheme in practice, we perform two numerical experiments. First, we examine the finite sample performance of an estimator of the roughness index α, introduced by Barndorff-Nielsen, Corcuera, and Podolskij [33] and Corcuera et al. [103]. This experiment enables us to assess how faithfully the hybrid scheme approximates the fine properties of the $\mathcal{BSS}$ process X. Second, we study Monte Carlo option pricing in the rough Bergomi stochastic volatility model of Bayer, Friz, and Gatheral [39]. We use the hybrid scheme to simulate the volatility process in this model, and we find that the resulting implied volatility smiles are indistinguishable from those simulated using a method that involves exact simulation of the volatility process. Thus, we are able to propose a solution to the problem of finding an efficient simulation scheme for the rough Bergomi model, left open in [39].

The rest of this chapter is organized as follows. In Section 7.1 we recall the rigorous definition of a $\mathcal{BSS}$ process and introduce our assumptions. We also introduce the hybrid scheme, state our main theoretical result concerning the asymptotics of the MSE, and discuss an extension of the scheme to a class of *truncated* $\mathcal{BSS}$ processes. Section 7.2 briefly discusses the implementation of the discretization scheme and presents the numerical experiments mentioned above. Finally, Section 7.3 contains the proofs of the theoretical and technical results.

7.1 ▪ The model and theoretical results

7.1.1 ▪ Brownian semistationary process

Let $(\Omega, \mathcal{F}, (\mathcal{F}_t)_{t\in\mathbb{R}}, \mathbb{P})$ be a filtered probability space, satisfying the usual conditions, supporting a (two-sided) standard Brownian motion $W = (W_t)_{t\in\mathbb{R}}$. We consider a $\mathcal{BSS}$ process

$$X_t = \int_{-\infty}^{t} g(t-s)\sigma_s \mathrm{d}W_s, \qquad t \in \mathbb{R}, \tag{7.3}$$

where $\sigma = (\sigma_t)_{t\in\mathbb{R}}$ is an $(\mathcal{F}_t)_{t\in\mathbb{R}}$-predictable process with locally bounded trajectories, which captures the stochastic volatility (intermittency) of X, and $g : (0,\infty) \to [0,\infty)$ is a Borel measurable kernel function.

To ensure that the integral (7.3) is well-defined, we assume that the kernel function g is square integrable, that is, $\int_0^\infty g(x)^2 \grave{u}\mathrm{d}x < \infty$. In fact, we will introduce more specific assumptions on g that will imply its square integrability. Throughout the chapter, we also assume that the process σ has finite second moments, $\mathbb{E}[\sigma_t^2] < \infty$ for all $t \in \mathbb{R}$, and that the process is covariance stationary, namely,

$$\mathbb{E}[\sigma_s] = \mathbb{E}[\sigma_t], \quad \mathrm{Cov}(\sigma_s, \sigma_t) = \mathrm{Cov}(\sigma_0, \sigma_{|s-t|}), \quad s, t \in \mathbb{R}.$$

These assumptions imply also that X is covariance stationary, that is,

$$\mathbb{E}[X_t] = 0, \quad \mathrm{Cov}(X_s, X_t) = \mathbb{E}[\sigma_0^2] \int_0^\infty g(x)g(x+|s-t|)\mathrm{d}x, \quad s, t \in \mathbb{R}.$$

However, the process X need not be *strictly* stationary as the dependence between the volatility process σ and the driving Brownian motion W may be time-varying.

7.1.2 ▪ Kernel function

As mentioned above, we consider a kernel function that satisfies $g(x) \propto x^\alpha$ for some $\alpha \in (-\frac{1}{2}, \frac{1}{2}) \setminus \{0\}$ when $x > 0$ is near zero. To make this idea rigorous and to allow for additional flexibility, we formulate our assumptions on g using the theory of regular variation [63] and, more specifically, slowly varying functions. To this end, recall that a measurable function $L : (0,1] \to [0,\infty)$ is *slowly varying* at 0 if, for any $t > 0$,

$$\lim_{x\downarrow 0} \frac{L(tx)}{L(x)} = 1.$$

Moreover, a function $f(x) = x^\beta L(x)$, $x \in (0,1]$, where $\beta \in \mathbb{R}$ and L is slowly varying at 0, is said to be *regularly varying* at 0, with β being the *index of regular variation*.

Remark 7.1.1. *Conventionally, slow and regular variations are defined at* ∞ [63, pages 6 and 17–18]. *However, L is slowly varying (resp., regularly varying) at* 0 *if and only if* $x \mapsto L(1/x)$ *is slowly varying (resp., regularly varying) at* ∞.

A key feature of slowly varying functions, which will be very important in the sequel, is that they can be sandwiched between polynomial functions as follows. If $\delta > 0$, and L is slowly varying at 0 and bounded away from 0 and ∞ on any interval $(u, 1]$, $u \in (0, 1)$, then there exist constants $\overline{C}_\delta \geq \underline{C}_\delta > 0$ such that

$$\underline{C}_\delta x^\delta \leq L(x) \leq \overline{C}_\delta x^{-\delta}, \quad x \in (0, 1]. \tag{7.4}$$

The inequalities above are an immediate consequence of the so-called *Potter bounds* for slowly varying functions; see [63, Theorem 1.5.6(ii)] and (7.28). Making δ very small therein, we see that slowly varying functions are asymptotically negligible in comparison with polynomially growing/decaying functions. Thus, by multiplying power functions and slowly varying functions, regular variation provides a flexible framework for constructing functions that behave asymptotically like power functions.

Our assumptions concerning the kernel function g are as follows:

(1) For some $\alpha \in (-\frac{1}{2}, \frac{1}{2}) \setminus \{0\}$,

$$g(x) = x^\alpha L_g(x), \quad x \in (0, 1],$$

where $L_g : (0, 1] \to [0, \infty)$ is continuously differentiable, slowly varying at 0, and bounded away from 0. Moreover, there exists a constant $C > 0$ such that the derivative L_g' of L_g satisfies

$$|L_g'(x)| \leq C(1 + x^{-1}), \quad x \in (0, 1].$$

(2) The function g is continuously differentiable on $(0, \infty)$, so that its derivative g' is ultimately monotonic and satisfies $\int_1^\infty g'(x)^2 \mathrm{d}x < \infty$.

(3) For some $\beta \in (-\infty, -\frac{1}{2})$,

$$g(x) = O(x^\beta), \quad x \uparrow \infty.$$

(Here and in the sequel, we use $f(x) = O(h(x))$, $x \to a$, to indicate that $\limsup_{x\to a} \left|\frac{f(x)}{h(x)}\right| < \infty$. Additionally, analogous notation is later used for sequences and computational complexity.) In view of the bound (7.4), these assumptions ensure that g is square integrable. It is worth pointing out that (1) accommodates functions L_g with $\lim_{x\downarrow 0} L_g(x) = \infty$, e.g., $L_g(x) = 1 - \log x$.

The assumption (1) influences the short-term behavior and roughness of the process X. A simple way to assess the roughness of X is to study the behavior of its *variogram* (also called the *second-order structure function* in the turbulence literature)

$$V_X(h) := \mathbb{E}[|X_h - X_0|^2], \quad h \geq 0,$$

as $h \downarrow 0$. Note that, by covariance stationarity,

$$V_X(|s - t|) = \mathbb{E}[|X_s - X_t|^2], \quad s, t \in \mathbb{R}.$$

Under our assumptions, we have the following characterization of the behavior of V_X near zero, which generalizes a result of Barndorff-Nielsen [29, page 9] and implies that X has a locally Hölder continuous modification. Therein, and in what follows, we write $a(x) \sim b(x)$, $x \to y$, to indicate that $\lim_{x\to y} \frac{a(x)}{b(x)} = 1$. The proof of this result is carried out in Section 7.3.1.

Proposition 7.1 (local behavior and continuity). *Suppose that* (1), (2), *and* (3) *hold.*

(i) *The variogram of X satisfies*

$$V_X(h) \sim \mathbb{E}[\sigma_0^2]\left(\frac{1}{2\alpha+1} + \int_0^\infty \left((y+1)^\alpha - y^\alpha\right)^2 \mathrm{d}y\right) h^{2\alpha+1} L_g(h)^2, \quad h \downarrow 0,$$

which implies that V_X is regularly varying at zero with index $2\alpha + 1$.

(ii) *The process X has a modification with locally φ-Hölder continuous trajectories for any $\varphi \in (0, \alpha + \frac{1}{2})$.*

Motivated by Proposition 7.1, we call α the *roughness index* of the process X. Ignoring the slowly varying factor $L_g(h)^2$ in (7.1), we see that the variogram $V(h)$ behaves like $h^{2\alpha+1}$ for small values of h, which is reminiscent of the scaling property of the increments of an fBm with Hurst index $H = \alpha + \frac{1}{2}$. Thus, the process X behaves locally like such an fBm, at least when it comes to second-order structure and roughness. (Moreover, the factor $\frac{1}{2\alpha+1} + \int_0^\infty ((y+1)^\alpha - y^\alpha)^2 \mathrm{d}y$ coincides with the normalization coefficient that appears in the Mandelbrot–Van Ness representation [243, Theorem 1.3.1] of an fBm with $H = \alpha + \frac{1}{2}$.)

Let us now look at two examples of a kernel function g that satisfies our assumptions.

Example 7.2 (the gamma kernel). The so-called *gamma kernel*

$$g(x) = x^\alpha \mathrm{e}^{-\lambda x}, \quad x \in (0, \infty),$$

with parameters $\alpha \in (-\frac{1}{2}, \frac{1}{2}) \setminus \{0\}$ and $\lambda > 0$, has been used extensively in the literature on $\mathcal{BSS}$ processes. It is particularly important in connection with statistical modeling of turbulence (see [103]), but it also provides a way to construct generalizations of Ornstein–Uhlenbeck (OU) processes with roughness that differs from the usual semimartingale case $\alpha = 0$, while mimicking the long-term behavior of an OU process. Moreover, $\mathcal{BSS}$ and $\mathcal{LSS}$ processes defined using the gamma kernel have interesting probabilistic properties; see [256]. An in-depth study of the gamma kernel can be found in [29]. Setting $L_g(x) := \mathrm{e}^{-\lambda x}$, which is slowly varying at 0 since $\lim_{x\downarrow 0} L_g(x) = 1$, it is evident that (1) holds. Since $g(x)$ decays exponentially fast to 0 as $x \uparrow \infty$, it is clear that (3) also holds. To verify (2), note that g satisfies

$$g'(x) = \left(\frac{\alpha}{x} - \lambda\right) g(x), \quad g''(x) = \left(\left(\frac{\alpha}{x} - \lambda\right)^2 - \frac{\alpha}{x^2}\right) g(x), \quad x \in (0, \infty),$$

where $\lim_{x\uparrow\infty}((\frac{\alpha}{x} - \lambda)^2 - \frac{\alpha}{x^2}) = \lambda^2 > 0$, so g' is ultimately increasing with

$$g'(x)^2 \leq (|\alpha| + \lambda)^2 g(x)^2, \quad x \in [1, \infty).$$

Thus, $\int_1^\infty g'(x)^2 \mathrm{d}x < \infty$ since g is square integrable.

Example 7.3 (power-law kernel). Consider the kernel function

$$g(x) = x^\alpha (1 + x)^{\beta - \alpha}, \quad x \in (0, \infty),$$

with parameters $\alpha \in (-\frac{1}{2}, \frac{1}{2}) \setminus \{0\}$ and $\beta \in (-\infty, -\frac{1}{2})$. The behavior of this kernel function near zero is similar to that of the gamma kernel, but $g(x)$ decays to zero polynomially as $x \uparrow \infty$, so it can be used to model long memory. In fact, it can be shown that if $\beta \in (-1, -\frac{1}{2})$, then the autocorrelation function of X is not integrable. Clearly, (1) holds with $L_g(x) := (1+x)^{\beta-\alpha}$, which is slowly varying at 0 since $\lim_{x\downarrow 0} L_g(x) = 1$. Moreover, note that we can write

$$g(x) = x^{\beta} K_g(x), \quad x \in (0,\infty),$$

where $K_g(x) := (1+x^{-1})^{\beta-\alpha}$ satisfies $\lim_{x\uparrow\infty} K_g(x) = 1$. Thus, (3) also holds. We can check (2) by computing

$$g'(x) = \left(\frac{\alpha+\beta x}{x(1+x)}\right) g(x),$$
$$g''(x) = \left(\left(\frac{\alpha+\beta x}{x(1+x)}\right)^2 + \frac{-\alpha - 2\alpha x - \beta x^2}{x^2(1+x)^2}\right) g(x), \quad x \in (0,\infty),$$

where $-\alpha - 2\alpha x - \beta x^2 \uparrow \infty$ when $x \uparrow \infty$ (as $\beta < -\frac{1}{2}$), so g' is ultimately increasing. Additionally, we note that

$$g'(x)^2 \le (|\alpha| + |\beta|)^2 g(x)^2, \quad x \in [1,\infty),$$

implying $\int_1^\infty g'(x)^2 \mathrm{d}x < \infty$ since g is square integrable.

7.1.3 ▪ Hybrid scheme

Let $t \in \mathbb{R}$, and consider discretizing X_t based on its integral representation (7.3) on the grid $\mathcal{G}_t^n := \{t, t-\frac{1}{n}, t-\frac{2}{n}, \ldots\}$ for $n \in \mathbf{N}$. To derive our discretization scheme, let us first note that if the volatility process σ does not vary too much, then it is reasonable to use the approximation

$$X_t = \sum_{k=1}^{\infty} \int_{t-\frac{k}{n}}^{t-\frac{k}{n}+\frac{1}{n}} g(t-s)\sigma_s \mathrm{d}W_s \approx \sum_{k=1}^{\infty} \sigma_{t-\frac{k}{n}} \int_{t-\frac{k}{n}}^{t-\frac{k}{n}+\frac{1}{n}} g(t-s)\mathrm{d}W_s, \tag{7.5}$$

that is, we keep σ constant in each discretization cell. (Here and in the sequel, "$\approx$" stands for an informal approximation used for purely heuristic purposes.) If k is "small," then due to (1) we may approximate

$$g(t-s) \approx (t-s)^{\alpha} L_g\left(\frac{k}{n}\right), \quad t-s \in \left[\frac{k-1}{n}, \frac{k}{n}\right] \setminus \{0\}, \tag{7.6}$$

as the slowly varying function L_g varies "less" than the power function $y \mapsto y^{\alpha}$ near zero; cf. (7.4). If k is "large," or at least $k \ge 2$, then choosing $b_k \in [k-1, k]$ provides an adequate approximation,

$$g(t-s) \approx g\left(\frac{b_k}{n}\right), \quad t-s \in \left[\frac{k-1}{n}, \frac{k}{n}\right], \tag{7.7}$$

by (2). Applying (7.6) to the first κ terms, where $\kappa = 1, 2, \ldots$, and (7.7) to the remaining terms in the approximating series in (7.5) yields

$$\begin{aligned}\sum_{k=1}^{\infty} \sigma_{t-\frac{k}{n}} \int_{t-\frac{k}{n}}^{t-\frac{k}{n}+\frac{1}{n}} g(t-s)\mathrm{d}W_s \approx{}& \sum_{k=1}^{\kappa} L_g\left(\frac{k}{n}\right) \sigma_{t-\frac{k}{n}} \int_{t-\frac{k}{n}}^{t-\frac{k}{n}+\frac{1}{n}} (t-s)^{\alpha}\mathrm{d}W_s \\ &+ \sum_{k=\kappa+1}^{\infty} g\left(\frac{b_k}{n}\right) \sigma_{t-\frac{k}{n}} \int_{t-\frac{k}{n}}^{t-\frac{k}{n}+\frac{1}{n}} \mathrm{d}W_s.\end{aligned} \tag{7.8}$$

For completeness, we also allow for $\kappa = 0$, in which case we require that $b_1 \in (0, 1]$ and interpret the first sum on the right-hand side of (7.8) as zero. To make numerical implementation feasible, we truncate the second sum on the right-hand side of (7.8) so that both sums have $N_n \geq \kappa + 1$ terms in total. Thus, we arrive at a discretization scheme for X_t, which we call a *hybrid scheme*, given by

$$X_t^n := \check{X}_t^n + \hat{X}_t^n,$$

where

$$\check{X}_t^n := \sum_{k=1}^{\kappa} L_g\left(\frac{k}{n}\right)\sigma_{t-\frac{k}{n}} \int_{t-\frac{k}{n}}^{t-\frac{k}{n}+\frac{1}{n}} (t-s)^\alpha \mathrm{d}W_s, \tag{7.9}$$

$$\hat{X}_t^n := \sum_{k=\kappa+1}^{N_n} g\left(\frac{b_k}{n}\right)\sigma_{t-\frac{k}{n}}\left(W_{t-\frac{k}{n}+\frac{1}{n}} - W_{t-\frac{k}{n}}\right), \tag{7.10}$$

and $\mathbf{b} := (b_k)_{k=\kappa+1}^{\infty}$ is a sequence of real numbers, evaluation points, that must satisfy $b_k \in [k-1, k] \setminus \{0\}$ for each $k \geq \kappa + 1$ but otherwise can be chosen freely.

As it stands, the discretization grid $\mathcal{G}_t^n$ depends on the time t, which may seem cumbersome with regard to sampling X_t^n simultaneously for different times t. However, note that whenever times t and t' are separated by a multiple of $\frac{1}{n}$, the corresponding grids $\mathcal{G}_t^n$ and $\mathcal{G}_{t'}^n$ will intersect. In fact, the hybrid scheme defined by (7.9) and (7.10) can be implemented efficiently, as we shall see in Section 7.2.1. Since

$$g\left(\frac{b_k}{n}\right) = g\left(t - \left(t - \frac{b_k}{n}\right)\right),$$

the degenerate case $\kappa = 0$ with $b_k = k$ for all $k \geq 1$ corresponds to the usual Riemann-sum discretization scheme of X_t with (Itô-type) forward sums from (7.10). Henceforth, we denote the associated sequence $(k)_{k=\kappa+1}^{\infty}$ by $\mathbf{b}_{\mathrm{FWD}}$, where the subscript "FWD" alludes to forward sums. However, including terms involving Wiener integrals of a power function given by (7.9), that is, having $\kappa \geq 1$, improves the accuracy of the discretization considerably, as we shall see. Having the leeway to select b_k within the interval $[k-1, k] \setminus \{0\}$, so that the function $g(t - \cdot)$ is evaluated at a point that does not necessarily belong to $\mathcal{G}_t^n$, leads additionally to a moderate improvement.

The truncation in the sum (7.10) entails that the stochastic integral (7.3) defining X is truncated at $t - \frac{N_n}{n}$. In practice, the value of the parameter N_n should be large enough to mitigate the effect of truncation. To ensure that the truncation point $t - \frac{N_n}{n}$ tends to $-\infty$ as $n \uparrow \infty$ in our asymptotic results, we introduce the following assumption:

(4) For some $\gamma > 0$,

$$N_n \sim n^{\gamma+1}, \quad n \uparrow \infty.$$

7.1.4 ▪ Asymptotic behavior of mean square error

We are now ready to state our main theoretical result, which gives a sharp description of the asymptotic behavior of the mean square error (MSE) of the hybrid scheme as $n \uparrow \infty$. We defer the proof of this result to Section 7.3.2.

Theorem 7.4 (asymptotics of mean square error). *Suppose that* (1), (2), (3), *and* (4) *hold, so that*

$$\gamma > -\frac{2\alpha+1}{2\beta+1}, \tag{7.11}$$

and that for some $\delta > 0$,

$$\mathbb{E}[|\sigma_s - \sigma_0|^2] = O\big(s^{2\alpha+1+\delta}\big), \quad s \downarrow 0. \tag{7.12}$$

Then for all $t \in \mathbb{R}$,

$$\mathbb{E}[|X_t - X_t^n|^2] \sim J(\alpha, \kappa, \mathbf{b})\mathbb{E}[\sigma_0^2]n^{-(2\alpha+1)}L_g(1/n)^2, \quad n \uparrow \infty, \tag{7.13}$$

where

$$J(\alpha, \kappa, \mathbf{b}) := \sum_{k=\kappa+1}^{\infty} \int_{k-1}^{k} (y^\alpha - b_k^\alpha)^2 \mathrm{d}y < \infty. \tag{7.14}$$

Remark 7.1.2. *Note that if $\alpha \in (-\frac{1}{2}, 0)$, then having*

$$\mathbb{E}[|\sigma_s - \sigma_0|^2] = O(s^\theta), \quad s \downarrow 0,$$

for all $\theta \in (0, 1)$, ensures that (7.12) holds. (Take, say, $\delta := \frac{1}{2}(1 - (2\alpha + 1)) > 0$ and $\theta := 2\alpha + 1 + \delta = \alpha + 1 \in (0, 1)$.)

When the hybrid scheme is used to simulate the $\mathcal{BSS}$ process X on an equidistant grid $\{0, \frac{1}{n}, \frac{2}{n}, \dots, \frac{\lfloor nT \rfloor}{n}\}$ for some $T > 0$ (see Section 7.2.1 for details of the implementation), the following consequence of Theorem 7.4 ensures that the covariance structure of the simulated process approximates that of the actual process X.

Corollary 7.5 (covariance structure). *Suppose that the assumptions of Theorem 7.4 hold. Then for any $s, t \in \mathbf{R}$, and $\varepsilon > 0$,*

$$|\mathbb{E}[X_t^n X_s^n] - \mathbb{E}[X_t X_s]| = O\big(n^{-(\alpha+\frac{1}{2})+\varepsilon}\big), \quad n \uparrow \infty.$$

Proof. Let $s, t \in \mathbf{R}$. Applying the Cauchy–Schwarz inequality, we get

$$\begin{aligned}|\mathbb{E}[X_t^n X_s^n] - \mathbb{E}[X_t X_s]| \leq \mathbb{E}[(X_t^n)^2]^{1/2}\mathbb{E}[|X_s - X_s^n|^2]^{1/2} \\ + \mathbb{E}[X_s^2]^{1/2}\mathbb{E}[|X_t - X_t^n|^2]^{1/2}.\end{aligned}$$

We have $\sup_{n\in\mathbf{N}} \mathbb{E}[(X_t^n)^2]^{1/2} < \infty$ since $\mathbb{E}[(X_t^n)^2] \to \mathbb{E}[X_t^2] < \infty$ as $n \uparrow \infty$, by Theorem 7.4. Moreover, Theorem 7.4 and the bound (7.4) imply that $\mathbb{E}[|X_s - X_s^n|^2]^{1/2} = O(n^{-(\alpha+\frac{1}{2})+\varepsilon})$ and $\mathbb{E}[|X_t - X_t^n|^2]^{1/2} = O(n^{-(\alpha+\frac{1}{2})+\varepsilon})$ for any $\varepsilon > 0$. □

In Theorem 7.4, the asymptotics of the MSE (7.13) are determined by the behavior of the kernel function g near zero, as specified in (1). The condition (7.11) ensures that the error from approximating g near zero is asymptotically larger than the error induced by the truncation of the stochastic integral (7.3) at $t - \frac{N_n}{n}$. In fact, a different kind of asymptotics of the MSE, where the truncation error becomes dominant, could be derived when (7.11) does not hold, under some additional assumptions, but we do not pursue this direction.

While the rate of convergence in (7.13) is fully determined by the roughness index α, which may seem discouraging at first, it turns out that the quantity $J(\alpha, \kappa, \mathbf{b})$, which we shall call the *asymptotic* MSE, can vary a lot, depending on how we choose κ and $\mathbf{b}$, and can have a substantial impact on the precision of the approximation of X. It is immediate from (7.14) that increasing κ will decrease $J(\alpha, \kappa, \mathbf{b})$. Moreover, for given α and κ, it is straightforward to choose $\mathbf{b}$ so that $J(\alpha, \kappa, \mathbf{b})$ is minimized, as shown in the following result.

Proposition 7.6 (optimal discretization). *Let $\alpha \in (-\frac{1}{2}, \frac{1}{2}) \setminus \{0\}$ and $\kappa \geq 0$. Among all sequences $\mathbf{b} = (b_k)_{k=\kappa+1}^{\infty}$ with $b_k \in [k-1, k] \setminus \{0\}$ for $k \geq \kappa + 1$, the function $J(\alpha, \kappa, \mathbf{b})$, and consequently the asymptotic MSE induced by the discretization, is minimized by the sequence $\mathbf{b}^*$ given by*

$$b_k^* = \left(\frac{k^{\alpha+1} - (k-1)^{\alpha+1}}{\alpha + 1} \right)^{1/\alpha}, \quad k \geq \kappa + 1.$$

Proof. Clearly, a sequence $\mathbf{b} = (b_k)_{k=\kappa+1}^{\infty}$ minimizes the function $J(\alpha, \kappa, \mathbf{b})$ if and only if b_k minimizes $\int_{k-1}^{k} (y^\alpha - b_k^\alpha)^2 \mathrm{d}y$ for any $k \geq \kappa + 1$. By standard L^2-space theory, $c \in \mathbb{R}$ minimizes the integral $\int_{k-1}^{k} (y^\alpha - c)^2 \mathrm{d}y$ if and only if the function $y \mapsto y^\alpha - c$ is orthogonal in L^2 to all constant functions. This is tantamount to

$$\int_{k-1}^{k} (y^\alpha - c) \mathrm{d}y = 0,$$

and computing the integral and solving for c yields

$$c = \frac{k^{\alpha+1} - (k-1)^{\alpha+1}}{\alpha + 1}.$$

Setting $b_k^* := c^{1/\alpha} \in (k-1, k)$ completes the proof. □

To understand how much increasing κ and using the optimal sequence $\mathbf{b}^*$ from Proposition 7.6 improves the approximation, we study numerically the asymptotic *root mean square error* (RMSE) $\sqrt{J(\alpha, \kappa, \mathbf{b})}$. In particular, we assess how much the asymptotic RMSE decreases relative to the RMSE of the forward Riemann-sum scheme ($\kappa = 0$ and $\mathbf{b} = \mathbf{b}_{\mathrm{FWD}}$) using the quantity

$$\text{reduction in asymptotic RMSE} = -\frac{\sqrt{J(\alpha, \kappa, \mathbf{b})} - \sqrt{J(\alpha, 0, \mathbf{b}_{\mathrm{FWD}})}}{\sqrt{J(\alpha, 0, \mathbf{b}_{\mathrm{FWD}})}} \cdot 100\%. \tag{7.15}$$

The results are presented in Figure 7.1. We find that employing the hybrid scheme with $\kappa \geq 1$ leads to a substantial reduction in the asymptotic RMSE relative to the forward Riemann-sum scheme when $\alpha \in (-\frac{1}{2}, 0)$. Indeed, when $\kappa \geq 1$, the asymptotic RMSE, as a function of α, does not blow up as $\alpha \to -\frac{1}{2}$, while with $\kappa = 0$ it does. This explains why the reduction in the asymptotic RMSE approaches 100% as $\alpha \to -\frac{1}{2}$. When $\alpha \in (0, \frac{1}{2})$, the improvement achieved using the hybrid scheme is more modest but still considerable. Figure 7.1 also highlights the importance of using the optimal sequence $\mathbf{b}^*$, instead of $\mathbf{b}_{\mathrm{FWD}}$, as evaluation points in the scheme, in particular when $\alpha \in (0, \frac{1}{2})$. Finally, we observe that increasing κ beyond 2 does not appear to lead to a significant further reduction. Indeed, in our numerical experiments, reported in Section 7.2.2 and 7.2.3, we observe that using $\kappa = 1, 2$ already leads to good results.

Remark 7.1.3. *It is nontrivial to evaluate the quantity $J(\alpha, \kappa, \mathbf{b})$ numerically. Computing the integral in* (7.14) *explicitly, we can approximate $J(\alpha, \kappa, \mathbf{b})$ by*

$$J_N(\alpha, \kappa, \mathbf{b}) := \sum_{k=\kappa+1}^{N} \left(\frac{k^{2\alpha+1} - (k-1)^{2\alpha+1}}{2\alpha + 1} - \frac{2b_k^\alpha \left(k^{\alpha+1} - (k-1)^{\alpha+1}\right)}{\alpha + 1} + b_k^{2\alpha} \right)$$

with some large $N \in \mathbf{N}$. This approximation is adequate when $\alpha \in (-\frac{1}{2}, 0)$, but its accuracy deteriorates when $\alpha \to \frac{1}{2}$. In particular, the singularity of the function $\alpha \mapsto J(\alpha, \kappa, \mathbf{b})$ at $\frac{1}{2}$ is difficult to capture using $J_N(\alpha, \kappa, \mathbf{b})$ with numerically feasible values of N. To overcome this numerical problem, we introduce a correction term in the case $\alpha \in (0, \frac{1}{2})$. The correction term

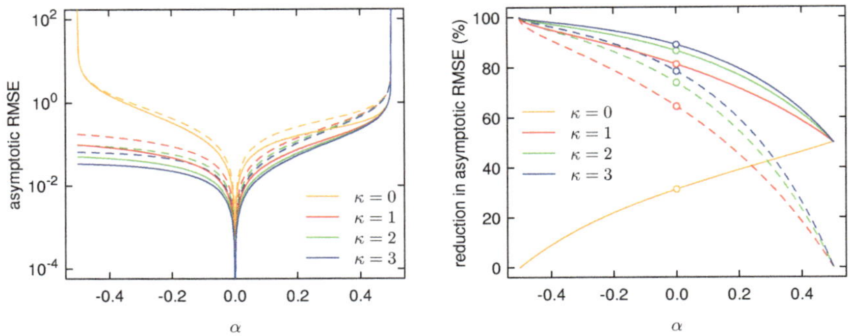

Figure 7.1: Left: The asymptotic RMSE given by $\sqrt{J(\alpha,\kappa,\mathbf{b})}$ as a function of $\alpha \in (-\frac{1}{2},\frac{1}{2})\backslash\{0\}$ for $\kappa = 0,1,2,3$ using $\mathbf{b} = \mathbf{b}^*$ of Proposition 7.6 (solid lines) and $\mathbf{b} = \mathbf{b}_{\mathrm{FWD}}$ (dashed lines). Right: Reduction in the asymptotic RMSE relative to the forward Riemann-sum scheme ($\kappa = 0$ and $\mathbf{b} = \mathbf{b}_{\mathrm{FWD}}$) given by the formula (7.15), plotted as a function of $\alpha \in (-\frac{1}{2},\frac{1}{2}) \setminus \{0\}$ for $\kappa = 0,1,2,3$ using $\mathbf{b} = \mathbf{b}^*$ (solid lines) and for $\kappa = 1,2,3$ using $\mathbf{b} = \mathbf{b}_{\mathrm{FWD}}$ (dashed lines). In all computations, we have used the approximations outlined in Remark 7.1.3 with $N = 1{,}000{,}000$.

can be derived informally as follows. By the mean value theorem, and since $b_k^* \approx k - \frac{1}{2}$ for large k, we have

$$(y^\alpha - b_k^\alpha)^2 = \alpha^2 \xi^{2\alpha-2}(y-b_k)^2 \approx \begin{cases} \alpha^2 k^{2\alpha-2}(y-k)^2, & \mathbf{b} = \mathbf{b}_{\mathrm{FWD}}, \\ \alpha^2 k^{2\alpha-2}(y-k+\frac{1}{2})^2, & \mathbf{b} = \mathbf{b}^*, \end{cases}$$

where $\xi = \xi(y,b_k) \in [k-1,k]$, for large k. Thus, for large N, we obtain

$$\begin{aligned} &J(\alpha,\kappa,\mathbf{b}) - J_N(\alpha,\kappa,\mathbf{b}) \\ &\quad = \sum_{k=N+1}^{\infty} \int_{k-1}^{k} (y^\alpha - b_k^\alpha)^2 \mathrm{d}y \\ &\quad \approx \begin{cases} \alpha^2 \sum_{k=N+1}^{\infty} k^{2\alpha-2} \int_{k-1}^{k} (y-k)^2 \mathrm{d}y, & \mathbf{b} = \mathbf{b}_{\mathrm{FWD}}, \\ \alpha^2 \sum_{k=N+1}^{\infty} k^{2\alpha-2} \int_{k-1}^{k} (y-k+\frac{1}{2})^2 \mathrm{d}y, & \mathbf{b} = \mathbf{b}^*, \end{cases} \\ &\quad = \begin{cases} \frac{\alpha^2}{3}\zeta(2-2\alpha, N+1), & \mathbf{b} = \mathbf{b}_{\mathrm{FWD}}, \\ \frac{\alpha^2}{12}\zeta(2-2\alpha, N+1), & \mathbf{b} = \mathbf{b}^*, \end{cases} \end{aligned}$$

where $\zeta(x,s) := \sum_{k=0}^{\infty} \frac{1}{(k+s)^x}$, $x > 1$, $s > 0$, is the Hurwitz zeta function, *which can be evaluated using accurate numerical algorithms.*

Remark 7.1.4. *Unlike the Fourier-based method of* [53], *the hybrid scheme does not require truncating the singularity of the kernel function g when $\alpha \in (-\frac{1}{2}, 0)$, which is beneficial to maintaining the accuracy of the scheme when α is near $-\frac{1}{2}$. Let us briefly analyze the effect of truncating the singularity of g on the approximation error* [53, pages 75–76]. *Consider, for any $\varepsilon > 0$, the modified $\mathcal{BSS}$ process*

$$\tilde{X}_t^\varepsilon := \int_{-\infty}^{t} g_\varepsilon(t-s)\sigma_s \mathrm{d}W_s, \quad t \in \mathbb{R},$$

defined using the truncated kernel function

$$g_\varepsilon(x) := \begin{cases} g(\varepsilon), & x \in (0, \varepsilon], \\ g(x), & x \in (\varepsilon, \infty). \end{cases}$$

Adapting the proof of Theorem 7.4 *in a straightforward manner, it is possible to show that, under* (1) *and* (3),

$$\begin{aligned} \mathbb{E}[|X_t - \tilde{X}_t^\varepsilon|^2] &= \mathbb{E}[\sigma_0^2] \int_0^\varepsilon \big(g(s) - g(\varepsilon)\big)^2 \mathrm{d}s \\ &\sim \underbrace{\left(\frac{1}{2\alpha+1} - \frac{2}{\alpha+1} + 1\right)}_{=:\tilde{J}(\alpha)} \mathbb{E}[\sigma_0^2]\varepsilon^{2\alpha+1} L_g(\varepsilon)^2, \quad \varepsilon \downarrow 0, \end{aligned}$$

for any $t \in \mathbb{R}$. *While the rate of convergence, as* $\varepsilon \downarrow 0$, *of the MSE that arises from replacing* g *with* g_ε *is analogous to the rate of convergence of the hybrid scheme, it is important to note that the factor* $\tilde{J}(\alpha)$ *blows up as* $\alpha \downarrow -\frac{1}{2}$. *In fact,* $\tilde{J}(\alpha)$ *is equal to the first term in the series that defines* $J(\alpha, 0, \mathbf{b}_{\mathrm{FWD}})$ *and*

$$\tilde{J}(\alpha) \sim J(\alpha, 0, \mathbf{b}_{\mathrm{FWD}}), \quad \alpha \downarrow -\frac{1}{2},$$

which indicates that the effect of truncating the singularity, in terms of the MSE, is similar to the effect of using the forward Riemann-sum scheme to discretize the process when α *is near* $-\frac{1}{2}$. *In particular, the truncation threshold* ε *would then have to be very small in order to keep the truncation error in check.*

7.1.5 ▪ Extension to truncated Brownian semistationary processes

It is useful to extend the hybrid scheme to a class of nonstationary processes that are closely related to $\mathcal{BSS}$ processes. This extension is important in connection with an application to the so-called rough Bergomi model, which we discuss in Section 7.2.3. More precisely, we consider processes of the form

$$Y_t = \int_0^t g(t-s)\sigma_s \mathrm{d}W_s, \quad t \geq 0, \tag{7.16}$$

where the kernel function g, volatility process σ, and driving Brownian motion W are as before. We call Y a *truncated Brownian semistationary* ($\mathcal{TBSS}$) process, as Y is obtained from the $\mathcal{BSS}$ process X by truncating the stochastic integral in (7.3) at 0. Of the preceding assumptions, only (1) and (2) are needed to ensure that the stochastic integral in (7.16) exists—in fact, of (2), only the requirement that g is differentiable on $(0, \infty)$ comes into play.

The $\mathcal{TBSS}$ process Y does not have covariance stationary increments, so we define its (time-dependent) variogram as

$$V_Y(h, t) := \mathbb{E}[|Y_{t+h} - Y_t|^2], \quad h, t \geq 0.$$

Extending Proposition 7.1, we can describe the behavior of $h \mapsto V_Y(h, t)$ near zero as follows. The existence of a locally Hölder continuous modification is then a straightforward consequence. We omit the proof of this result, as it would be a straightforward adaptation of the proof of Proposition 7.1.

Proposition 7.7 (local behavior and continuity). *Suppose that* (1) *and* (2) *hold.*

(i) *The variogram of Y satisfies, for any $t \geq 0$,*

$$V_Y(h,t) \sim \mathbb{E}[\sigma_0^2]\left(\frac{1}{2\alpha+1} + \mathbf{1}_{(0,\infty)}(t)\int_0^\infty \left((y+1)^\alpha - y^\alpha\right)^2 \mathrm{d}y\right) h^{2\alpha+1} L_g(h)^2,$$

as $h \downarrow 0$, which implies that $h \mapsto V_Y(h,t)$ is regularly varying at zero with index $2\alpha + 1$.

(ii) *The process Y has a modification with locally φ-Hölder continuous trajectories for any $\varphi \in (0, \alpha + \frac{1}{2})$.*

Note that while the increments of Y are not covariance stationary, the asymptotic behavior of $V_Y(h,t)$ is the same as that of $V_X(h)$ as $h \downarrow 0$ (see Proposition 7.1) for any $t > 0$. Thus, the increments of Y (apart from increments starting at time 0) are locally like the increments of X.

We define the hybrid scheme to discretize Y_t, for any $t \geq 0$, as

$$Y_t^n := \check{Y}_t^n + \hat{Y}_t^n, \tag{7.17}$$

where

$$\check{Y}_t^n := \sum_{k=1}^{\min\{\lfloor nt \rfloor, \kappa\}} L_g\left(\frac{k}{n}\right)\sigma_{t-\frac{k}{n}} \int_{t-\frac{k}{n}}^{t-\frac{k}{n}+\frac{1}{n}} (t-s)^\alpha \mathrm{d}W_s,$$

$$\hat{Y}_t^n := \sum_{k=\kappa+1}^{\lfloor nt \rfloor} g\left(\frac{b_k}{n}\right)\sigma_{t-\frac{k}{n}}\left(W_{t-\frac{k}{n}+\frac{1}{n}} - W_{t-\frac{k}{n}}\right).$$

In effect, we simply drop the summands in (7.9) and (7.10) that correspond to integrals and increments on the negative real line. We make remarks on the implementation of this scheme in Section 7.2.1.

The MSE of the hybrid scheme for the $\mathcal{TBSS}$ process Y has the following asymptotic behavior as $n \uparrow \infty$, which is, in fact, identical to the asymptotic behavior of the MSE of the hybrid scheme for $\mathcal{BSS}$ processes. We omit the proof of this result, which would be a simple modification of the proof of Theorem 7.4.

Theorem 7.8 (asymptotics of mean square error). *Suppose that* (1) *and* (2) *hold, and that for some $\delta > 0$,*

$$\mathbb{E}[|\sigma_s - \sigma_0|^2] = O\left(s^{2\alpha+1+\delta}\right), \quad s \downarrow 0.$$

Then for all $t > 0$,

$$\mathbb{E}[|Y_t - Y_t^n|^2] \sim J(\alpha,\kappa,\mathbf{b})\mathbb{E}[\sigma_0^2] n^{-(2\alpha+1)} L_g(1/n)^2, \quad n \uparrow \infty,$$

where $J(\alpha,\kappa,\mathbf{b})$ is as in Theorem 7.4.

Remark 7.1.5. *Under the assumptions of Theorem 7.8, the conclusion of Corollary 7.5 holds mutatis mutandis. In particular, the covariance structure of the discretized $\mathcal{TBSS}$ process approaches that of Y when $n \uparrow \infty$.*

7.2 ▪ Implementation and numerical experiments

7.2.1 ▪ Practical implementation

Simulating the $\mathcal{BSS}$ process X on the equidistant grid $\{0, \frac{1}{n}, \frac{2}{n}, \dots, \frac{\lfloor nT \rfloor}{n}\}$ for some $T > 0$ using the hybrid scheme entails generating

$$X^n_{\frac{i}{n}}, \quad i = 0, 1, \dots, \lfloor nT \rfloor. \tag{7.18}$$

Provided that we can simulate the random variables

$$W^n_{i,j} := \int_{\frac{i}{n}}^{\frac{i+1}{n}} \left(\frac{i+j}{n} - s\right)^{\alpha} \mathrm{d}W_s, \qquad i = -N_n, -N_n + 1, \dots, \lfloor nT \rfloor - 1, \quad j = 1, \dots, \kappa, \tag{7.19}$$

$$W^n_i := \int_{\frac{i}{n}}^{\frac{i+1}{n}} \mathrm{d}W_s, \qquad i = -N_n, -N_n + 1, \dots, \lfloor nT \rfloor - 1, \tag{7.20}$$

$$\sigma^n_i := \sigma_{\frac{i}{n}}, \qquad i = -N_n, -N_n + 1, \dots, \lfloor nT \rfloor - 1,$$

we can compute (7.18) via the formula

$$X^n_{\frac{i}{n}} = \underbrace{\sum_{k=1}^{\kappa} L_g\left(\frac{k}{n}\right)\sigma^n_{i-k} W^n_{i-k,k}}_{=\check{X}^n_{\frac{i}{n}}} + \underbrace{\sum_{k=\kappa+1}^{N_n} g\left(\frac{b^*_k}{n}\right)\sigma^n_{i-k} W^n_{i-k}}_{=\hat{X}^n_{\frac{i}{n}}}. \tag{7.21}$$

In order to simulate (7.19) and (7.20), it is instrumental to note that the $\kappa+1$-dimensional random vectors

$$\mathbf{W}^n_i := \left(W^n_i, W^n_{i,1}, \dots, W^n_{i,\kappa}\right), \quad i = -N_n, -N_n + 1, \dots, \lfloor nT \rfloor - 1,$$

are independent and identically distributed (i.i.d.) according to a multivariate Gaussian distribution with mean zero and covariance matrix Σ given by

$$\Sigma_{1,1} = \frac{1}{n}, \qquad \Sigma_{1,j} = \Sigma_{j,1} = \frac{(j-1)^{\alpha+1} - (j-2)^{\alpha+1}}{(\alpha+1)n^{\alpha+1}},$$

$$\Sigma_{j,j} = \frac{(j-1)^{2\alpha+1} - (j-2)^{2\alpha+1}}{(2\alpha+1)n^{2\alpha+1}},$$

for $j = 2, \dots, \kappa + 1$, and

$$\Sigma_{j,k} = \frac{1}{(\alpha+1)n^{2\alpha+1}}\Bigg((j-1)^{\alpha+1}(k-1)^{\alpha}{}_2F_1\left(-\alpha, 1, \alpha+2, \frac{j-1}{k-1}\right) - (j-2)^{\alpha+1}(k-2)^{\alpha}{}_2F_1\left(-\alpha, 1, \alpha+2, \frac{j-2}{k-2}\right)\Bigg), \tag{7.22}$$

for $j, k = 2, \dots, \kappa + 1$ such that $j < k$, where ${}_2F_1$ stands for the *Gauss hypergeometric function*; see [37, page 56] for the definition. (When $k < j$, set $\Sigma_{j,k} = \Sigma_{k,j}$.) For the convenience of the reader, we provide a proof of (7.22) in Section 7.3.3.

Thus, $(\mathbf{W}^n_i)_{i=-N_n}^{\lfloor nT \rfloor - 1}$ can be generated by taking independent draws from the multivariate Gaussian distribution $N_{\kappa+1}(\mathbf{0}, \Sigma)$. If the volatility process σ is independent of W, then $(\sigma^n_i)_{i=-N_n}^{\lfloor nT \rfloor - 1}$ can be generated separately, possibly using exact methods. (Exact methods are available, e.g., for Gaussian processes, as mentioned in the introduction, and diffusions; see [59].) In the case where σ depends on W, simulating $(\mathbf{W}_i^n)_{i=-N_n}^{\lfloor nT \rfloor-1}$ and $(\sigma_i^n)_{i=-N_n}^{\lfloor nT \rfloor-1}$ is less straightforward. That said, if σ is driven by a standard Brownian motion Z, correlated with W, say, one could rely on a factor decomposition

$$Z_t := \varrho W_t + \sqrt{1-\varrho^2} W_t^{\perp}, \quad t \in \mathbb{R}, \tag{7.23}$$

where $\varrho \in [-1, 1]$ is the correlation parameter, and $(W_t^{\perp})_{t\in[0,T]}$ is a standard Brownian motion independent of W. Then one would first generate $(\mathbf{W}_i^n)_{i=-N_n}^{\lfloor nT \rfloor-1}$, use (7.23) to generate $\big(Z_{\frac{i+1}{n}} - Z_{\frac{i}{n}}\big)_{i=-N_n}^{\lfloor nT \rfloor-1}$, and employ some appropriate approximate method to produce $(\sigma_i^n)_{i=-N_n}^{\lfloor nT \rfloor-1}$ thereafter. This approach has, however, the caveat that it induces an additional approximation error not quantified in Theorem 7.4.

Remark 7.2.1. *In the case of the $\mathcal{TBSS}$ process Y, introduced in Section* 7.1.5, *the observations $Y_{\frac{i}{n}}^n$, $i = 0, 1, \ldots, \lfloor nT \rfloor$, given by the hybrid scheme* (7.17) *can be computed via*

$$Y_{\frac{i}{n}}^n = \sum_{k=1}^{\min\{i,\kappa\}} L_g\left(\frac{k}{n}\right) \sigma_{i-k}^n W_{i-k,k}^n + \sum_{k=\kappa+1}^{i} g\left(\frac{b_k^*}{n}\right) \sigma_{i-k}^n W_{i-k}^n, \tag{7.24}$$

using random vectors $\mathbf{W}_i^n$, $i = 0, 1, \ldots, \lfloor nT \rfloor - 1$, and random variables σ_i^n, $i = 0, 1, \ldots, \lfloor nT \rfloor - 1$.

In the hybrid scheme, it typically suffices to take κ to be at most 3. Thus, in (7.21), the first sum $\check{X}_{\frac{i}{n}}^n$ requires only a negligible computational effort. By contrast, the number of terms in the second sum $\hat{X}_{\frac{i}{n}}^n$ increases as $n \uparrow \infty$. It is then useful to note that

$$\hat{X}_{\frac{i}{n}}^n = \sum_{k=1}^{N_n} \Gamma_k \Xi_{i-k} = (\Gamma \star \Xi)_i,$$

where

$$\Gamma_k := \begin{cases} 0, & k = 1, \ldots, \kappa, \\ g\left(\frac{b_k^*}{n}\right), & k = \kappa+1, \kappa+2, \ldots, N_n, \end{cases}$$
$$\Xi_k := \sigma_k^n W_k^n, \quad k = -N_n, -N_n + 1, \ldots, \lfloor nT \rfloor - 1,$$

and $\Gamma \star \Xi$ stands for the discrete convolution of the sequences Γ and Ξ. It is well known that the discrete convolution can be evaluated efficiently using a fast Fourier transform (FFT). The computational complexity of simultaneously evaluating $(\Gamma \star \Xi)_i$ for all $i = 0, 1, \ldots, \lfloor nT \rfloor$ using an FFT is $O(N_n \log N_n)$ (see [234, pages 79–80]), which under (4) translates to $O(n^{\gamma+1} \log n)$. The computational complexity of the entire hybrid scheme is then $O(n^{\gamma+1} \log n)$, provided that $(\sigma_i^n)_{i=-N_n}^{\lfloor nT \rfloor-1}$ is generated using a scheme with complexity not exceeding $O(n^{\gamma+1} \log n)$. As a comparison, we mention that the complexity of an exact simulation of a stationary Gaussian process using circulant embeddings is $O(n \log n)$ [21, page 316], whereas the complexity of the Cholesky factorization is $O(n^3)$ [21, page 312].

Remark 7.2.2. *With $\mathcal{TBSS}$ processes, the computational complexity of the hybrid scheme via* (7.24) *is $O(n \log n)$.*

Figure 7.2 presents examples of trajectories of the $\mathcal{BSS}$ process X using the hybrid scheme with $\kappa = 1, 2$ and $\mathbf{b} = \mathbf{b}^*$. We choose the kernel function g to be the gamma kernel (Example 7.2) with $\lambda = 1$. We also discretize X using the Riemann-sum scheme, $\kappa = 0$ with $\mathbf{b} \in \{\mathbf{b}_{\mathrm{FWD}}, \mathbf{b}^*\}$ (that is, the forward Riemann-sum scheme and its counterpart with optimally chosen evaluation

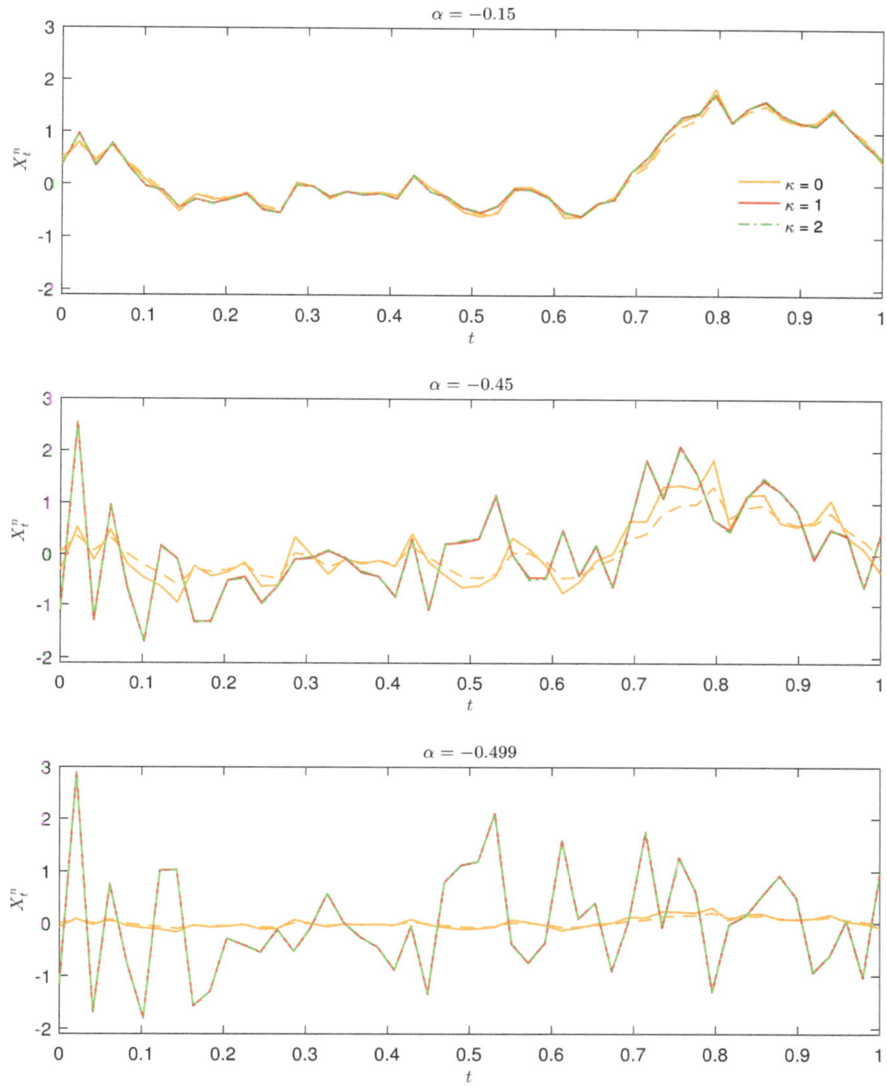

Figure 7.2: Discretized trajectories of a $\mathcal{BSS}$ process, where g is the gamma kernel (Example 7.2), $\lambda = 1$, and $\sigma_t = 1$ for all $t \in \mathbb{R}$. Trajectories consisting of $n = 50$ observations on $[0, 1]$ were generated with the hybrid scheme ($\kappa = 1, 2$ and $\mathbf{b} = \mathbf{b}^*$) and Riemann-sum scheme ($\kappa = 0$ and $\mathbf{b} = \mathbf{b}^*$ (solid lines), $\mathbf{b} = \mathbf{b}_{\text{FWD}}$ (dashed lines)), using the same innovations for the driving Brownian motion in all cases and $N_n = \lfloor 50^{1.5} \rfloor = 353$. The simulated processes were normalized to have unit (stationary) variance.

points). We can make two observations: First, we see how the roughness parameter α controls the regularity properties of the trajectories of X—as we decrease α, the trajectories of X become increasingly rough. Second, and more importantly, we see how the simulated trajectories coming from the Riemann-sum and hybrid schemes can be rather different, even though we use the same innovations for the driving Brownian motion. In fact, the two variants of the hybrid scheme ($\kappa = 1, 2$) yield almost identical trajectories, while the Riemann-sum scheme ($\kappa = 0$)

produces trajectories that are comparatively smoother; this difference becomes more apparent as α approaches $-\frac{1}{2}$. Indeed, in the extreme case with $\alpha = -0.499$, both variants of the Riemann-sum scheme break down and yield anomalous trajectories with very little variation, while the hybrid scheme continues to produce accurate results. The fact that the hybrid scheme is able to reproduce the fine properties of rough $\mathcal{BSS}$ processes, even for values of α very close to $-\frac{1}{2}$, is backed up by a further experiment reported in the following section.

7.2.2 ▪ Estimation of the roughness parameter

Suppose that we have observations $X_{\frac{i}{m}}$, $i = 0, 1, \ldots, m$, of the $\mathcal{BSS}$ process X, given by (7.3), for some $m \in \mathbf{N}$. References [33] and [103] discuss how the roughness index α can be estimated consistently as $m \uparrow \infty$. The method is based on the *change-of-frequency* (COF) statistics

$$\mathrm{COF}(X, m) = \frac{\sum_{k=5}^{m} \left| X_{\frac{k}{m}} - 2X_{\frac{k-2}{m}} + X_{\frac{k-4}{m}} \right|^2}{\sum_{k=3}^{m} \left| X_{\frac{k}{m}} - 2X_{\frac{k-1}{m}} + X_{\frac{k-2}{m}} \right|^2}, \quad m \geq 5,$$

which compare the realized quadratic variations of X, using second-order increments, with two different lag lengths. Reference [103] has shown that under some assumptions on the process X, which are similar to (1), (2), and (3), albeit slightly more restrictive, it holds that

$$\hat{\alpha}(X, m) := \frac{\log\left(\mathrm{COF}(X, m)\right)}{2 \log 2} - \frac{1}{2} \xrightarrow{\mathbb{P}} \alpha, \quad m \uparrow \infty. \tag{7.25}$$

An in-depth study of the finite sample performance of this COF estimator can be found in [49].

To examine how well the hybrid scheme reproduces the fine properties of the $\mathcal{BSS}$ process in terms of regularity/roughness, we apply the COF estimator to discretized trajectories of X, where the kernel function g is again the gamma kernel (Example 7.2) with $\lambda = 1$, generated using the hybrid scheme with $\kappa = 1, 2, 3$ and $\mathbf{b} = \mathbf{b}^*$. We consider the case where the volatility process satisfies $\sigma_t = 1$, that is, the process X is Gaussian. This allows us to quantify and control for the intrinsic bias and noisiness, measured in terms of standard deviation, of the estimation method itself, by initially applying the estimator to trajectories that have been simulated using an exact method based on the Cholesky factorization. We then study the behavior of the estimator when applied to a discretized trajectory, while decreasing the step size of the discretization scheme. More precisely, we simulate $\hat{\alpha}(X^n, m)$, where $m = 500$, and X_n is the hybrid scheme for X with $n = ms$ and $s \in \{1, 2, 5\}$. This means that we compute $\hat{\alpha}(X^n, m)$ using m observations obtained by subsampling every sth observation in the sequence $X^n_{\frac{i}{n}}$, $i = 0, 1, \ldots, n$. As a comparison, we repeat these simulations, substituting the hybrid scheme with the Riemann-sum scheme, using $\kappa = 0$ with $\mathbf{b} \in \{\mathbf{b}_{\mathrm{FWD}}, \mathbf{b}^*\}$.

The results are presented in Figure 7.3. We observe that the intrinsic bias of the estimator with $m = 500$ observations is negligible, and hence the bias of the estimates computed from discretized trajectories is then attributable to approximation error arising from the respective discretization scheme, where positive (resp., negative) bias indicates that the simulated trajectories are smoother (resp., rougher) than those of the process X. Concentrating first on the baseline case $s = 1$, we note that the hybrid scheme produces essentially unbiased results when $\alpha \in (-\frac{1}{2}, 0)$, while there is moderate bias when $\alpha \in (0, \frac{1}{2})$, which disappears when passing from $\kappa = 1$ to $\kappa = 3$, even for values of α very close to $\frac{1}{2}$. (The largest value of α considered in our simulations is $\alpha = 0.49$; one would expect the performance to weaken as α approaches $\frac{1}{2}$ (Figure 7.1), but this range of parameter values seems to be of limited practical interest.) The standard deviations exhibit a similar pattern. The corresponding results for the Riemann-sum scheme are clearly

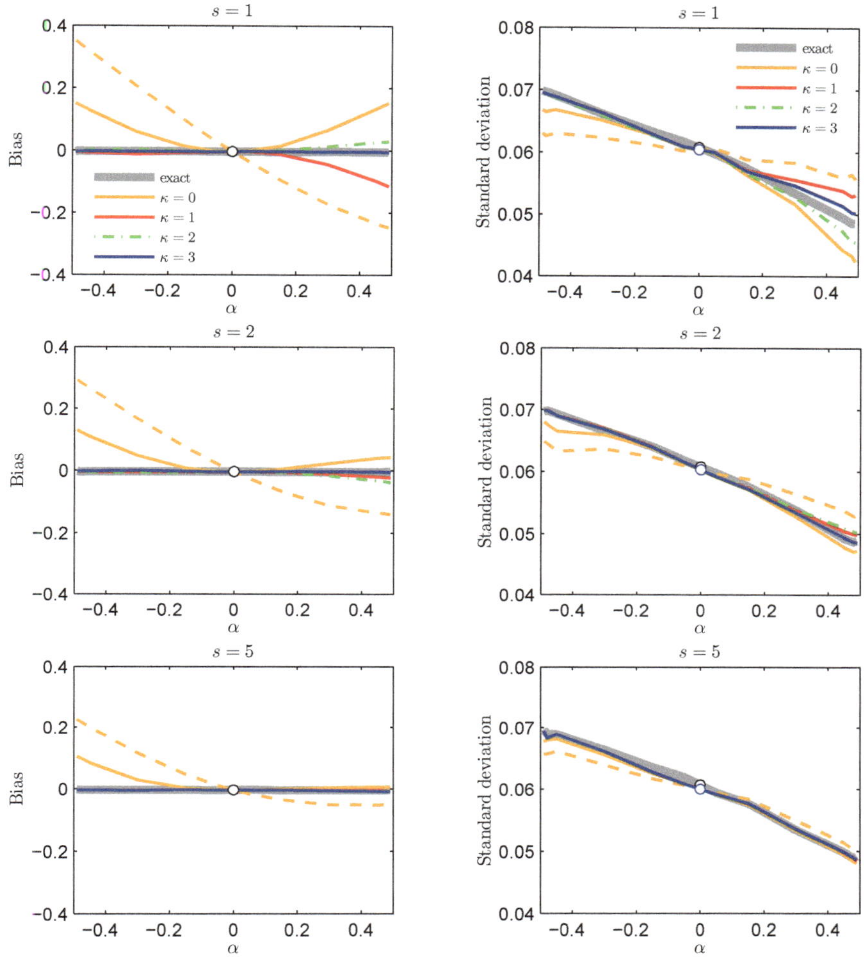

Figure 7.3: Bias and standard deviation of the COF estimator (7.25) of the roughness index α, when applied to discretized trajectories of a $\mathcal{BSS}$ process with the gamma kernel (Example 7.2), $\lambda = 1$ and $\sigma_t = 1$ for all $t \in \mathbb{R}$. Trajectories were generated using an exact method based on the Cholesky factorization, the hybrid scheme ($\kappa = 1, 2, 3$ and $\mathbf{b} = \mathbf{b}^*$), and Riemann-sum scheme ($\kappa = 0$ and $\mathbf{b} = \mathbf{b}^*$ (solid lines), $\mathbf{b} = \mathbf{b}_{\text{FWD}}$ (dashed lines)). In the experiment, $n = ms$ observations were generated, where $m = 500$ and $s \in \{1, 2, 5\}$, on $[0, 1]$ using $N_n = \lfloor n^{1.5} \rfloor$. Every sth observation was then subsampled, resulting in $m = 500$ observations that were used to compute the estimate $\hat{\alpha}(X^n, m)$ of the roughness index α. Number of Monte Carlo replications: 10,000.

inferior, exhibiting significant bias, while using optimal evaluation points ($\mathbf{b} = \mathbf{b}^*$) improves the situation slightly. In particular, the bias in the case $\alpha \in (-\frac{1}{2}, 0)$ is positive, indicating too smooth discretized trajectories, which is connected with the failure of the Riemann-sum scheme with α near $-\frac{1}{2}$, illustrated in Figure 7.2. With $s = 2$ and $s = 5$, the results improve with both schemes. Notably, in the case $s = 5$, the performance of the hybrid scheme even with $\kappa = 1$ is on par with the exact method. However, the improvements with the Riemann-sum scheme are more meager, as considerable bias persists when α is near $-\frac{1}{2}$.

7.2.3 ▪ Option pricing under rough volatility

As another experiment, we study Monte Carlo option pricing in the *rough Bergomi* (rBergomi) model of Bayer, Friz, and Gatheral [39]. In the rBergomi model, the logarithmic spot variance of the price of the underlying is modeled by a rough Gaussian process, which is a special case of (7.16). By virtue of the rough volatility process, the model fits well to observed implied volatility smiles [39, pages 893–895].

More precisely, the price of the underlying asset in the rBergomi model with time horizon $T > 0$ is defined, under an equivalent martingale measure identified with $\mathbb{P}$, as

$$S_t := S_0 \exp\left(\int_0^t \sqrt{v_s}\mathrm{d}Z_s - \frac{1}{2}\int_0^t v_s\mathrm{d}s\right), \quad t \in [0,T],$$

using the spot variance process

$$v_t := \xi_t^0 \exp\Bigg(\underbrace{\eta\sqrt{2\alpha+1}\int_0^t (t-s)^\alpha \mathrm{d}W_s}_{=:Y_t} - \frac{\eta^2}{2}t^{2\alpha+1}\Bigg), \quad t \in [0,T].$$

Above, $S_0 > 0$, $\eta > 0$, and $\alpha \in (-\frac{1}{2}, 0)$ are deterministic parameters, and Z is a standard Brownian motion given by

$$Z_t := \varrho W_t + \sqrt{1-\varrho^2}W_t^\perp, \quad t \in [0,T], \tag{7.26}$$

where $\varrho \in (-1,1)$ is the correlation parameter, and $(W_t^\perp)_{t\in[0,T]}$ is a standard Brownian motion independent of W. The process $(\xi_t^0)_{t\in[0,T]}$ is the so-called *forward variance curve* [39, page 891], which we assume here to be flat, $\xi_t^0 = \xi > 0$ for all $t \in [0,T]$.

Using Monte Carlo simulation, we aim to compute the price of a European call option struck at $K > 0$ with maturity T, which is given by

$$\mathrm{C}(S_0, K, T) := \mathbb{E}[(S_T - K)_+]. \tag{7.27}$$

The approach suggested by [39] involves sampling the Gaussian processes Z and Y on a discrete time grid using exact simulation and then approximating S and v using Euler discretization. We modify this approach by using the hybrid scheme to simulate Y, instead of the computationally more costly exact simulation. As the hybrid scheme involves simulating increments of the Brownian motion W driving Y, we can conveniently simulate the increments of Z, needed for the Euler discretization of S, using the representation (7.26).

We map the option price $\mathrm{C}(S_0, K, T)$ to the corresponding Black–Scholes implied volatility $\mathrm{IV}(S_0, K, T)$; see, e.g., [163]. Reparameterizing the implied volatility using the log-strike $k := \log(K/S_0)$ allows us to drop the dependence on the initial price, so we will abuse notation slightly and write $\mathrm{IV}(k, T)$ for the corresponding implied volatility. Figure 7.4 displays implied volatility smiles obtained from the rBergomi model using the hybrid and Riemann-sum schemes to simulate Y, as discussed above, and compares these to the smiles obtained using an exact simulation of Y via Cholesky factorization. The parameter values are given in Table 7.1. They have been adopted from [39] where it is demonstrated that they result in realistic volatility smiles. We consider two different maturities: "short" $T = 0.041$, and "long" $T = 1$.

We observe that the Riemann-sum scheme ($\kappa = 0$, $\mathbf{b} \in \{\mathbf{b}_{\mathrm{FWD}}, \mathbf{b}^*\}$) is able capture the shape of the implied volatility smile but not its level. Alas, the method even breaks down with more extreme log-strikes (the prices are so low that the root-finding algorithm used to compute the implied volatility returns zero). In contrast, the hybrid scheme with $\kappa = 1, 2$ and $\mathbf{b} = \mathbf{b}^*$ yields implied volatility smiles that are indistinguishable from the benchmark smiles obtained using exact simulation. Further, there is no discernible difference between the smiles obtained using

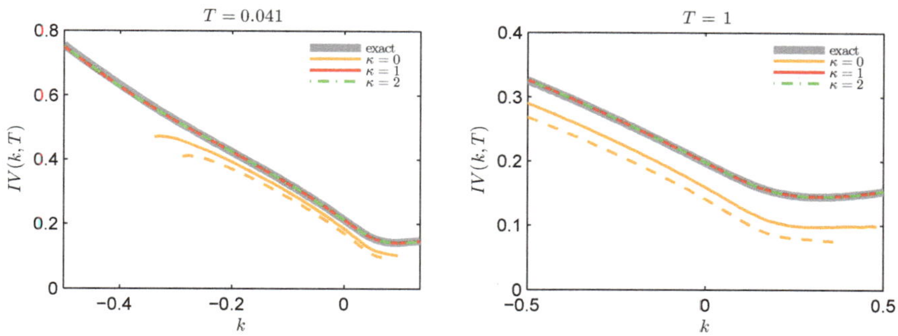

Figure 7.4: Implied volatility smiles corresponding to the option price (7.27), computed using Monte Carlo simulation (500 time steps, 1,000,000 replications), with two maturities: $T = 0.041$ (left) and $T = 1$ (right). The spot variance process v was simulated using an exact method, the hybrid scheme ($\kappa = 1, 2$ and $\mathbf{b} = \mathbf{b}^*$), and Riemann-sum scheme ($\kappa = 0$ and $\mathbf{b} = \mathbf{b}^*$ (solid lines), $\mathbf{b} = \mathbf{b}_{\text{FWD}}$ (dashed lines)). The parameter values used in the rBergomi model are given in Table 7.1.

Table 7.1: Parameter values used in the rBergomi model.

S_0	ξ	η	α	ϱ
1	0.235^2	1.9	−0.43	−0.9

$\kappa = 1$ and $\kappa = 2$. As in the previous section, we observe that the hybrid scheme is indeed capable of producing very accurate trajectories of $\mathcal{TBSS}$ processes, particularly in the case $\alpha \in (-\frac{1}{2}, 0)$, even when $\kappa = 1$.

7.3 ▪ Proofs

Throughout the proofs, we rely on two useful inequalities. The first one is the Potter bound for slow variation at 0, which follows immediately from the corresponding result for slow variation at ∞ [63, Theorem 1.5.6]. Namely, if $L : (0, 1] \to (0, \infty)$ is slowly varying at 0 and bounded away from 0 and ∞ on any interval $(u, 1]$, $u \in (0, 1)$, then for any $\delta > 0$ there exists a constant $C_\delta > 0$ such that

$$\frac{L(x)}{L(y)} \leq C_\delta \max\left\{\left(\frac{x}{y}\right)^{\delta}, \left(\frac{x}{y}\right)^{-\delta}\right\}, \quad x, y \in (0, 1]. \tag{7.28}$$

The second one is the elementary inequality

$$|x^\alpha - y^\alpha| \leq |\alpha|(\min\{x, y\})^{\alpha-1}|x - y|, \quad x, y \in (0, \infty), \quad \alpha \in (-\infty, 1), \tag{7.29}$$

which can be easily shown using the mean value theorem. Additionally, we use the following variant of Karamata's theorem for regular variation at 0. Its proof is similar to that of the usual Karamata's theorem for regular variation at ∞ [63, Proposition 1.5.10].

Lemma 7.9 (Karamata's theorem). *If $\alpha \in (-1, \infty)$ and $L : (0, 1] \to [0, \infty)$ is slowly varying at 0, then*

$$\int_0^y x^\alpha L(x)\mathrm{d}x \sim \frac{1}{\alpha + 1} y^{\alpha+1} L(y), \quad y \to 0.$$

7.3.1 ▪ Proof of Proposition 7.1

***Proof of Proposition* 7.1.** (i) By the covariance stationarity of the volatility process σ, we may express the variogram $V(h)$ for any $h \geq 0$ as

$$\begin{aligned} V(h) = \mathbb{E}[|X_h - X_0|^2] &= \int_{-\infty}^{h} \big(g(h-u) - g(-u)\mathbf{1}_{(-\infty,0)}(u)\big)^2 \mathbb{E}[\sigma_u^2]\mathrm{d}u \\ &= \mathbb{E}[\sigma_0^2]\bigg(\int_0^h g(x)^2\mathrm{d}x + \int_0^\infty (g(x+h) - g(x))^2\mathrm{d}x\bigg). \end{aligned} \tag{7.30}$$

Invoking (1) and Lemma 7.9, we find that

$$\int_0^h g(x)^2\mathrm{d}x \sim \frac{1}{2\alpha+1} h^{2\alpha+1} L_g(h)^2, \quad h \downarrow 0. \tag{7.31}$$

We may clearly assume that $h < 1$, which allows us to work with the decomposition

$$\int_0^\infty (g(x+h) - g(x))^2\mathrm{d}x = A_h + A'_h,$$

where

$$A_h := \int_0^{1-h} (g(x+h) - g(x))^2\mathrm{d}x, \quad A'_h := \int_{1-h}^\infty (g(x+h) - g(x))^2\mathrm{d}x.$$

According to (2), there exists $M > 1$ such that $x \mapsto |g'(x)|$ is nonincreasing on $[M, \infty)$. Thus, using the mean value theorem, we deduce that

$$|g(x+h) - g(x)| = |g'(\xi)|h \leq \begin{cases} \sup_{y \in (1-h,M]} |g'(y)|h, & x \in (1-h, M), \\ |g'(x)|h, & x \in [M, \infty), \end{cases}$$

where $\xi = \xi(x, h) \in [x, x+h]$. It follows then that

$$\limsup_{h\downarrow 0} \frac{A'_h}{h^2} \leq (M-1) \sup_{y \in [1,M]} g'(y)^2 + \int_1^\infty g'(x)^2\mathrm{d}x < \infty,$$

which in turn implies that

$$A'_h = O(h^2), \quad h \downarrow 0. \tag{7.32}$$

Making a substitution $y = \frac{x}{h}$, we obtain

$$\begin{aligned} A_h = \int_0^{1-h} (g(x+h) - g(x))^2\mathrm{d}x &= h \int_0^{1/h-1} \big(g(h(y+1)) - g(hy)\big)^2\mathrm{d}y \\ &= h^{2\alpha+1} L_g(h)^2 \int_0^\infty G_h(y)\mathrm{d}y, \end{aligned}$$

where

$$G_h(y) := \left((y+1)^\alpha \frac{L_g(h(y+1))}{L_g(h)} - y^\alpha \frac{L_g(hy)}{L_g(h)}\right)^2 \mathbf{1}_{(0,1/h-1)}(y), \quad y \in (0, \infty).$$

By the definition of slow variation at 0,

$$\lim_{h\downarrow 0} G_h(y) = \big((y+1)^\alpha - y^\alpha\big)^2, \qquad y \in (0, \infty).$$

We shall show below that the functions G_h, $h \in (0,1)$, have an integrable dominant. Thus, by the dominated convergence theorem,

$$A_h \sim h^{2\alpha+1} L_g(h)^2 \int_0^\infty \left((y+1)^\alpha - y^\alpha\right)^2 \mathrm{d}y, \quad h \downarrow 0. \tag{7.33}$$

Since $\alpha < \frac{1}{2}$, we have $\lim_{h\downarrow 0} \frac{A_h'}{h^{2\alpha+1}L_g(h)^2} = 0$ by (7.4) and (7.32), so we get from (7.31) and (7.33)

$$\begin{aligned}\int_0^h g(x)^2 \mathrm{d}x &+ \int_0^\infty (g(x+h)-g(x))^2 \mathrm{d}x \\ &\sim \left(\frac{1}{2\alpha+1} + \int_0^\infty \left((y+1)^\alpha - y^\alpha\right)^2 \mathrm{d}y\right) h^{2\alpha+1} L_g(h)^2, \quad h \downarrow 0,\end{aligned}$$

which, together with (7.30), implies the assertion.

It remains to justify the use of the dominated convergence theorem to deduce (7.33). For any $y \in (0,1]$, we have, by the Potter bound (7.28) and the elementary inequality $(u+v)^2 \le 2u^2 + 2v^2$,

$$\begin{aligned} G_h(y) &\le 2(y+1)^{2\alpha} \left(\frac{L_g(h(y+1))}{L_g(h)}\right)^2 + 2y^{2\alpha}\left(\frac{L_g(hy)}{L_g(h)}\right)^2 \\ &\le 2C_{\delta_1}^2 \left((y+1)^{2(\alpha+\delta_1)} + y^{2(\alpha-\delta_1)}\right),\end{aligned}$$

where we choose $\delta_1 \in (0, \alpha + \frac{1}{2})$ to ensure that $2(\alpha - \delta_1) > -1$. Consider then $y \in [1,\infty)$. By adding and subtracting the term $(y+1)^\alpha \frac{L_g(hy)}{L_g(h)}$ and using again the inequality $(u+v)^2 \le 2u^2 + 2v^2$, we get

$$\begin{aligned} G_h(y) &= \Bigg((y+1)^\alpha \frac{L_g(h(y+1))}{L_g(h)} - (y+1)^\alpha \frac{L_g(hy)}{L_g(h)} \\ &\quad + (y+1)^\alpha \frac{L_g(hy)}{L_g(h)} - y^\alpha \frac{L_g(hy)}{L_g(h)}\Bigg)^2 \mathbf{1}_{(0,1/h-1)}(y) \\ &\le 2(y+1)^{2\alpha}\left(\frac{L_g(h(y+1)) - L_g(hy)}{L_g(h)} \mathbf{1}_{(0,1/h-1)}(y)\right)^2 \\ &\quad + 2\left((y+1)^\alpha - y^\alpha\right)^2 \left(\frac{L_g(hy)}{L_g(h)}\mathbf{1}_{(0,1/h-1)}(y)\right)^2.\end{aligned}$$

We recall that $\underline{L_g} := \inf_{x\in(0,1]} L_g(x) > 0$ by (1), so

$$\begin{aligned}&\left|\frac{L_g(h(y+1)) - L_g(hy)}{L_g(h)}\mathbf{1}_{(0,1/h-1)}(y)\right| \\ &\qquad \le \frac{1}{\underline{L_g}} |L_g(h(y+1)) - L_g(hy)| \mathbf{1}_{(0,1/h-1)}(y).\end{aligned}$$

Using the mean value theorem and the bound for the derivative of L_g from (1), we observe that

$$|L_g(h(y+1)) - L_g(hy)| = |L_g'(\xi)||h(y+1) - hy| \le hC\left(1 + \frac{1}{\xi}\right) \le C\left(h + \frac{1}{y}\right),$$

where $\xi = \xi(y,h) \in [hy, h(y+1)]$. Noting that the constraint $y < \frac{1}{h} - 1$ is equivalent to $h < \frac{1}{y+1}$, we further obtain

$$\begin{aligned}\left|\frac{L_g(h(y+1)) - L_g(hy)}{L_g(h)}\mathbf{1}_{(0,1/h-1)}(y)\right| &\le \frac{C}{\underline{L_g}}\left(h + \frac{1}{y}\right)\mathbf{1}_{(0,1/h-1)}(y) \\ &\le \frac{C}{\underline{L_g}}\left(\frac{1}{y+1} + \frac{1}{y}\right) \le \frac{C}{\underline{L_g}} \frac{3}{y+1},\end{aligned}$$

as $y \geq 1$, which we then use to deduce that

$$2(y+1)^{2\alpha}\left(\frac{L_g(h(y+1)) - L_g(hy)}{L_g(h)}\mathbf{1}_{(0,1/h-1)}(y)\right)^2 \leq \frac{18C^2}{\underline{L}_g^2}(y+1)^{2(\alpha-1)}.$$

Additionally, we observe that, by (7.28) and (7.29),

$$2\big((y+1)^{\alpha} - y^{\alpha}\big)^2\left(\frac{L_g(hy)}{L_g(h)}\mathbf{1}_{(0,1/h-1)}(y)\right)^2 \leq 2C_{\delta_2}^2\alpha^2 y^{2(\alpha-1+\delta_2)},$$

where we choose $\delta_2 \in (0, \frac{1}{2} - \alpha)$, ensuring that $2(\alpha - 1 + \delta_2) < -1$. We may finally define a function

$$G(y) := \begin{cases} 2C_{\delta_1}^2\left((y+1)^{2(\alpha+\delta_1)} + y^{2(\alpha-\delta_1)}\right), & y \in (0,1], \\ \frac{18C^2}{\underline{L}_g^2}(y+1)^{2(\alpha-1)} + 2C_{\delta_2}^2\alpha^2 y^{2(\alpha-1+\delta_2)}, & y \in (1,\infty), \end{cases}$$

which satisfies $0 \leq G_h(y) \leq G(y)$ for any $y \in (0,\infty)$ and $h \in (0,1)$, and is integrable on $(0,\infty)$ with the aforementioned choices of δ_1 and δ_2.

(ii) To show existence of the modification, we need a localization procedure that involves an ancillary process

$$F_t := \int_1^{\infty} g'(s)^2\sigma_{t-s}^2 \mathrm{d}s, \quad t \in \mathbb{R}.$$

We check first that F is locally bounded under (1) and (2), which is essential for localization. To this end, let $T \in (0,\infty)$, and for any $t \in [-T,T]$ write

$$F_t = F_t^{\flat} + F_t^{\sharp},$$

where

$$F_t^{\flat} := \int_1^{M+2T} g'(s)^2\sigma_{t-s}^2 \mathrm{d}s, \quad F_t^{\sharp} := \int_{M+2T}^{\infty} g'(s)^2\sigma_{t-s}^2 \mathrm{d}s,$$

and $M > 1$ is such that $x \mapsto |g'(x)|$ is nonincreasing on $[M,\infty)$, as in the proof of (i).

Since g' is continuous on $(0,\infty)$ and σ locally bounded, for any $t \in [-T,T]$ we have

$$0 \leq F_t^{\flat} \leq (M+2T-1) \sup_{y\in[1,M+2T]} g'(y)^2 \sup_{u\in[-M-3T,T-1]} \sigma_u^2 < \infty.$$

Further, when $t \in [-T,T]$,

$$F_t^{\sharp} = \int_{-\infty}^{t-(M+2T)} g'(t-u)^2\sigma_u^2 \mathrm{d}u,$$

where $g'(t-u)^2 \leq g'(-T-u)^2$ since the arguments satisfy

$$t - u \geq -T - u \geq -T - \big(t - (M+2T)\big) \geq M.$$

Thus,

$$0 \leq F_t^{\sharp} \leq \int_{-\infty}^{-(M+T)} g'(-T-u)^2\sigma_u^2 \mathrm{d}u \leq \int_1^{\infty} g'(s)^2\sigma_{-T-s}^2 \mathrm{d}s < \infty$$

for any $t \in [-T,T]$ almost surely, as we have

$$\mathbb{E}\left[\int_1^{\infty} g'(s)^2\sigma_{-T-s}^2 \mathrm{d}s\right] = \int_1^{\infty} g'(s)^2\mathbb{E}[\sigma_{-T-s}^2]\mathrm{d}s = \mathbb{E}[\sigma_0^2]\int_1^{\infty} g'(s)^2 \mathrm{d}s < \infty,$$

where we change the order of expectation and integration by relying on Tonelli's theorem, and where the final equality follows from the covariance stationarity of σ. So, we can conclude that F is indeed locally bounded.

Now let $m \in \mathbf{N}$, and for localization define a sequence of stopping times

$$\tau_{m,n} := \inf\{t \in [-m,\infty) : F_t > n \text{ or } |\sigma_t| > n\}, \quad n \in \mathbf{N},$$

that satisfies $\tau_{m,n} \uparrow \infty$ almost surely as $n \uparrow \infty$ since both F and σ are locally bounded. (We follow the usual convention that $\inf \varnothing = \infty$.) Consider now the modified $\mathcal{BSS}$ process

$$X_t^{m,n} := \int_{-\infty}^{t} g(t-s)\sigma_{\min\{s,\tau_{m,n}\}}\mathrm{d}W_s, \quad t \in [-m,\infty),$$

that coincides with X on the stochastic interval $[\![-m,\tau_{m,n}]\!]$. The process $X^{m,n}$ satisfies the assumptions of [32, Lemma 1], so for any $p > 0$ there exists a constant $\hat{C}_p > 0$ such that

$$\mathbb{E}[|X_s^{m,n} - X_t^{m,n}|^p] \leq \hat{C}_p V(|s-t|)^{p/2}, \quad s,t \in [-m,\infty). \tag{7.34}$$

Using the upper bound in (7.4), we can deduce from (i) that for any $\delta > 0$ there are constants $\tilde{C}_\delta > 0$ and $\underline{h}_\delta > 0$ such that

$$V(h) \leq \tilde{C}_\delta h^{2\alpha+1-\delta}, \quad h \in (0,\underline{h}_\delta). \tag{7.35}$$

Applying (7.35) to (7.34), we get

$$\begin{aligned}\mathbb{E}[|X_s^{m,n} &- X_t^{m,n}|^p] \\ &\leq \hat{C}_p\tilde{C}_\delta^{p/2}|s-t|^{1+p(\alpha+\frac{1}{2}-\frac{\delta}{2}-\frac{1}{p})}, \quad s,t \in [-m,\infty), \quad |s-t| < \underline{h}_\delta.\end{aligned}$$

We may note that $p(\alpha + \frac{1}{2} - \frac{\delta}{2} - \frac{1}{p}) > 0$ for small enough δ and large enough p and, in particular,

$$\frac{p(\alpha + \frac{1}{2} - \frac{\delta}{2} - \frac{1}{p})}{p} \uparrow \alpha + \frac{1}{2},$$

as $\delta \downarrow 0$ and $p \uparrow \infty$. Thus it follows from the Kolmogorov–Chentsov theorem [218, Theorem 3.22] that $X^{m,n}$ has a modification with locally φ-Hölder continuous trajectories for any $\varphi \in (0,\alpha+\frac{1}{2})$. Moreover, a modification of X on $\mathbf{R}$, having locally φ-Hölder continuous trajectories for any $\varphi \in (0,\alpha+\frac{1}{2})$, can then by constructed from these modifications of $X^{m,n}$, $m \in \mathbf{N}$, $n \in \mathbf{N}$, by letting first $n \uparrow \infty$ and then $m \uparrow \infty$. □

7.3.2 ▪ Proof of Theorem 7.4

As preparation, we shall first establish an auxiliary result that deals with the asymptotic behavior of certain integrals of regularly varying functions.

Lemma 7.10. *Suppose that $L : (0,1] \to [0,\infty)$ is bounded away from 0 and ∞ on any set of the form $(u,1]$, $u \in (0,1)$, and slowly varying at 0. Moreover, let $\alpha \in (-\frac{1}{2},\infty)$ and $k \geq 1$. If $b \in [k-1,k] \setminus \{0\}$, then*

(i)
$$\lim_{n\uparrow\infty}\int_{k-1}^{k}\left(x^\alpha\frac{L(x/n)}{L(1/n)} - b^\alpha\frac{L(b/n)}{L(1/n)}\right)^2\mathrm{d}x = \int_{k-1}^{k}(x^\alpha - b^\alpha)^2\mathrm{d}x < \infty,$$

(ii)
$$\lim_{n\uparrow\infty}\int_{k-1}^{k}x^{2\alpha}\left(\frac{L(x/n)}{L(1/n)} - \frac{L(b/n)}{L(1/n)}\right)^2\mathrm{d}x = 0.$$

Proof. We only prove (i) as (ii) can be shown similarly. By the definition of slow variation at 0, the function

$$f_n(x) := \left(x^\alpha \frac{L(x/n)}{L(1/n)} - b^\alpha \frac{L(b/n)}{L(1/n)} \right)^2, \quad x \in [k-1, k] \setminus \{0\},$$

satisfies $\lim_{n\uparrow\infty} f_n(x) = (x^\alpha - b^\alpha)^2$ for any $x \in [k-1,k]\setminus\{0\}$. In view of the dominated convergence theorem, it suffices to find an integrable dominant for the functions f_n, $n \in \mathbf{N}$. The construction of the dominant is quite similar to the one seen in the proof of Proposition 7.1, but we provide the details for the convenience of the reader.

Using the Potter bound (7.28) and the inequality $(u+v)^2 \leq 2u^2 + 2v^2$, we find that for any $x \in [k-1,k] \setminus \{0\}$,

$$\begin{aligned} 0 \leq f_n(x) &\leq 2x^{2\alpha} \left(\frac{L(x/n)}{L(1/n)} \right)^2 + 2b^{2\alpha} \left(\frac{L(b/n)}{L(1/n)} \right)^2 \\ &\leq 2C_\delta^2 \left(x^{2\alpha} \max\left\{x^\delta, x^{-\delta}\right\}^2 + b^{2\alpha} \max\left\{b^\delta, b^{-\delta}\right\}^2 \right) =: f(x), \end{aligned}$$

where we choose $\delta \in (0, \alpha + \frac{1}{2})$. When $k \geq 2$, we have $x \geq 1$ and $b \geq 1$, so

$$f(x) = 2C_\delta^2 \left(x^{2(\alpha+\delta)} + b^{2(\alpha+\delta)} \right)$$

is a bounded function of x on $[k-1,k]$. When $k = 1$, we have $x \leq 1$ and $b \leq 1$, implying that

$$f(x) = 2C_\delta^2 \left(x^{2(\alpha-\delta)} + b^{2(\alpha-\delta)} \right),$$

where $2(\alpha - \delta) > -1$ with our choice of δ, so f is an integrable function on $(0, 1]$. □

***Proof of Theorem* 7.4.** Let $t \in \mathbb{R}$ be fixed. It will be convenient to write X_t^n as

$$X_t^n = \sum_{k=1}^{\kappa} \int_{t-\frac{k}{n}}^{t-\frac{k-1}{n}} (t-s)^\alpha L_g\left(\frac{k}{n}\right) \sigma_{t-\frac{k}{n}} \mathrm{d}W_s + \sum_{k=\kappa+1}^{N_n} \int_{t-\frac{k}{n}}^{t-\frac{k-1}{n}} g\left(\frac{b_k}{n}\right) \sigma_{t-\frac{k}{n}} \mathrm{d}W_s.$$

Moreover, we introduce an ancillary approximation of X_t, namely,

$$\check{X}_t^n = \sum_{k=1}^{N_n} \int_{t-\frac{k}{n}}^{t-\frac{k-1}{n}} g(t-s) \sigma_{t-\frac{k}{n}} \mathrm{d}W_s + \int_{-\infty}^{t-\frac{N_n}{n}} g(t-s) \sigma_s \mathrm{d}W_s.$$

By Minkowski's inequality, we have

$$\begin{aligned} \mathbb{E}\left[|X_t^n - X_t|^2\right]^{\frac{1}{2}} &\geq \mathbb{E}\left[|X_t^n - \check{X}_t^n|^2\right]^{\frac{1}{2}} - \mathbb{E}\left[|\check{X}_t^n - X_t|^2\right]^{\frac{1}{2}}, \\ \mathbb{E}\left[|X_t^n - X_t|^2\right]^{\frac{1}{2}} &\leq \mathbb{E}\left[|X_t^n - \check{X}_t^n|^2\right]^{\frac{1}{2}} + \mathbb{E}\left[|\check{X}_t^n - X_t|^2\right]^{\frac{1}{2}}, \end{aligned}$$

which together, after taking squares, imply that

$$E_n\left(1 - 2\sqrt{\frac{E_n'}{E_n}} + \frac{E_n'}{E_n}\right) \leq \mathbb{E}\left[|X_t^n - X_t|^2\right] \leq E_n\left(1 + 2\sqrt{\frac{E_n'}{E_n}} + \frac{E_n'}{E_n}\right), \tag{7.36}$$

where

$$E_n := \mathbb{E}\big[|X_t^n - \bar{X}_t^n|^2\big], \quad E_n' := \mathbb{E}\big[|X_t - \bar{X}_t^n|^2\big].$$

Using the Itô isometry, and recalling that σ is covariance stationary, we obtain

$$\begin{aligned} E_n' &= \sum_{k=1}^{N_n} \int_{t-\frac{k}{n}}^{t-\frac{k-1}{n}} g(t-s)^2 \mathbb{E}\big[\big(\sigma_{t-\frac{k}{n}} - \sigma_s\big)^2\big] \mathrm{d}s \\ &\le \sup_{u\in(0,\frac{1}{n}]} \mathbb{E}\big[|\sigma_u - \sigma_0|^2\big] \int_0^\infty g(s)^2 \mathrm{d}s \end{aligned}$$

and

$$\begin{aligned} E_n &= \sum_{k=1}^{\kappa} \int_{t-\frac{k}{n}}^{t-\frac{k-1}{n}} \left((t-s)^\alpha L_g\left(\frac{k}{n}\right) - g(t-s)\right)^2 \mathbb{E}\big[\sigma^2_{t-\frac{k}{n}}\big] \mathrm{d}s \\ &\quad + \sum_{k=\kappa+1}^{n} \int_{t-\frac{k}{n}}^{t-\frac{k-1}{n}} \left(g\left(\frac{b_k}{n}\right) - g(t-s)\right)^2 \mathbb{E}\big[\sigma^2_{t-\frac{k}{n}}\big] \mathrm{d}s \\ &\quad + \sum_{k=n+1}^{N_n} \int_{t-\frac{k}{n}}^{t-\frac{k-1}{n}} \left(g\left(\frac{b_k}{n}\right) - g(t-s)\right)^2 \mathbb{E}\big[\sigma^2_{t-\frac{k}{n}}\big] \mathrm{d}s \\ &\quad + \int_{-\infty}^{t-\frac{N_n}{n}} g(t-s)^2 \mathbb{E}[\sigma_s^2] \mathrm{d}s \\ &= \mathbb{E}[\sigma_0^2](D_n + D_n' + D_n'' + D_n'''), \end{aligned}$$

where

$$\begin{aligned} D_n &:= \sum_{k=1}^{\kappa} \int_{\frac{k-1}{n}}^{\frac{k}{n}} \left(s^\alpha L_g\left(\frac{k}{n}\right) - g(s)\right)^2 \mathrm{d}s, \\ D_n' &:= \sum_{k=\kappa+1}^{n} \int_{\frac{k-1}{n}}^{\frac{k}{n}} \left(g\left(\frac{b_k}{n}\right) - g(s)\right)^2 \mathrm{d}s, \\ D_n'' &:= \sum_{k=n+1}^{N_n} \int_{\frac{k-1}{n}}^{\frac{k}{n}} \left(g\left(\frac{b_k}{n}\right) - g(s)\right)^2 \mathrm{d}s, \\ D_n''' &:= \int_{\frac{N_n}{n}}^{\infty} g(s)^2 \mathrm{d}s. \end{aligned}$$

(We may assume without loss of generality that $N_n > n > \kappa$, as this will be the case for large enough n.) In what follows, we study the asymptotic behavior of the terms D_n, D_n', D_n'', and D_n''' separately, showing that D_n, D_n'', and D_n''' are negligible in comparison with D_n', and that D_n' gives rise to the convergence rate given in Theorem 7.4.

Let us analyze the terms D_n''', D_n'', and D_n first. By (3) and (4), we have

$$D_n''' = O\left(\left(\frac{N_n}{n}\right)^{2\beta+1}\right) = O\big(n^{\gamma(2\beta+1)}\big), \quad n \uparrow \infty. \tag{7.37}$$

Regarding the term D_n'', recall that by (2) there is $M > 1$ such that $x \mapsto |g'(x)|$ is nonincreasing on $[M,\infty)$. So, by the mean value theorem we have

$$\left|g\left(\frac{b_k}{n}\right)-g(s)\right| = |g'(\xi)|\left|\frac{b_k}{n}-s\right| \le \begin{cases} \frac{1}{n}\sup_{y\in[1,M]}|g'(y)|, & \frac{k-1}{n}<M,\\ \frac{1}{n}|g'\left(\frac{k-1}{n}\right)|, & \frac{k-1}{n}\ge M,\end{cases}$$

where $\xi=\xi(\frac{b_k}{n},s)\in[\frac{k-1}{n},\frac{k}{n}]$. Thus,

$$\limsup_{n\uparrow\infty} n^2 D_n'' \le (M-1)\sup_{y\in[1,M]} g'(y)^2 + \int_1^\infty g'(s)^2 ds < \infty,$$

which implies

$$D_n'' = O(n^{-2}), \quad n\uparrow\infty. \tag{7.38}$$

To analyze the behavior of D_n, we substitute $y=ns$ and invoke (1), yielding

$$\begin{aligned} D_n &= \sum_{k=1}^{\kappa}\int_{k-1}^{k}\left(\left(\frac{y}{n}\right)^{\alpha}L_g\left(\frac{k}{n}\right)-g\left(\frac{y}{n}\right)\right)^2\frac{dy}{n}\\ &= n^{-(2\alpha+1)}L_g(1/n)^2\sum_{k=1}^{\kappa}\int_{k-1}^{k}y^{2\alpha}\left(\frac{L_g(k/n)}{L_g(1/n)}-\frac{L_g(y/n)}{L_g(1/n)}\right)^2 dy,\end{aligned}$$

where, by Lemma 7.10(ii), we have

$$\lim_{n\uparrow\infty}\int_{k-1}^{k}y^{2\alpha}\left(\frac{L_g(k/n)}{L_g(1/n)}-\frac{L_g(y/n)}{L_g(1/n)}\right)^2 dy = 0$$

for any $k=1,\ldots,\kappa$. Thus, we find that

$$\lim_{n\uparrow\infty}\frac{D_n}{n^{-(2\alpha+1)}L_g(1/n)^2} = 0. \tag{7.39}$$

Analyzing the asymptotic behavior of the term D_n' is more delicate. By (1), and substituting $y=ns$, we can write

$$\begin{aligned} D_n' &= \sum_{k=\kappa+1}^{n}\int_{k-1}^{k}\left(g\left(\frac{b_k}{n}\right)-g\left(\frac{y}{n}\right)\right)^2\frac{dy}{n}\\ &= n^{-(2\alpha+1)}\sum_{k=\kappa+1}^{n}\int_{k-1}^{k}\left(b_k^{\alpha}L_g\left(\frac{b_k}{n}\right)-y^{\alpha}L_g\left(\frac{y}{n}\right)\right)^2 dy\\ &= n^{-(2\alpha+1)}L_g(1/n)^2\sum_{k=\kappa+1}^{n}A_{n,k},\end{aligned}$$

where

$$A_{n,k} := \int_{k-1}^{k}\left(y^{\alpha}\frac{L_g(y/n)}{L_g(1/n)}-b_k^{\alpha}\frac{L_g(b_k/n)}{L_g(1/n)}\right)^2 dy.$$

Let us study the asymptotic behavior of the sum $\sum_{k=\kappa+1}^{n}A_{n,k}$ as $n\uparrow\infty$. By Lemma 7.10, we have, for any $k\in\mathbf{N}$,

$$\lim_{n\uparrow\infty}A_{n,k} = \int_{k-1}^{k}(y^{\alpha}-b_k^{\alpha})^2 dy < \infty.$$

To be able to then deduce, using the dominated convergence theorem, that

$$\lim_{n\uparrow\infty}\sum_{k=\kappa+1}^{n} A_{n,k} = \sum_{k=1}^{\infty}\int_{k-1}^{k}(y^{\alpha}-b_k^{\alpha})^2 \mathrm{d}y = J(\alpha,\kappa,\mathbf{b}) < \infty, \tag{7.40}$$

we seek a sequence $(A_k)_{k=\kappa+1}^{\infty} \subset [0,\infty)$ such that

$$0 \le A_{n,k} \le A_k, \quad k=\kappa+1,\dots,n, \quad n\in\mathbf{N},$$

and that $\sum_{k=\kappa+1}^{\infty} A_k < \infty$. Let us assume without loss of generality that $\kappa = 0$. Clearly, we may set $A_1 := \sup_{n\in\mathbf{N}} A_{n,1} < \infty$. Consider now $k \ge 2$. The construction of A_k in this case parallels some arguments seen in the proof of Proposition 7.1, but we provide the details for the sake of clarity. By adding and subtracting $b_k^{\alpha}\frac{L_g(y/n)}{L_g(1/n)}$ and using the inequality $(u+v)^2 \le 2u^2+2v^2$, we get

$$\begin{aligned} A_{n,k} &= \left(y^{\alpha}\frac{L_g(y/n)}{L_g(1/n)} - b_k^{\alpha}\frac{L_g(y/n)}{L_g(1/n)} + b_k^{\alpha}\frac{L_g(y/n)}{L_g(1/n)} - b_k^{\alpha}\frac{L_g(b_k/n)}{L_g(1/n)}\right)^2 \\ &\le 2\int_{k-1}^{k}(y^{\alpha}-b_k^{\alpha})^2\left(\frac{L_g(y/n)}{L_g(1/n)}\right)^2 \mathrm{d}y \\ &\quad + 2b_k^{2\alpha}\int_{k-1}^{k}\left(\frac{L_g(y/n)-L_g(b_k/n)}{L_g(1/n)}\right)^2 \mathrm{d}y =: I_{n,k} + I'_{n,k}. \end{aligned} \tag{7.41}$$

Recall that $\underline{L}_g := \inf_{x\in(0,1]} L_g(x) > 0$, so by the estimates $b_k^{2\alpha} \le \max\{k^{2\alpha},(k-1)^{2\alpha}\} \le 2(k-1)^{2\alpha}$, valid when $\alpha < \frac{1}{2}$, we obtain

$$I'_{n,k} \le \frac{4}{\underline{L}_g^2}(k-1)^{2\alpha}\int_{k-1}^{k}\Big(L_g(y/n)-L_g(b_k/n)\Big)^2 \mathrm{d}y.$$

Note that, thanks to (1) and the mean value theorem,

$$\begin{aligned} |L_g(y/n)-L_g(b_k/n)| &= |L'_g(\xi)|\left|\frac{y}{n}-\frac{b_k}{n}\right| \\ &\le \frac{C(1+\xi^{-1})}{n} \le \frac{C}{n}+\frac{C}{k-1} \le \frac{2C}{k-1}, \end{aligned}$$

where $\xi = \xi(y/n, b_k/n) \in [\frac{k-1}{n},\frac{k}{n}]$ and where the final inequality follows since $k-1<n$. Thus,

$$I'_{n,k} \le \frac{16C^2}{\underline{L}_g^2}(k-1)^{2(\alpha-1)}. \tag{7.42}$$

Moreover, the Potter bound (7.28) and inequality (7.29) imply

$$I_{n,k} \le 2\alpha^2 C_\delta^2\int_{k-1}^{k}(\min\{y,b_k\})^{2(\alpha-1)}y^{2\delta}\mathrm{d}y \le 2^{1+2\delta}\alpha^2C_\delta^2(k-1)^{2(\alpha-1+\delta)}, \tag{7.43}$$

where we choose $\delta \in (0,\frac{1}{2}-\alpha)$. Applying the bounds (7.42) and (7.43) to (7.41) shows that

$$A_{n,k} \le 2^{1+2\delta}\alpha^2C_\delta^2(k-1)^{2(\alpha-1+\delta)} + \frac{16C^2}{\underline{L}_g^2}(k-1)^{2(\alpha-1)} =: A_k,$$

where $2(\alpha-1)<-1$ and $2(\alpha-1+\delta)<-1$ with our choice of δ, so that $\sum_{k=1}^{\infty}A_k<\infty$. Thus, we have shown (7.40), which in turn implies that

$$D'_n \sim J(\alpha,\kappa,(b_k)_{k=\kappa+1}^{\infty})n^{-(2\alpha+1)}L_g(1/n)^2, \quad n\uparrow\infty. \tag{7.44}$$

We will now use the obtained asymptotic relations, (7.37), (7.38), (7.39), and (7.44) to complete the proof. To this end, it will be convenient to introduce a relation $x_n \gg y_n$ for any sequences $(x_n)_{n=1}^{\infty}$ and $(y_n)_{n=1}^{\infty}$ of positive real numbers that satisfy $\lim_{n\uparrow\infty} \frac{x_n}{y_n} = \infty$. By (7.39), we have $D_n' \gg D_n$. Since $2\alpha+1 < 2$, we find that also $D_n' \gg D_n''$, in view of (7.38). The assumption $\gamma > -\frac{2\alpha+1}{2\beta+1}$ is equivalent to $-(2\alpha+1) > \gamma(2\beta+1)$, so by the estimate (7.4) for slowly varying functions, we have $D_n' \gg D_n'''$. It then follows that $E_n \sim \mathbb{E}[\sigma_0^2]D_n'$ as $n \uparrow \infty$. Further, the condition (7.12) implies that $E_n \gg E_n'$. In view of (7.36), we finally find that $\mathbb{E}[|X_t^n - X_t|^2] \sim \mathbb{E}[\sigma_0^2]D_n'$ as $n \uparrow \infty$, which completes the proof. □

7.3.3 ▪ Proof of equation (7.22)

In order to prove (7.22), we rely on the following integral identity for the Gauss hypergeometric function ${}_2F_1$.

Lemma 7.11. *For all $\alpha \in (-1,\infty)$ and $0 \le a < b$,*

$$\int_0^a (a-x)^\alpha (b-x)^\alpha \mathrm{d}x = \frac{a^{\alpha+1}b^\alpha}{\alpha+1}\,{}_2F_1\Big(-\alpha,1,\alpha+2,\frac{a}{b}\Big).$$

Proof. In the case $a = 0$, the asserted identity becomes trivial, so we may assume that $a > 0$. Substituting $y = \frac{x}{a}$, we get

$$\int_0^a (a-x)^\alpha (b-x)^\alpha \mathrm{d}x = a^{\alpha+1}b^\alpha \int_0^1 (1-y)^\alpha \Big(1-\frac{a}{b}y\Big)^\alpha \mathrm{d}y.$$

Using Euler's formula [37, page 59], we find that

$$\int_0^1 (1-y)^\alpha \Big(1-\frac{a}{b}y\Big)^\alpha \mathrm{d}y = \frac{\Gamma(1)\Gamma(\alpha+1)}{\Gamma(\alpha+2)}\,{}_2F_1\Big(-\alpha,1,\alpha+2,\frac{a}{b}\Big),$$

where Γ denotes the gamma function, observing that this step is valid since $\alpha+2 > 1 > 0$ and $0 < \frac{a}{b} < 1$ under our assumptions. By the connection between the gamma function Γ and the beta function B (see, e.g., [37, page 9]), we further obtain

$$\frac{\Gamma(1)\Gamma(\alpha+1)}{\Gamma(\alpha+2)} = \mathrm{B}(1,\alpha+1) = \int_0^1 (1-x)^\alpha \mathrm{d}x = \frac{1}{\alpha+1},$$

concluding the proof. □

***Proof of equation* (7.22).** Let $\alpha \in (-\frac{1}{2},\frac{1}{2}) \setminus \{0\}$, and let $j,k = 2,\ldots,\kappa+1$ be such that $j<k$. By the Itô isometry, we have

$$\begin{aligned}\Sigma_{j,k} = \mathbb{E}[W_{0,j-1}^n W_{0,k-1}^n] &= \int_0^{\frac{1}{n}} \Big(\frac{j-1}{n}-s\Big)^\alpha \Big(\frac{k-1}{n}-s\Big)^\alpha \mathrm{d}s \\ &= \frac{1}{n^{2\alpha+1}} \int_0^1 (j-1-x)^\alpha (k-1-x)^\alpha \mathrm{d}x,\end{aligned}$$

where we have substituted $x = ns$. The second integral above can now be expressed as

$$
\begin{aligned}
\int_0^1 & (j-1-x)^\alpha (k-1-x)^\alpha \mathrm{d}x \\
&= \int_0^{j-1} (j-1-x)^\alpha (k-1-x)^\alpha \mathrm{d}x - \int_0^{j-2} (j-2-x)^\alpha (k-2-x)^\alpha \mathrm{d}x \\
&= \frac{1}{\alpha+1}\bigg((j-1)^{\alpha+1}(k-1)^\alpha {}_2F_1\bigg(-\alpha, 1, \alpha+2, \frac{j-1}{k-1}\bigg) \\
&\quad - (j-2)^{\alpha+1}(k-2)^\alpha {}_2F_1\bigg(-\alpha, 1, \alpha+2, \frac{j-2}{k-2}\bigg)\bigg),
\end{aligned}
$$

where the second equality follows from Lemma 7.11, which is applicable to both integrals on the second line as $0 < j-1 < k-1$ and $0 \leq j-2 < k-2$. □

Acknowledgments

We would like to thank Heidar Eyjolfsson and Emil Hedevang for useful discussions regarding simulation of $\mathcal{BSS}$ processes, and Ulises Márquez for assistance with symbolic computation. Our research has been supported by CREATES (DNRF78), funded by the Danish National Research Foundation, by Aarhus University Research Foundation (project "Stochastic and Econometric Analysis of Commodity Markets"), and by the Academy of Finland (project 258042).

Chapter 8
Asymptotics

Christian Bayer, Peter K. Friz, Masaaki Fukasawa, Antoine Jacquier

8.1 ▪ Introduction

8.1.1 ▪ Option pricing

We are interested in the asymptotic evaluation of option prices. Ignoring rates and dividends, the underlying (S_t), with today's spot price S_0, is thus modeled as a positive martingale, with normalized log-price $X_t := \log(S_t/S_0)$. Denote by $\mathrm{P}(k,\theta)$ the normalized put option price with log-moneyness k and time-to-maturity θ under the model at time $t = 0$, given by

$$\mathrm{P}(k,\theta) := \mathbb{E}\left[\left(\mathrm{e}^k - \mathrm{e}^{X_\theta}\right)_+\right]. \tag{8.1}$$

Denote by $\sigma_{\mathrm{imp}}(k,\theta)$ the Black–Scholes implied volatility at time 0 for log-moneyness k and time-to-maturity θ, that is,

$$\mathrm{P}(k,\theta) = \mathrm{P}_{\mathrm{BS}}(k,\theta,\sigma_{\mathrm{imp}}(k,\theta)),$$

where $\mathrm{P}_{\mathrm{BS}}(k,\theta,\sigma)$ is the normalized Black–Scholes put option price with volatility σ:

$$\mathrm{P}_{\mathrm{BS}}(k,\theta,\sigma) = \mathrm{e}^k\Phi(f_+) - \Phi(f_-), \qquad \text{with} \qquad f_\pm = \frac{k}{\sigma\sqrt{\theta}} \pm \frac{\sigma\sqrt{\theta}}{2}. \tag{8.2}$$

We leave the analogous definitions of the call option C_{BS} to the reader. We remark that, in practice, (S_t) may only be specified in terms of an Itô integral/equation that a priori only exhibits S as a local martingale with a potential integrability issue for call prices. This explains why many authors start with put prices.[44]

8.1.2 ▪ A class of rough volatility models

We now consider two-factor rough volatility models. To this end, we employ a two-dimensional standard Brownian motion $(W,\overline{W})$, introduce a correlation parameter $\varrho \in [-1,1]$, and set $\overline{\varrho} :=$

[44]This issue is only hidden, though. For instance, large deviation based pricing of in-the-money puts is done via put-call parity and out-of-money calls, so that integrability is required afterward to deal with all strikes or, equivalently, the full smile.

$\sqrt{1-\varrho^2}$ and

$$\widetilde{W}_t := \overline{\varrho}\,\overline{W}_t + \varrho W_t \qquad \text{and} \qquad \widehat{W}_t := (K * \dot{W})(t) = \int_0^t K(t,s)\mathrm{d}W_s, \tag{8.3}$$

with (prototypical) kernel $K = K^H$, with $H \in (0, \frac{1}{2}]$, given by

$$K^H(t,s) = \sqrt{2H}(t-s)^{H-\frac{1}{2}} \qquad \text{for } t > 0, \tag{8.4}$$

with prefactor chosen such that $\widehat{W}_1$ has unit variance. Let us further agree that $K(t,s) = 0$ for $s \le t$; such kernels are said to be of *Volterra* type. Note that $\widetilde{W}$ is a standard Brownian motion, correlated with W (with correlation ϱ), whereas $\widehat{W}$ is a Riemann–Liouville fractional Brownian motion (fBm) with Hurst index H. Prototypical rough volatility dynamics are given by

$$\begin{cases} \dfrac{\mathrm{d}S_t}{S_t} &= v_t \mathrm{d}\widetilde{W}_t, \\ \quad v_t &= \sigma\left(\widehat{W}_t, t\right), \end{cases} \qquad \text{or, equivalently,} \qquad \begin{cases} \mathrm{d}X_t &= -\dfrac{1}{2}V_t\mathrm{d}t + \sqrt{V_t}\,\mathrm{d}\widetilde{W}_t, \\ \quad V_t &= \sigma^2\left(\widehat{W}_t, t\right), \end{cases} \tag{8.5}$$

with the *rough stochastic volatility* (resp., *variance*) process v (resp., V) given as an explicit function of the fBm via some continuous "volatility function" $\sigma = \sigma(x,t)$. We insist that $S = S_0 e^X$ is a continuous local martingale, whereas the (nontradable) instantaneous volatility process v_t is, generically for $H < \frac{1}{2}$, not even a semimartingale.[45]

Martingality

A sufficient condition for $(S_t : 0 \le t \le T)$ to be a martingale can be given by Novikov's criterion $\mathbb{E}\left[\exp\left\{\frac{1}{2}\int_0^T v_t^2(\omega)\mathrm{d}t\right\}\right] < \infty$. Unfortunately, already the *rough Stein–Stein model*, of affine linear form

$$v_t(\omega) = \sigma_0 + \sigma_0' \widehat{W}_t(\omega) \tag{8.6}$$

is a borderline case (Novikov only applies for small enough σ_0') and it does not apply at all to the *rough Bergomi model*, with lognormal variance given by

$$V_t := v_t^2(\omega) = \xi(t)\exp\left\{\eta\widehat{W}_t(\omega) - \frac{\eta^2}{2}t^{2H}\right\}, \tag{8.7}$$

with (deterministic) forward variance curve $\xi(t) = \mathbb{E}[V_t]$. However, martingality holds true in these examples when $\varrho \le 0$, as was understood in [231, 269] in the classical setting of diffusive scaling, $H = \frac{1}{2}$, and recently [162, 182] in the rough regime $H < \frac{1}{2}$.

Spot vol and spot skew

Given the model specification (8.5), the importance of *spot vol*

$$v_0 = \sigma(0,0) := \sigma_0,$$

assumed strictly positive, should be evident for short-time consideration. Granted some regularity, the *spot skew*,

$$\sigma_0' := \partial_x \sigma(x,0)|_{x=0}, \tag{8.8}$$

[45] In general, v_t will not enjoy finite quadratic variation.

will also play an important role. Indeed, we shall see, repeatedly and from different perspectives, that the short-dated implied volatility (resp., variance) skew behaves like $t^{H-\frac{1}{2}}$ times

$$\frac{\sqrt{2H}\varrho\sigma_0'}{(H+\frac{1}{2})(H+\frac{3}{2})\sigma_0} \qquad \left(\text{resp., } \frac{2\sqrt{2H}\varrho\sigma_0'}{(H+\frac{1}{2})(H+\frac{3}{2})}\right). \tag{8.9}$$

Example 8.1 (rough Stein–Stein). This applies directly to the specification (8.6).

Example 8.2 (rough Bergomi). From $\sigma(x,t) = \xi(t)e^{\eta x-\eta^2 t^{2H}/2}$, as specified in (8.7), we see that $\sigma_0'/\sigma_0 = \eta$ has a direct interpretation as vol-of-vol parameter η. The above implied volatility skew formula simplifies to $t^{H-\frac{1}{2}}$ times

$$\frac{\sqrt{2H}\varrho\eta}{(H+\frac{1}{2})(H+\frac{3}{2})}. \tag{8.10}$$

Example 8.3 (rough Heston). The *rough Heston model*, covered in Chapter 4, commonly employs the kernel $\widetilde{K}_H(t,s) = (t-s)^{H-\frac{1}{2}}/\Gamma(H+\frac{1}{2})$. While its instantaneous stochastic variance $V_t \equiv v_t^2$ of Itô-Volterra dynamics

$$V_t = V_0 + \left(\widetilde{K}_H * \left(\lambda(\theta - V) + \nu\sqrt{V}\dot{W}\right)\right)(t)$$

does not quite fit into the above framework, it can be approximated as

$$V_t \approx V_0 + \nu\sqrt{V_0}\widetilde{K}_H * \dot{W} \approx \left(\sigma_0 + \sigma_0'\widehat{W}_t\right)^2,$$

where we expanded to first order in $\widehat{W}_t \approx t^H$ and so construct an approximate rough Stein–Stein model with $\sigma_0 = \sqrt{V_0}$ and $2\sigma_0'\sqrt{2H} = \nu/\Gamma(H+\frac{1}{2})$. Inserting this into (8.9) yields the implied volatility (resp., implied variance) skew formula: $t^{H-\frac{1}{2}}$ times

$$\frac{\varrho\nu}{2\sqrt{V_0}\Gamma(H+5/2)} \qquad \left(\text{resp., } \frac{\varrho\nu}{\Gamma(H+5/2)}\right). \tag{8.11}$$

In contrast to (8.10), these prefactors do not vanish as $H \downarrow 0$. Mathematically, this is the simple consequence of being in a situation where $\sigma_0'\sqrt{2H}$ has a nonzero limit. But one can also regard σ_0' as a vol-of-vol parameter, which the rough Heston scale effectively prescribes to be divergent as $H \downarrow 0$.

Structure of this article

In Section 8.2.1 we briefly review the pioneering work of Alòs, León, and Vives [11] (see also [10]) that provides a direct route to at-the-money (ATM), short-dated implied volatility skew asymptotics

$$\left.\frac{\partial}{\partial k}\right|_{k=0} \sigma_{\text{imp}}(k,t) \sim t^{H-\frac{1}{2}} \times (\text{const}).$$

Section 8.3 considers the central limit theorem (CLT) regime, with nonzero log-strike $k_t = xt^{\gamma}$ for $\gamma = \frac{1}{2}$, which naturally leads to the asymptotic formula for the discrete skew

$$\mathcal{S}_{\text{BS}} := \frac{\sigma_{\text{imp}}(k_t,t) - \sigma_{\text{imp}}(k_t',t)}{k_t - k_t'} \sim t^{H-\frac{1}{2}} \times (\text{const})_{x,x'}$$

with $k_t' = x't^\gamma$. In Section 8.4 we see how large deviation leads to a similar formula, but now further away from the money, with $\gamma = \frac{1}{2} - H$, for $H \in (0, \frac{1}{2}]$. We also comment on the intermediate moderate deviation regime, $\gamma \in (\frac{1}{2} - H, \frac{1}{2})$. At last, in Section 8.5 we consider the behavior as H tends to zero and use CLT-based methods to analyze the square-root blow-up of the implied volatility skew.

8.2 ▪ At-the-money short-time skew

8.2.1 ▪ Malliavin calculus

In [11], Alòs, León, and Vives pioneered asymptotic results for rough stochastic volatility. An exhaustive review of this approach and its application to (classical and rough) stochastic volatility models can be found in the recent monograph of Alòs and Lorite [12]. Their framework is that of a general stochastic volatility model (potentially with jumps as well), for which they derive a generalized Hull–White formula (originating in [7]) for a European call option price as follows.

Proposition 8.4. *Assume that* $v \in L^2([0,T], \mathbb{D}_W^{1,2})$, *where* $\mathbb{D}_W^{1,2}$ *denotes the domain of the Malliavin derivative operator with respect to* W. *Then*

$$\mathbb{E}_0\left[\left(S_T - \mathrm{e}^k\right)_+\right] = \mathbb{E}_0\left[\widetilde{\mathrm{C}}_{\mathrm{BS}}\left(0, X_0, \overline{V}_0\right)\right] + \frac{\varrho}{2}\mathbb{E}_0\left[\int_0^T \mathfrak{F}(u, X_u, \overline{V}_u)\mathrm{d}u\right],$$

where C_{BS} *denotes the Black–Scholes call price introduced above and*

$$\widetilde{\mathrm{C}}_{\mathrm{BS}}(t,x,\sigma) := \mathrm{e}^x \mathrm{C}_{\mathrm{BS}}(k - x, T - t, \sigma) = \mathrm{e}^x \Phi\left(\tilde{d}_+(t)\right) - \mathrm{e}^k \Phi\left(\tilde{d}_-(t)\right),$$

with $\tilde{d}_\pm(t) := \frac{x-k}{\sigma\sqrt{T-t}} \pm \frac{\sigma\sqrt{T-t}}{2}$ *and* $\overline{V}_u := \sqrt{\frac{1}{T-u}\int_u^T v_r^2 \mathrm{d}r}$. *Furthermore,*

$$\mathfrak{F}(u, X_u, \overline{V}_u) := \partial_x G(u, X_u, \overline{V}_u)\Lambda_u,$$

with $G(u, X_u, v) := (\partial_{xx} - \partial_x)\widetilde{\mathrm{C}}_{\mathrm{BS}}(u, x, v)$ *and* $\Lambda_u := v_u \int_u^T \left(\mathrm{D}_u^W v_r^2\right)\mathrm{d}r$.

The key tool for proving this formula is the anticipating Itô formula due to Nualart and Pardoux [253]. The term $\mathrm{D}_t^W v_r^2$ corresponds to the Malliavin derivative (with respect to the Brownian motion W) of the instantaneous variance process v^2. Taking derivative with respect to the time t log-stock price yields almost directly the following expression for the ATM implied volatility skew:

$$\left.\frac{\partial\sigma_{\mathrm{imp}}(k,T)}{\partial k}\right|_{k=X_0} = -\left.\frac{\mathbb{E}_t\left[\int_0^T \left(\partial_x \mathfrak{F}(u, X_u, \overline{V}_u) - \frac{1}{2}\mathfrak{F}(u, X_u, \overline{V}_u)\right)\mathrm{d}u\right]}{\partial_\sigma \widetilde{\mathrm{C}}_{\mathrm{BS}}(0, k, \sigma_{\mathrm{imp}}(k,T))}\right|_{k=X_0}$$

A subtle analysis of the short-time behavior of the right-hand side then provides us with the corresponding behavior of the implied volatility skew. The following assumption is needed.

Assumption 8.5. *The process* v *has right-continuous paths, belongs to* $\mathbb{L}^{2,4}$, *and is bounded below almost surely by some strictly positive constant. Furthermore, there exists* $\delta \in (-\frac{1}{2}, 0)$ *such that, for all* $0 < s < r < T$, *for some* $C > 0$,

$$\mathbb{E}\left[(\mathrm{D}_s v_r)^2\right] \le C(r-s)^{2\delta},$$
$$\mathbb{E}\left[(\mathrm{D}_\theta \mathrm{D}_s v_r)^2\right] \le C(r-s)^{2\delta}(r-\theta)^{-2\delta}.$$

Theorem 8.6. *If Assumption* 8.5 *holds and* $\lim_{T\downarrow 0}\sup_{s,r,u\in[0,T]}\mathbb{E}\left[(v_s v_r - v_u^2)^2\right] = 0$*, and there exists* $v_0' \in \mathcal{F}_0$ *such that*

$$\lim_{T\downarrow 0}\left\{\frac{1}{T^{\delta+2}}\int_0^T\int_s^T \mathbb{E}_0[\mathrm{D}_s v_r]\mathrm{d}r\mathrm{d}s\right\} =: v_0',$$

then

$$\lim_{T\downarrow 0}\left.\frac{\partial_k \sigma_{\mathrm{imp}}(k,T)}{T^{\delta}}\right|_{k=X_0} = \frac{\varrho v_0'}{v_0}. \tag{8.12}$$

Remark 8.2.1. *The* lim sup *assumption in the theorem may not always be straightforward to check. However, it is implied by the condition that the spot volatility* $(v_t)_{t\geq 0}$ *converges to* $v_0 = \sigma_0 > 0$ *uniformly in time and in* L^4*. It is clear that it holds for both the rough Stein–Stein and the rough Bergomi models.*

Example 8.7. In the rough Stein–Stein model (8.6) equipped with the Riemann–Liouville kernel (8.4), $\mathrm{d}_s v_r = \sqrt{2H}\sigma_0'(r-s)^{H-\frac{1}{2}}$ for $s < r$, and therefore the first bound in Assumption 8.5 holds with $\delta = H - \frac{1}{2}$ and $C = 2H(\sigma_0')^2$, and the second one holds trivially as the second derivative is null. It is straightforward to compute the limit in Theorem 8.6 (it is, in fact, independent of T) as

$$v_0' := \frac{1}{T^{H+\frac{3}{2}}}\int_0^T\int_s^T \mathbb{E}_0[\mathrm{D}_s v_r]\mathrm{d}r\mathrm{d}s = \frac{\sqrt{2H}\sigma_0'}{(H+\frac{1}{2})(H+\frac{3}{2})}.$$

As announced in (8.9), we get from the above theorem that

$$\lim_{T\downarrow 0}\left.\frac{\partial_k \sigma_{\mathrm{imp}}(k,T)}{T^{H-\frac{1}{2}}}\right|_{k=X_0} = \frac{\varrho\sqrt{2H}\sigma_0'}{\sigma_0(H+\frac{1}{2})(H+\frac{3}{2})}.$$

Example 8.8. In the rough Bergomi model (8.7) with the kernel (8.4),

$$v_t = \sqrt{\xi(t)}\exp\left\{\frac{\eta}{2}\widehat{W}_t - \frac{\eta^2}{4}t^{2H}\right\},$$

we can compute the Malliavin derivatives

$$\mathrm{D}_s v_r = \frac{\eta}{2}\sqrt{2H}(r-s)^{H-\frac{1}{2}}v_r \qquad \text{and} \qquad \mathrm{D}_\theta \mathrm{D}_s v_r = \frac{\eta^2 H}{2}(r-s)^{H-\frac{1}{2}}(r-\theta)^{H-\frac{1}{2}}v_r,$$

for $\theta, s < r$, and therefore, since $\mathbb{E}[v_r^2] = \xi(r)$,

$$\begin{aligned}\mathbb{E}\left[(\mathrm{D}_s v_r)^2\right] &= \frac{\eta^2 H\xi(r)}{2}(r-s)^{2H-1},\\ \mathbb{E}\left[(\mathrm{D}_\theta \mathrm{D}_s v_r)^2\right] &= \frac{\eta^4 H^2\xi(r)}{4}(r-s)^{2H-1}(r-\theta)^{2H-1},\end{aligned}$$

showing the validity of Assumption 8.5 with $\delta = H - \frac{1}{2}$. Now the limit in Theorem 8.6 can be computed as

$$
\begin{aligned}
v_0' &:= \lim_{T\downarrow 0} \frac{1}{T^{H+\frac{3}{2}}} \int_0^T \int_s^T \mathbb{E}_0[\mathrm{D}_s v_r]\mathrm{d}r\mathrm{d}s \\
&= \frac{\eta\sqrt{2H}}{2} \lim_{T\downarrow 0} \frac{1}{T^{H+\frac{3}{2}}} \int_0^T \int_s^T (r-s)^{H-\frac{1}{2}} \mathbb{E}_0[v_r]\mathrm{d}r\mathrm{d}s \\
&= \frac{\eta\sqrt{2H}}{2} \lim_{T\downarrow 0} \frac{1}{T^{H+\frac{3}{2}}} \int_0^T \int_s^T (r-s)^{H-\frac{1}{2}} \sqrt{\xi(r)}\mathrm{d}r\mathrm{d}s \\
&\overset{\text{(Fubini)}}{=} \frac{\eta\sqrt{2H}}{2} \lim_{T\downarrow 0} \frac{1}{T^{H+\frac{3}{2}}} \int_0^T \left\{ \int_0^r (r-s)^{H-\frac{1}{2}} \mathrm{d}s \right\} \sqrt{\xi(r)}\mathrm{d}r \\
&= \frac{\eta\sqrt{2H}}{2(H+\frac{1}{2})} \lim_{T\downarrow 0} \frac{1}{T^{H+\frac{3}{2}}} \int_0^T r^{H+\frac{1}{2}} \sqrt{\xi(r)}\mathrm{d}r \\
&= \frac{\eta\sqrt{2H}\sqrt{\xi(0)}}{2(H+\frac{1}{2})(H+\frac{3}{2})},
\end{aligned}
$$

assuming $\xi(0) > 0$. Therefore, the limiting skew (8.12) reads

$$
\lim_{T\downarrow 0} \left. \frac{\partial_k \sigma_{\text{imp}}(k,T)}{T^{H-\frac{1}{2}}} \right|_{k=X_0} = \frac{\varrho\eta\sqrt{2H}}{2(H+\frac{1}{2})(H+\frac{3}{2})},
$$

as was announced in (8.10).

Remark 8.2.2. *In the setup* (8.5), *it is proved in* [11] *that, if* $\sigma(\cdot, t) = \sigma(\cdot) \in C_b^1(\mathbb{R})$, *then Assumption* 8.5 *and Theorem* 8.6 *hold with spot volatility* $v_0 = \sigma_0$ *and*

$$
v_0' = \frac{\sigma_0'}{(H+\frac{1}{2})(H+\frac{3}{2})},
$$

with σ_0' *defined as in* (8.8) *and* $\delta = H - \frac{1}{2}$.

8.2.2 ▪ From Bergomi–Guyon to diamonds

Let us include a pointer to Chapter 9, Sections 9.2.1 and 9.4.3. There, the Bergomi–Guyon expansion [58] is revisited from a "diamond" perspective. More specifically, in a generic stochastic volatility model, one has

$$
\left. \frac{\partial}{\partial k} \right|_{k=0} (T\sigma_{\text{imp}}^2(k,T)) \sim \frac{X \diamond M}{M}
$$

where X denotes the log-price, $M_t(T) = \int_t^T \mathbb{E}_t[V_u]\mathrm{d}u$ with stochastic variance process V, and $\diamond$ is the diamond product defined in Chapter 9. When given in forward variance form, a quick computation in the rough Heston model (Section 9.4.3 in Chapter 9) gives

$$
\left. \frac{\partial}{\partial k} \right|_{k=0} \sigma_{\text{BS}}^2(k,T) \sim \frac{\varrho\nu}{\Gamma(H+\frac{5}{2})} T^{H-\frac{1}{2}}
$$

in agreement with (8.11).

8.3 ▪ Central limit regime

8.3.1 ▪ Short-time vs. small volatility of volatility

We first consider asymptotics where a (rescaled) limit law of X_θ is normal or lognormal. Mainly, we focus on short-time asymptotics ($\theta \downarrow 0$), for which the limit is the Gaussian Bachelier model. An example of the lognormal limit is the so-called small volatility of volatility (vol-of-vol) asymptotics, where the limit is the Black–Scholes model. Note also that under the short-time asymptotics, the (rescaled) Black–Scholes model converges to the Bachelier model. First, we recall the simplest version of the martingale central limit theorem.

Theorem 8.9. *Let $(M^n)_{n\geq 1}$ be a sequence of one-dimensional continuous local martingales with $M_0^n = 0$. Let $(T_n)_{n\geq 1}$ be a sequence of random times, such that $(\langle M^n\rangle_{T_n})_{n\geq 1}$ converges to a constant $\Sigma > 0$ in probability. Then $(M^n_{T_n})_{n\geq 1}$ converges to $\mathcal{N}(0,\Sigma)$ in law.*

This one-dimensional continuous version is quite easy to prove: by the Dambis–Dubins–Schwarz time-change, we can express $M_t^n = B^n_{\langle M^n\rangle_t}$ for some standard Brownian motion B^n. Since Σ is deterministic, the law of $(B^n, \langle M^n\rangle_{T_n})$ converges to that of (B,Σ). Then the continuous mapping theorem implies the result. This time-change device appears, e.g., in [273]. The general case of vector-valued local martingales, also in the presence of jumps, is detailed in [209].

Now, let M be the local martingale part of the log-price X. This martingale CLT implies that if

$$\langle M\rangle_\theta = \int_0^\theta v_t^2 \,\mathrm{d}t$$

converges to a positive deterministic value Σ under some asymptotics, then X_θ converges to $\mathcal{N}(-\Sigma/2,\Sigma)$ in law, and so the put price $\mathrm{P}(k,\theta)$ given in (8.1) converges to its Black–Scholes counterpart $\mathrm{P}_{\mathrm{BS}}(k,\theta,\sqrt{\Sigma/\theta})$ from (8.2). The small vol-of-vol asymptotics $v_t = \sigma(\varepsilon\widehat{W}_t, t)$, as ε tends to zero, is the simplest example. Indeed, in this case, v_t converges to $\sigma(0,t)$ as ε tends to zero, and therefore $\langle M\rangle_\theta$ converges to $\int_0^\theta \sigma(0,t)^2 \mathrm{d}t =: \Sigma > 0$. If, furthermore, $\sigma_0 := \sigma(0,\cdot)$ does not depend on t, then $\Sigma = \sigma_0^2\theta$, and therefore the put price $\mathrm{P}(k,\theta)$ converges to $\mathrm{P}_{\mathrm{BS}}(k,\theta,\sigma_0)$. More precise results using other perturbation methods and a refinement of the CLT by a martingale expansion can be found in [151].

A related approach [159, 160, 161], based on martingale expansion in spirit, was developed for specific fractional stochastic volatility models, where the perturbation is only viewed through the mean reversion speed. It turns out that the shape of the asymptotic formulas heavily depends on the *speed of mean reversion*. In particular, assuming that the option's maturity is of the same order as the price process's characteristic diffusion time, the power-law explosion of the implied volatility skew is only recovered in the case of slow mean reversion for $H < \frac{1}{2}$. On the other hand, in the case of fast mean reversion, the asymptotic skew is independent of H for $H < \frac{1}{2}$. We refer the reader to [160, Section 6] for a nice summary thereof.

Alternatively, one can consider applying the theorem to the local martingale S itself (assuming it is so) instead of the local martingale part of X. Under the short-time asymptotics, letting $M_t := \theta^{-\frac{1}{2}}(S_{t\theta}/S_0 - 1)$, we have

$$\langle M\rangle_1 = \frac{1}{\theta S_0^2}\int_0^\theta S_u^2 v_u^2 \,\mathrm{d}u \to v_0^2,$$

as θ tends to zero. As a result, $\theta^{-\frac{1}{2}}(S_\theta/S_0 - 1) \to \mathcal{N}(0, v_0^2)$, and so,

$$\frac{1}{\sqrt{\theta}}\mathrm{P}\left(\sqrt{\theta}z,\theta\right) = \mathbb{E}\left[\left(\frac{\mathrm{e}^{\sqrt{\theta}z}-1}{\sqrt{\theta}} - M_1\right)_+\right] \longrightarrow \mathbb{E}\left[\left(z - \mathcal{N}(0,v_0^2)\right)_+\right],$$

which corresponds to the Bachelier put option price.

8.3.2 ▪ Asymptotic expansions for short time

We now consider a region of log-moneyness close to ATM (situated at the origin). More specifically, we scale the log-moneyness as $k = \sqrt{\theta}z$ and let θ tend to zero. With this scaling the model converges to the Bachelier model in law (by the CLT) and we can consider an Edgeworth-type expansion around this limit normal model. The theorem below is a mild adaptation of [154, Theorem 2.1]. Let V denote the variance process of the form $V_t = \sigma(\widehat{W}_t, t)^2$ as in (8.5), where $\widehat{W}_t = \int_0^t K(t,s)\mathrm{d}W_s$ is the continuous Gaussian Volterra process with variance $\mathfrak{s}^2(t) := \int_0^t K(t,s)^2\mathrm{d}s$. We impose the following conditions, inspired by [276].

Assumption 8.10. *Set* $\xi(t) := \mathbb{E}[V_t]$.

(i) *The asset price process S as specified in* (8.5) *is a martingale.*

(ii) *The random variable* $\frac{1}{\mathfrak{s}(t)}\left(\frac{V_t}{\xi(t)} - 1\right)$ *is uniformly integrable in* $t \in [0,1]$.

(iii) *The couple*

$$\left(\frac{1}{\sqrt{t}}\left(\frac{S_t}{S_0} - 1\right), \frac{1}{\mathfrak{s}(t)}\left(\frac{V_t}{\xi(t)} - 1\right)\right)$$

converges in law as t tends to zero to a pair of random variables (γ, δ).

(iv) *There exists* $0 \le H < \frac{1}{2}$ *such that* $\mathfrak{s}(t) = t^H L(t)$ *for some slowly varying function L, and* $\lim_{t\downarrow 0}\mathfrak{s}(t) = 0$.

This assumption covers our prototypical kernel K^H, as detailed in Example 8.12, and coincides with the assumptions of [154], except that the reference assumes $L \equiv 1$. It is also reminiscent of the more general kernels studied in [50].

Theorem 8.11. *Under Assumption 8.10, the Black–Scholes implied volatility at maturity t and log-moneyness $z\sqrt{t}$ satisfies*

$$\sigma_{\text{imp}}\left(z\sqrt{t}, t\right) = \sqrt{\overline{\xi}(t)}\left(1 + \alpha(z)\mathfrak{s}(t)\right) + o(\mathfrak{s}(t)),$$

as t tends to zero, where

$$\begin{aligned}
\overline{\xi}(t) &:= \frac{1}{t}\int_0^t \xi(s)\mathrm{d}s, \\
\alpha(z) &:= \frac{1}{2}\int_0^1\int_{-\infty}^{\infty} u^H\mathbb{E}\left[\delta \mid \gamma = \sqrt{u}z + \sqrt{(1-u)\xi(0)}w\right]\varphi(w)\,\mathrm{d}w\,\mathrm{d}u,
\end{aligned} \tag{8.13}$$

where φ denotes the density of the standard normal distribution.

Proof. We sketch the proof from [154]. Consider the family of martingales Y^t indexed by $0 < t \le 1$ defined by

$$Y_u^t := \frac{1}{\sqrt{t}}\left(\frac{S_{tu}}{S_0} - 1\right), \quad u \in [0,1].$$

We now re-express the (normalized) put option price with log-moneyness $z\sqrt{t}$ as

$$\frac{\mathbb{E}\left[\left(S_0\mathrm{e}^{z\sqrt{t}} - S_t\right)_+\right]}{\sqrt{t}S_0} = \mathbb{E}\left[(\Delta - Y_1^t)_+\right], \qquad \text{with } \Delta = \frac{\mathrm{e}^{z\sqrt{t}} - 1}{\sqrt{t}}.$$

We re-express the price in terms of the solution $p = p(u, x)$ of the Bachelier equation

$$\partial_u p(u,x) + \frac{1}{2}\xi(tu)\partial_x^2 p(u,x) = 0, \qquad \text{with} \qquad p(1,x) = (\Delta - x)_+.$$

Indeed, by Itô's formula,

$$\begin{aligned}\mathbb{E}\big[(\Delta - Y_1^t)_+\big] &= \mathbb{E}\big[p(1, Y_1^t)\big] \\ &= p(0,0) + \frac{1}{2}\int_0^1 \mathbb{E}\left[\partial_x^2 p(u, Y_u^t)\left(\left(1 + \sqrt{t}Y_u^t\right)^2 V_{tu} - \xi(tu)\right)\right] \mathrm{d}u.\end{aligned}$$

The above formula can now be expanded in $t \downarrow 0$ using the following ingredients:

- The explicit solution of the Bachelier equation,

$$p(u,x) = (\Delta - x)\Phi\left(\frac{\Delta - x}{\sqrt{w_{u,1}}}\right) + \sqrt{w_{u,1}}\varphi\left(\frac{\Delta - x}{\sqrt{w_{u,1}}}\right), \qquad w_{u,1} := \frac{1}{t}\int_{tu}^t \xi(s)\mathrm{d}s,$$

  in terms of the standard normal cumulative distribution function Φ and density φ as well as the similar formula for $\partial_x^2 p(u,x)$.

- The convergence in law (using the structural assumption on $\mathfrak{s}$)

$$\left(Y_u^t, \frac{V_{tu} - \xi(tu)}{\mathfrak{s}(t)}\right) \overset{t\downarrow 0}{\longrightarrow} \left(\sqrt{u}\boldsymbol{\gamma}, \xi(0)u^H\boldsymbol{\delta}\right),$$

  for the random variables $(\boldsymbol{\gamma}, \boldsymbol{\delta})$ introduced in Assumption 8.10, for fixed u.

- By the martingale CLT, we know that $\boldsymbol{\gamma} \sim \mathcal{N}(0, \xi(0))$.

- Since $\boldsymbol{\gamma}$ is normal, we trivially have, for arbitrary integrable g,

$$\mathbb{E}\big[g(\boldsymbol{\gamma},\boldsymbol{\delta})\big] = \int_{-\infty}^{\infty} \mathbb{E}\big[g(\boldsymbol{\gamma},\boldsymbol{\delta}) \mid \boldsymbol{\gamma} = w\big] \frac{1}{\sqrt{\xi(0)}}\varphi\left(\frac{w}{\sqrt{\xi(0)}}\right)\mathrm{d}w.$$

Applying these facts, we finally obtain the asymptotic expansion

$$\begin{aligned}&\frac{\mathbb{E}\left[\left(S_0 e^{\sqrt{t}z} - S_t\right)_+\right]}{\sqrt{t}S_0} \\ &= \Delta\Phi\left(\frac{\Delta}{\sqrt{\xi(t)}}\right) + \sqrt{\xi(t)}\varphi\left(\frac{\Delta}{\sqrt{\xi(t)}}\right)\left[1 + \alpha(z)\mathfrak{s}(t) + \frac{1}{2}z\sqrt{t}\right] + o(\mathfrak{s}(t)).\end{aligned}$$

In order to obtain an expansion for the implied volatility, we next compare the above put price expansion against a put price expansion for the Black–Scholes model. Of course, the same expansion holds for the Black–Scholes model, with $\mathfrak{s}(t) = \sqrt{t}$ and $\boldsymbol{\delta} \equiv 0$, implying $\alpha \equiv 0$ and

$$\frac{\mathrm{P}_{\mathrm{BS}}\left(S_0 e^{\sqrt{t}z}, t, \sigma\right)}{\sqrt{t}S_0} = \Delta\Phi\left(\frac{\Delta}{\sigma}\right) + \sigma\varphi\left(\frac{\Delta}{\sigma}\right)\left[1 + \frac{1}{2}\sqrt{t}z\right] + o(\sqrt{t}).$$

Inserting the ansatz $\sigma = \sigma_0 + a\mathfrak{s}(t)$ into the expansion for the Black–Scholes price, we arrive at the conclusion. □

Let us now come back to our structural assumption $V_t = \sigma(\widehat{W}_t, t)^2$. The following formal discussion was made rigorous in the rough Bergomi and log-modulated rough Bergomi in [42]. Formally, under suitable regularity and boundedness assumptions on σ, we expect that

$$\left(\frac{1}{\sqrt{t}}\left(\frac{S_t}{S_0}-1\right), \frac{1}{\mathfrak{s}(t)}\left(\frac{\sigma(\widehat{W}_t,t)^2}{\xi(t)}-1\right)\right) \approx \left(\sqrt{\xi(0)}\frac{\widetilde{W}_t}{\sqrt{t}}, 2\frac{\sigma_0'}{\sigma_0}\frac{\widehat{W}_t}{\mathfrak{s}(t)}\right) \stackrel{t\downarrow 0}{\longrightarrow} \mathcal{N}(0,\Sigma), \tag{8.14}$$

where σ_0' is defined as in (8.8) and where we recall that $\widetilde{W}_t = \varrho W_t + \overline{\varrho}\,\overline{W}_t$ and

$$\Sigma := \begin{pmatrix} \sigma_0^2 & 2\varrho\partial_x\sigma_0\mathcal{K} \\ 2\varrho\partial_x\sigma_0\mathcal{K} & 4\left(\frac{\sigma_0'}{\sigma_0}\right)^2 \end{pmatrix}, \qquad \text{where} \qquad \mathcal{K} := \lim_{t\downarrow 0}\frac{\int_0^t K(t,s)\mathrm{d}s}{\mathfrak{s}(t)\sqrt{t}}. \tag{8.15}$$

Simple calculations for two-dimensional normal random variables imply that

$$\mathbb{E}[\boldsymbol{\delta}|\boldsymbol{\gamma}=x] = \frac{\Sigma_{1,2}}{\Sigma_{1,1}}x = \frac{2\varrho\sigma_0'\mathcal{K}}{\sigma_0^2}x,$$

with $\boldsymbol{\delta}$ and $\boldsymbol{\gamma}$ defined as in Assumption 8.10(iii), from which we further conclude that

$$\alpha(z) = \frac{\varrho\sigma_0'\mathcal{K}}{(H+\frac{3}{2})\sigma_0^2}z. \tag{8.16}$$

Plugging this into the asymptotic expansion for the implied volatility, using that (under reasonable assumptions on σ),

$$\lim_{t\downarrow 0}\overline{\xi}(t) = \lim_{t\downarrow 0}\left\{\frac{1}{t}\int_0^t \mathbb{E}\left[\sigma(\widehat{W}_s,s)^2\right]\mathrm{d}s\right\} = \sigma_0^2,$$

we arrive, at least for nonzero $\varrho\sigma_0'$, at the *skew formula*

$$\frac{\sigma_{\text{imp}}(\sqrt{t}z,t)-\sigma_{\text{imp}}(\sqrt{t}z',t)}{\sqrt{t}z-\sqrt{t}z'} \sim \frac{\varrho\sigma_0'\mathcal{K}}{(H+\frac{3}{2})\sigma_0}\frac{\mathfrak{s}(t)}{\sqrt{t}}, \tag{8.17}$$

as t tends to zero [154].

Example 8.12. For the rough Bergomi model with the Riemann–Liouville kernel $K^H(t,s) = \sqrt{2H}(t-s)^{H-\frac{1}{2}}$, the above computations can be justified [154, Remark 2.5]. Indeed, we have

$$\mathfrak{s}(t) = t^H \qquad \text{and} \qquad \mathcal{K} := \lim_{t\downarrow 0}\frac{\int_0^t K^H(t,s)\mathrm{d}s}{\sqrt{t}\mathfrak{s}(t)} = \frac{\sqrt{2H}}{H+\frac{1}{2}}.$$

For this kernel and general σ, the asymptotic equivalence (8.17) specializes to

$$\frac{\sigma_{\text{imp}}(\sqrt{t}z,t)-\sigma_{\text{imp}}(\sqrt{t}z',t)}{\sqrt{t}(z-z')} \sim \frac{\sqrt{2H}\varrho\sigma_0'}{(H+\frac{1}{2})(H+\frac{3}{2})\sigma_0}t^{H-\frac{1}{2}}.$$

For the specific choice $\sigma(x,t)^2 := \xi(t)\exp\left(\eta x - \frac{1}{2}\eta^2 t^{2H}\right)$, we obtain

$$\frac{\sigma_{\text{imp}}(\sqrt{t}z,t)-\sigma_{\text{imp}}(\sqrt{t}z',t)}{\sqrt{t}(z-z')} \sim \frac{\sqrt{2H}\varrho\eta}{(H+\frac{1}{2})(2H+3)}t^{H-\frac{1}{2}}.$$

8.3.3 ▪ The Edgeworth expansion

In the proof of Theorem 8.11, we gave an expansion for put option prices. Under additional assumptions we can prove higher-order expansions for a more general payoff function. The leading term of such an expansion is a Bachelier price of the payoff, which is the expectation of the payoff under the limit centered Normal random variable. As an expansion around the latter, one can consider the Gram–Charlier expansion, that is, the expansion of a density function by orthogonal polynomials with respect to the limit Normal distribution. The orthogonal polynomials are the probabilist's Hermite polynomials defined as

$$H_n(x, \Sigma) = \exp\left(\frac{\Sigma}{2}x^2\right)(-1)^n \frac{\mathrm{d}^n}{\mathrm{d}x^n} \exp\left(-\frac{\Sigma}{2}x^2\right), \tag{8.18}$$

for $n \geq 0$, where Σ is the variance of the limit centered Normal distribution. The *Gram–Charlier* expansion is a series expansion and requires the existence of a square integrable density with respect to some centered Normal distribution, which is too restrictive in most applications because it requires in particular a Gaussian tail. In many situations of interest, there is an asymptotic variable; this may be the number $N \to \infty$ of independent and identically distributed (i.i.d.) realizations in elementary Law of Large Numbers and CLT theory, a fast mean-reverting parameter $\varepsilon \downarrow 0$ as in [150], or in fact $t \downarrow 0$. Generally speaking, the *Edgeworth expansion* is a rearrangement of the Gram–Charlier series so that it is verified not as a series expansion but as an asymptotic expansion with respect to this asymptotic variable, reflecting the structure of the distribution in question.

The classical proof [60] (see also the reprint [61, Sections 7–9, Chapter 2]) for the validity of the Edgeworth expansion for sums of i.i.d. random variables consists of error estimates in an expansion of the characteristic function and in the Fourier inversion. The first step is essentially a Taylor expansion and requires the existence of finite moments up to some order. The second step requires some smoothness of the distribution to be expanded in order to ensure that the expansion error from the former step is negligible in the Fourier inversion.

The first step is naturally extended to functionals of dependent processes which converge in law to a Normal distribution. Indeed, in this central limit regime with the short-time asymptotics for rough (or nonrough) volatility models, an Itô–Taylor-type expansion is possible, which in turn provides an asymptotic expansion of the characteristic function. By the Fourier inversion, we then obtain an approximate density. To estimate the corresponding approximation error, we need some smoothness of the distribution. Under stochastic volatility models, the marginal distribution of the log-price is mixed normal, that is, a normal distribution with random mean and variance, except in the case of extreme leverage parameter $\varrho = \pm 1$. Indeed, for example, under the current framework, X_t is normally distributed conditionally on W, with conditional mean

$$-\frac{1}{2}\int_0^t V_s \mathrm{d}s + \varrho \int_0^t \sqrt{V_s}\mathrm{d}W_s$$

and conditional variance

$$\overline{\varrho}^2 \int_0^t V_s \mathrm{d}s.$$

In particular, X_t admits a density if $\overline{\varrho} \neq 0$ (i.e., $\varrho \notin \{-1, +1\}$), which serves as a smoothness condition required in the abovementioned Fourier inversion step.

A general first-order expansion result as a refinement of the martingale CLT is given by [276, 277]. The required smoothness comes in the form of suitably nondegeneracy of Malliavin's covariance matrix. A second-order expansion for stochastic volatility models under the short-time asymptotics is available in [130]. Here we only state a result for the rough Bergomi model.

Theorem 8.13. *Assume $\xi(\cdot)$ continuous at 0, and let $\zeta(t) := \sqrt{t\overline{\xi}(t)}$, with $\overline{\xi}$ as in* (8.13). *Under* (8.7) *with $\overline{\varrho} \neq 0$, for any $t > 0$, the random variable $\frac{X_t}{\zeta(t)}$ admits a density p_t with*

$$\sup_{x\in\mathbb{R}} \left(1 + |x|^2\right)^{\alpha} |p_t(x) - q_t(x)| = o\left(t^{2H}\right),$$

as t tends to zero for any nonnegative integer α, where

$$\begin{aligned} q_t(x) &= \varphi\left(x + \frac{\zeta(t)}{2}\right)\left\{1 + \lambda(t)\left[H_3\left(x + \frac{\zeta(t)}{2}\right) - \zeta(t)H_2\left(x + \frac{\zeta(t)}{2}\right)\right]\right\} \\ &\quad + \frac{\lambda(t)^2}{2}H_6(x)\varphi(x) + \kappa t^{2H} H_4(x)\varphi(x), \\ \lambda(t) &= \frac{\varrho\eta\sqrt{2H}}{2\zeta(t)^3}\int_0^t\int_0^u (u-s)^{H-\frac{1}{2}}\sqrt{\xi(s)}\,\mathrm{d}s\,\xi(u)\,\mathrm{d}u, \\ \kappa &= \frac{2H\eta^2}{(2H+1)^2}\left\{\frac{1+2\varrho^2}{4(H+1)} + \varrho^2\beta\left(H+\frac{3}{2}, H+\frac{3}{2}\right)\right\}, \end{aligned}$$

with $H_n(x) := H_n(x, 1)$ denoting the Hermite polynomials from (8.18), *φ the standard normal density, and β the Beta function.*

From this density expansion follows an expansion of the European option price with a general bounded payoff. In particular, we can derive a second-order expansion of put option prices, which in turn yields a second-order expansion of the implied volatility as follows.

Corollary 8.14. *For any $k \in \mathbb{R}$,*

$$\begin{aligned} &\sigma_{\text{imp}}\left(k\sqrt{t}, t\right) \\ &= \overline{\xi}(t)^{\frac{1}{2}} + \lambda(t)k + 3\lambda(t)^2\left[\frac{\overline{\xi}(t)^{\frac{1}{2}}}{2} - \frac{k^2}{\overline{\xi}(t)^{\frac{1}{2}}}\right] + \kappa t^{2H}\left[\frac{k^2}{\overline{\xi}(t)^{\frac{1}{2}}} - \overline{\xi}(t)^{\frac{1}{2}}\right] + o\left(t^{2H}\right), \end{aligned}$$

as t tends to zero. Moreover, again for small enough t,

$$\begin{aligned} \left.\frac{\partial}{\partial k}\right|_{k=0} \sigma_{\text{BS}}(k,t) &= \frac{\lambda(t)}{\sqrt{t}} + o\left(t^{2H-\frac{1}{2}}\right), \\ \left.\frac{\partial^2}{\partial k^2}\right|_{k=0} \sigma_{\text{BS}}(k,t) &= \frac{2\kappa}{\overline{\xi}(t)^{\frac{1}{2}}}t^{2H-1} - \frac{6\lambda(t)^2}{t\overline{\xi}(t)^{\frac{1}{2}}} + o\left(t^{2H-1}\right). \end{aligned} \tag{8.19}$$

Figure 8.1 provides a numerical comparison between the Monte Carlo values and the asymptotic expansion under realistic parameters. The skew and curvature formulae (8.19) are derived by combining the Edgeworth expansion of the distribution function and an expression of the derivative of the implied volatility in terms of a digital option price. We refer the reader to [130] for precise details. Since

$$\lambda(t) \sim \frac{\sqrt{2H}\varrho\eta}{2(H+\frac{1}{2})(H+\frac{3}{2})}t^H,$$

the result is consistent to what we observed in Example 8.12. We also note that the small-time Edgeworth expansion under the rough Heston model is given by [137] using an explicit formula of the characteristic function.

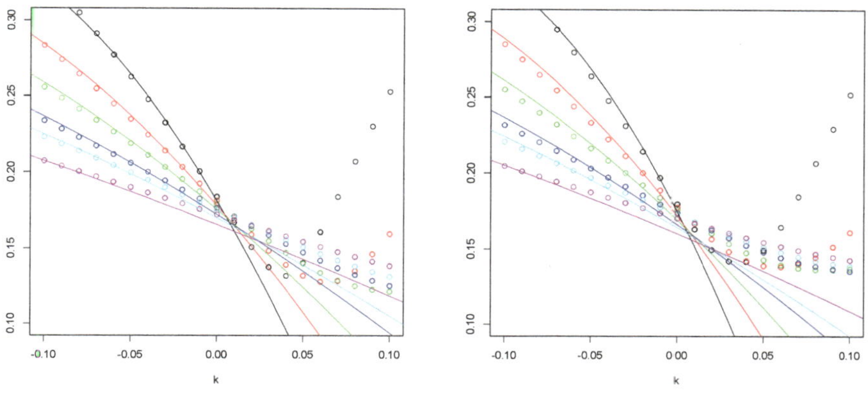

Figure 8.1: The Black–Scholes implied volatility under the rough Bergomi model with $\xi \equiv .04$ and $(H, \varrho, \eta) = (.05, -.9, 2.3)$ (left) or $(H, \varrho, \eta) = (.07, -.9, 1.9)$ (right). The points are computed by the Monte Carlo and the curves are by the asymptotic formula given in Corollary 8.14. The different colors are for different time-to-maturities: black for $t = .02$, red for $t = .05$, green for $t = .1$, blue for $t = .2$, cyan for $t = .3$, and magenta for $t = .6$. Note that the sets of parameters are those calibrated from option data in [39].

8.4 ▪ Short-time asymptotics, large deviation regime

In the setup (8.5) introduced above, with the Gaussian processes $\widetilde{W}$ and $\widehat{W}$ in (8.3) and the prototypical kernel K^H in (8.4), denote the spot variance $v_0 = \sigma_0^2 = f^2(0,0)$.

8.4.1 ▪ Large deviations for prices and implied volatilities

For the two-dimensional Brownian motion $(W, \overline{W})$ from (8.3), denote by $\mathcal{H}$ the associated Cameron–Martin space and

$$\|\mathrm{h}\|^2_{\mathcal{H}} := \left\| \left(h, \overline{h}\right) \right\|^2_{\mathcal{H}} = \int_0^1 \left(\dot{h}_t^2 + \dot{\overline{h}}_t^2 \right) \mathrm{d}t.$$

Schilder's theorem asserts that the family $\{\varepsilon \mathrm{W} : \varepsilon > 0\}$, viewed as random variables with values in pathspace $C([0,1], \mathbb{R}^2)$, satisfies a large deviation principle (LDP) with speed ε^2 and good rate function $I(\mathbf{x}) = \frac{1}{2}\|\mathbf{x}\|^2_{\mathcal{H}}$ when $\mathbf{x} \in \mathcal{H}$, and $+\infty$ otherwise. Intuitively, this means that for any reasonable set $A \subset C([0,1], \mathbb{R}^2)$ and small $\varepsilon > 0$,

$$\mathbb{P}(\varepsilon \mathrm{W} \in A) \approx \exp\left\{ -\frac{I(A)}{\varepsilon^2} \right\}, \qquad \text{with } I(A) := \inf\{I(\mathbf{x}) : \mathbf{x} \in A\},$$

(The precise definition bounds liminf and limsup of $\varepsilon^2 \log \mathbb{P}(\varepsilon \mathrm{W} \in A)$ from above and below by $I(A)$, for open and closed sets, respectively.) The *contraction principle* [113, Section 4.2.1] then asserts that LDPs are stable under continuous maps. As a trivial example, consider the map $(W, \overline{W}) \mapsto \sigma W_1$, in which case Schilder, combined with Brownian scaling $W_t \overset{\Delta}{=} \varepsilon W_1$, with $t = \varepsilon^2$, shows that $\{W_t : t > 0\}$ satisfies an LDP with speed t and good rate (a.k.a. energy) function $k \mapsto k^2/\sigma^2$, which leads to (elementary, Gaussian) asymptotics, whenever $\sigma, k > 0$, as $\varepsilon \downarrow 0$,

$$\mathbb{P}(\sigma W_t > k) = \exp\left\{ -\frac{1}{2t}\left(\frac{k^2}{\sigma^2} + o(1) \right) \right\} \approx \exp\left\{ -\frac{k^2}{2\sigma^2 t} \right\}.$$

(These can be viewed as out-of-money short-time asymptotics of Bachelier binary call option prices.) This logic applies, mutatis mutandis, to rough volatility. To this end, recall the precise setting of Section 8.1.2, with fBm motion $\widehat{W}_t = K^H * \dot{W}$. Note that $\widehat{W} = W$ for Hurst parameter $H = \frac{1}{2}$, and the same notational logic applies to

$$\widehat{\varepsilon} := \varepsilon^{2H}, \qquad \widehat{X}^\varepsilon := \varepsilon^{2H-1} X^\varepsilon,$$

with small noise log-price process

$$X_t^\varepsilon := \varepsilon \int_0^t \sigma\left(\varepsilon^{2H}\widehat{W}_s, \varepsilon^2 s\right) \mathrm{d}\widetilde{W}_s - \frac{\varepsilon^2}{2} \int_0^t \sigma^2\left(\varepsilon^{2H}\widehat{W}_s, \varepsilon^2 s\right) \mathrm{d}s \tag{8.20}$$

so that $X_t \overset{\Delta}{=} X_1^\varepsilon$ (equality in law). The rescaled process $\widehat{X}^\varepsilon$ can be seen to satisfy an LDP with speed $\widehat{\varepsilon}^2 := \varepsilon^{4H}$, equivalently obtained from the process given by

$$\int_0^t \sigma\left(\widehat{\varepsilon}\widehat{W}_s, 0\right) \widehat{\varepsilon}\, \mathrm{d}\widetilde{W}_s.$$

This reduction is valid for continuous σ, such that $\sigma(\cdot, u) \to \sigma(\cdot, 0)$ uniformly on compacts, as $u \downarrow 0$. This applies in particular to the rough Bergomi model, and from (8.7) we see that $\sigma^2(x, 0) = \sigma_0^2 \mathrm{e}^{\eta x}$ with spot variance $\sigma_0^2 = \xi(0)$.

Pretending that the *Itô map*, by which we here mean the stochastic integration map

$$(\widehat{\varepsilon}W, \widehat{\varepsilon}\overline{W}) \mapsto \int_0^{\cdot} \sigma\left(\widehat{\varepsilon}\widehat{W}_s, 0\right) \widehat{\varepsilon}\, \mathrm{d}(\overline{\varrho}\,\overline{W}_s + \varrho W_s) =: \varphi(\widehat{\varepsilon}W, \widehat{\varepsilon}\overline{W}), \tag{8.21}$$

is continuous, the contraction principle readily gives an LDP for $\widehat{X}_1^\varepsilon \overset{\Delta}{=} t^{H-\frac{1}{2}} X_t$ with speed $\widehat{\varepsilon}^2 = t^{2H}$ and rate function

$$\Lambda(x) := \inf\left\{ \frac{1}{2} \|(h, \overline{h})\|_{\mathcal{H}}^2 : \varphi_1(h, \overline{h}) := \int_0^1 \sigma\left(\widehat{h}, 0\right) \mathrm{d}(\overline{\varrho}\overline{h} + \varrho h) = x \right\}, \tag{8.22}$$

where $\widehat{h}_t = (K^H * \dot{h})_t$. This was first achieved by [138], using exponentially good Riemann-sum approximation to the above stochastic integral, under a (global) Hölder assumption on σ that rules out the rough Bergomi model. This restriction was removed by several authors with different methods, as detailed in Section 8.4.1. One can see that Λ in (8.22) is increasing (resp., decreasing) on $x > 0$ (resp., $x < 0$) (cf. [149, Prop. 6.1]), continuous [138], and smooth near zero [40, 142], provided $\sigma(\cdot, 0)$ is smooth (near zero) with nondegenerate spot-vol $\sigma_0 := \sigma(0, 0) > 0$, as is the case for rough Bergomi.

Consider log-moneyness $k_t = x t^{\frac{1}{2}-H} > 0$. As before, the LDP gives (of-out-money) binary call price asymptotics,

$$\mathbb{P}\left(X_t > x t^{\frac{1}{2}-H}\right) = \mathbb{P}\left(t^{H-\frac{1}{2}} X_t > x\right) \approx \exp\left\{ -\frac{\Lambda(x)}{t^{2H}} \right\}. \tag{8.23}$$

Matching exponents with (when $H < \frac{1}{2}$ moderately OTM) Bachelier prices,

$$\mathbb{P}\left(\sigma W_t > x t^{\frac{1}{2}-H}\right) = \mathbb{P}(\sigma W_{t^{2H}} > x) \approx \exp\left\{ -\frac{x^2}{2\sigma^2 t^{2H}} \right\},$$

then gives an effective (normal implied) volatility such that, as $t \downarrow 0$,

$$\sigma^2_{\text{normal}}\left(x t^{\frac{1}{2}-H}, t\right) \sim \frac{x^2}{2\Lambda(x)} =: \Sigma^2(x) > 0, \tag{8.24}$$

which also holds true for $x < 0$ by considering (out-of-money) binary puts. This "rough BBF formula" [55, 138] remains valid with the usual Black–Scholes implied volatility, σ_{imp}, the reason being that large deviation asymptotics for binary and classical (out-of-money) options are essentially identical. Indeed, if one sandwiches an out-of-money put payoff $(\mathrm{e}^{k_t} - \mathrm{e}^{X_t})_+$ by step functions, the put analogue of (8.23), i.e., $\mathbb{P}(X_t < k_t = xt^{\frac{1}{2}-H}) \approx \exp(-\Lambda(x)/t^{2H})$, readily leads to, as $t \downarrow 0$,

$$\mathrm{P}\left(xt^{\frac{1}{2}-H}, t\right) = \mathbb{E}\left[\left(\mathrm{e}^{xt^{\frac{1}{2}-H}} - \mathrm{e}^{X_t}\right)_+\right] = \exp\left\{-\frac{\Lambda(x) + o(1)}{t^{2H}}\right\} \approx \exp\left\{-\frac{\Lambda(x)}{t^{2H}}\right\}.$$

The corresponding result for ($x > 0$, i.e., out-of-money) call prices,

$$\mathrm{C}\left(xt^{\frac{1}{2}-H}, t\right) = \mathbb{E}\left[\left(\mathrm{e}^{X_t} - \mathrm{e}^{xt^{\frac{1}{2}-H}}\right)_+\right] \approx \exp\left\{-\frac{\Lambda(x)}{t^{2H}}\right\}, \tag{8.25}$$

holds true on a mild integrability assumption, namely existence of pth moments of e^{X_t} for *some* $p > 1$. (Note that martingality of the price process $\exp\{X\}$, shown in [162] for the rough Bergomi model, already entails existence of a first moment; negative correlation, $\varrho < 0$, is conjectured to yield existence of (some) higher moments, as is the case in classical lognormal volatility models (e.g., $\sqrt{V_t} = \sigma_0 \exp(\eta W_t)$).

Remark 8.4.1. *Note that $k_t = xt^{\frac{1}{2}-H}$ tends to zero as t tends to zero. The Caravenna–Corbetta formula* [83, (2.41)]*, an extension of Benaim and Friz* [46]*, gives*

$$\sigma^2_{\mathrm{BS}}(k_t, t) \sim \frac{k_t^2}{2t \log\left(\frac{k_t}{\mathrm{C}(k_t,t)}\right)} \sim \frac{x^2}{2\Lambda(x)} = \Sigma^2(x) \tag{8.26}$$

where we used the precise log-asymptotic equivalence of (8.25)*, in agreement with expected* (8.24)*.*

Remark 8.4.2. *Both option price and implied volatility asymptotics derived here are consistent with those in the rough Heston model as in Theorem* 4.4 *in Chapter* 4*.*

In Theorem 8.11 we gave a CLT based expansion of $\sigma_{\text{imp}}(x\sqrt{t}, t)$; formally subtracting the expansion for distinct x, x' led us to the skew formula (8.17). Proceeding in the same fashion with LDP-based formula (8.24), written as

$$\sigma_{\text{imp}}\left(xt^{\frac{1}{2}-H}, t\right) = \Sigma(x) + o(1),$$

gives the discrete skew formula

$$\mathcal{S}_{\mathrm{BS}}(t; x, x') := \frac{\sigma_{\text{imp}}\left(xt^{\frac{1}{2}-H}, t\right) - \sigma_{\text{imp}}\left(x't^{\frac{1}{2}-H}, t\right)}{xt^{\frac{1}{2}-H} - x't^{\frac{1}{2}-H}} \sim \frac{\Sigma(x) - \Sigma(x')}{x - x'} t^{H-\frac{1}{2}} \tag{8.27}$$

as t tends to zero. Note that $|xt^{\frac{1}{2}-H}| \gg |x\sqrt{t}|$, as $t \downarrow 0$; in this sense the large deviation approach deals with a larger strike regime than the CLT regime discussed in Section 8.3. However, this "gap" closes as H tends to zero!

Remark 8.4.3 (moderate deviations). *The intermediate regime with log-moneyness $xt^{\frac{1}{2}-H+\beta}$ for $\beta \in (0, H)$ is known as the moderate deviation regime, which effectively interpolates between LDP ($\beta = 0$) and CLT ($\beta = H$) regimes. A formal application of* (8.24) *gives, as t tends to zero,*

$$\sigma_{\text{imp}}\left(xt^{\frac{1}{2}-H+\beta}, t\right) \approx \Sigma\left(xt^{\beta}\right) \approx \Sigma(0) + \Sigma'(0)xt^{\beta} + \dots,$$

in terms of spot vol $\Sigma(0) = \sigma_0$ *and short-dated ATM skew* $\Sigma'(0)$. *These are higher-order moderate deviation expansions, more informative than first order (of Gaussian) moderate deviations which amount to* $\sigma_{\rm imp}(xt^{\frac{1}{2}-H+\beta}, t) \approx \Sigma(0)$. *A quick formal derivation of the latter can be based on* (8.23), *replacing x by* xt^β *with Gaussian (quadratic) approximation* $\Lambda(xt^\beta) \approx \Lambda''(0)x^2t^{2\beta}/2 = x^2t^{2\beta}/(2\sigma_0^2)$ *from which*

$$\mathbb{P}\left[X_t > xt^{\frac{1}{2}-H+\beta}\right] \approx \exp\left\{-\frac{x^2}{2\sigma_0^2 t^{2(H-\beta)}}\right\}. \tag{8.28}$$

Moderate deviations asymptotics were introduced in a diffusive option pricing context in [145], *already as higher-order expansion. The corresponding results under rough volatility were obtained in* [40], *a unifying perspective was given in* [142, 143]. *Classical (first order, Gaussian) moderate deviation in this context were derived in* [211], *also* [182, 210], *in a general setting under pathwise uniqueness (an open problem for rough Heston), and agrees with the result of formal affine calculations in* [134, *Section* 3.6].

Under the standing assumption that spot vol is strictly positive, it is seen in [40, 142] that Σ is differentiable (near 0), so that for x', x near 0, the slope appearing above is well approximated by $\Sigma'(0)$. It is clear from (8.24) that (short-dated) implied volatilities are directly related to the energy function $\Lambda = \Lambda(x)$. The latter is zero ATM ($x = 0$), nonnegative; granted sufficient smoothness (cf. references above!), it must be of the form

$$\Lambda(x) = \Lambda''(0)\frac{x^2}{2} + \Lambda'''(0)\frac{x^3}{3!} + O(x^4).$$

With spot vol σ_0 and spot skew σ_0', as defined in (8.8), we can compute [40]

$$\Lambda''(0) = \frac{1}{\sigma_0^2}, \quad \Lambda'''(0) = -\frac{6\varrho\sigma_0'\langle K^H 1, 1\rangle}{\sigma_0^4},$$

which in turn gives the short-dated smile expansion

$$\Sigma(k) = \sigma_0 + \frac{\varrho\sigma_0'\langle K^H 1, 1\rangle}{\sigma_0}k + O(k^2).$$

Using the explicit form of the Riemann–Liouville kernel, K^H from (8.4), we compute the prefactor of $t^{H-\frac{1}{2}}$ in (8.27), near the money, as

$$\Sigma'(0) = \frac{\varrho\sigma_0'\langle K^H 1, 1\rangle}{\sigma_0} = \frac{\sqrt{2H}\varrho\sigma_0'}{(H+\frac{1}{2})(H+\frac{3}{2})\sigma_0}, \tag{8.29}$$

which agrees with the right-hand side factor of the skew formula (8.17), derived earlier with CLT and Edgeworth methods.

Large deviations for stochastic integrals

In (8.21) we encountered

$$\int_0^\cdot \sigma\left(\widehat{W}_s, 0\right) \mathrm{d}\left(\overline{\varrho}\,\overline{W}_s + \varrho W_s\right) = \varphi\left(W, \overline{W}\right)$$

and highlighted the importance of an LDP for φ applied to $(\varepsilon W, \varepsilon\overline{W})$. In [138] Forde and Zhang used exponentially good approximations and the *extended contraction principle* [113, Theorem

4.2.23], noting that their (global) Hölder assumption on σ rules out the rough Bergomi model. This was overcome with different methods by several authors. In [38], Bayer et al. use rough paths and regularity structures to see that $\varphi(W,\overline{W}) = \Phi(\mathbf{W})$ for some lifted noise $\mathbf{W}$ of rough path type and continuous Φ, such that one can employ general facts on large deviations for Gaussian models (in the terminology of Hairer), followed by the classical contraction principle. This requires no growth restriction but does require local smoothness (depending on H), which is fine for rough Bergomi. (As noted after (8.20), the "t^{2H} term" in the rough Bergomi volatility specification (8.7) can be neglected for large deviations.) In [38, Section 5] the authors also obtained LDPs for stochastic Volterra dynamics.

Around the same time, large deviations were derived for the rough Bergomi model [172, 210], followed by Gulisashvili [181] who relaxed the conditions of [138], also via the extended contraction principle, to a mild local regularity condition on σ satisfied by the rough Bergomi model.

In [210], and again later in [200], the authors use *Garcia's theorem* [158] (see also [156] for some generalizations). In the context of adapted continuous processes, it states that for a sequence of such processes $(Y^\varepsilon)_{\varepsilon\geq 0}$, and a sequence of uniformly exponentially tight[46] semimartingales $(Z^\varepsilon)_{\varepsilon\geq 0}$, an LDP for the sequence $(Y^\varepsilon, Z^\varepsilon)_{\varepsilon>0}$, with rate function Λ, implies an LDP for the sequence of triples $(Y^\varepsilon, Z^\varepsilon, \int_0^\cdot Y_s^\varepsilon \mathrm{d}Z_s^\varepsilon)_{\varepsilon>0}$ with rate function

$$\Lambda^\sharp(\mathrm{y},\mathrm{z},\varphi) = \Lambda(\mathrm{y},\mathrm{z}), \quad \text{if} \begin{cases} \varphi = \displaystyle\int_0^\cdot \mathrm{y}(t)\mathrm{d}\mathrm{z}(t), \\ \mathrm{z} \in \mathrm{FV}, \end{cases}$$

and is infinite otherwise, where FV denotes the space of continuous paths of finite variations on $[0,1]$. The LDP for the sequence $(\int_0^1 Y_s^\varepsilon \mathrm{d}Z_s^\varepsilon)_{\varepsilon>0}$ then follows from the contraction principle. This applies in particular to the indefinite stochastic integral

$$\int_0^\cdot \sigma\left(\widehat{\varepsilon}\widehat{W}_t, \widehat{\varepsilon}^2 t\right)\widehat{\varepsilon}\mathrm{d}\left(\overline{\varrho}\,\overline{W}_t + \varrho W_t\right),$$

perhaps noting that the integrator enjoys an LDP thanks to Schilder, while the LDP for the integrand comes from the contraction principle, with rate function

$$I(y) := \inf\left\{\frac{1}{2}\|h\|_{H^1}^2 : y = \sigma(\widehat{h}_\cdot, 0)\right\},$$

and so entails the LDP for $\widehat{X}_1^\varepsilon \triangleq t^{H-\frac{1}{2}}X_t$ with speed $\widehat{\varepsilon}^2 = t^{2H}$ and rate function (8.22).

In [88] Cellupica and Pacchiarotti employ, in the uncorrelated setting $\varrho = 0$, a result of [89] which gives, in some generality, an LDP for $(\widehat{W}, X)$ provided one has LDPs for $\widehat{W}$ and X conditionally on $\widehat{W}$; for $\varrho \neq 0$ this direct approach fails, and again approximations are used. Let us also mention [185], which considers small-time asymptotics for implied volatilities in a stochastic volatility model with instantaneous volatility $v_t = |X_t|$ for a self-similar Gaussian process X in terms of the Karhunen–Loeve decomposition of the latter (see also [184] for extreme strike asymptotics obtained in a similar setting).

A different route was followed in [211], using the weak convergence approach pioneered by Boué and Dupuis [68] and developed further by Budhiraja and Dupuis [75] (see also the excellent

[46]The sequence $(Z^\varepsilon)_{\varepsilon\geq 0}$ is called uniformly exponentially tight if for any $c > 0$, there exists $K_c > 0$ such that

$$\limsup_{\varepsilon\downarrow 0} \varepsilon \log\left(\sup_{U:\sup_{t\in[0,1]\geq 0}|U_t|\leq 1} \mathbb{P}\left(\sup_{t\in[0,1]}\left|\int_0^t U_{s-}\mathrm{d}Z_s^\varepsilon\right| \geq K_{c,t}\right)\right) \leq -c,$$

where the outer sup is taken over the space of simple, real-valued, adapted processes.

monograph [76]). This method relies on the equivalence between the LDP and the Laplace principle. Recall that a family $\{X^\varepsilon\}_{\varepsilon>0}$ of continuous processes, viewed as random variables with values in $\mathcal{X} = C([0,T],\mathbb{R}^n)$, is said to satisfy the Laplace principle with speed ε^{-2} and rate function $I : \mathcal{X} \to [0,+\infty]$ if for all continuous bounded maps $F : \mathcal{X} \to \mathbb{R}$,

$$\lim_{\varepsilon\downarrow 0} -\varepsilon^2 \log \mathbb{E}\left[\exp\left\{-\frac{F(X^\varepsilon)}{\varepsilon^2}\right\}\right] = \inf_{x\in\mathcal{X}} \left\{I(x) + F(x)\right\}. \tag{8.30}$$

Assume now that X^ε is the image under some Itô map, say $X^\varepsilon = \mathcal{G}^\varepsilon(W)$, for some m-dimensional Brownian motion W. This allows one to compute the left-hand side prelimit with the Boué-Dupuis formula [68, Thm. 3.1], which states that

$$-\log \mathbb{E}\left[e^{-f(W)}\right] = \inf_{v\in\mathcal{A}} \mathbb{E}\left[\frac{1}{2}\int_0^\cdot |v_s|^2 \mathrm{d}s + f(W + \Im v)\right],$$

where $\mathcal{A} := \left\{v : \Omega\times[0,1] \to \mathbb{R}^m \text{ progressively measurable}, \mathbb{E}\int_0^T |v_t|^2 \mathrm{d}t < +\infty\right\}$ for any bounded measurable functional f, and $(\Im v)(t) := \int_0^t v_s \mathrm{d}s$. For fixed $\varepsilon > 0$, take $f : W \mapsto (F \circ \mathcal{G}^\varepsilon)(W)$. The crux of the method is then as follows: one can show that there exists a sequence of controls $(v^\varepsilon)_{\varepsilon>0} \in \mathcal{A}$ such that both sequences $(v^\varepsilon)_{\varepsilon>0}$ and $\left(\mathcal{G}^\varepsilon\left(W + \varepsilon^{-1}\Im v^\varepsilon\right)\right)_{\varepsilon>0}$ converge in law to v^0 and $\mathcal{G}^0(\Im v^0)$, respectively, for some measurable mapping $\mathcal{G}^0$ (this is formulated as an abstract condition in [68]). From this, one can conclude that the Laplace principle (8.30) holds with

$$I(\varphi) = \frac{1}{2}\inf\left\{\int_0^\cdot |v_s|^2 \mathrm{d}s : v \in \mathcal{A} \text{ and } \mathcal{G}^0(\Im v) = \varphi\right\}.$$

As an example, consider rescaled Brownian motion $X^\varepsilon = \varepsilon W$ so that Φ is the identity map and $\mathcal{G}^\varepsilon(W) := \varepsilon W$. Then, for any sequence of controls $(v^\varepsilon)_{\varepsilon>0} \in \mathcal{A}$,

$$\mathcal{G}^\varepsilon\left(W + \varepsilon^{-1}\Im v^\varepsilon\right) = \varepsilon W + \Im v^\varepsilon \longrightarrow \Im v^0$$

in law as ε tends to zero. Taking $\mathcal{G}^0$ as the identity, the Boué–Dupuis theorem applies, showing that the sequence $(\mathcal{G}^\varepsilon(W) = \varepsilon W)_{\varepsilon>0}$ satisfies a large deviations principle with rate function $I(\cdot)$, and we recover Schilder's theorem.

This generic approach has been applied successfully to derive large (and moderate) deviations in a variety of models without appealing to their particular features (continuous-time diffusions [93]; multiscale systems [77, 125, 271]; and SDEs driven by infinite-dimensional Brownian motions [78], Poisson random measures [74], or both [79] (including stochastic PDEs)), and has been applied in mathematical finance for (nonrough) stochastic volatility models [212]. This was further extended in [211] for general multidimensional stochastic Volterra systems, relaxing some of the classical assumptions on the coefficients (such as Lipschitz continuity and sublinear growth, which are particularly applicable to the setting in [138]), the rough Bergomi model, and the rough Heston one (under the assumption—yet to be proved—of pathwise uniqueness). Compared to the more classical Freidlin–Wentzell approach, this weak convergence method has clear benefits: it is the only available route for pathwise moderate deviations, provides a usually simpler rate function representation, and is more universal, not requiring model-tailored proofs.

8.4.2 ▪ Laplace method and refined LDP

We now refine our asymptotic option price formulae. As before, we assume dynamics of the form (8.5) with rough volatility process $\sigma(\widehat{W}_t, t)$. Recall also the energy function Λ given in (8.22), which can be seen to be differentiable, at least for small x provided $\sigma_0 > 0$, which we always assume. Set $\sigma_x^2 := 2\Lambda(x)/\Lambda'(x)^2$, and note the consistency of this notation, when

$x = 0$, with spot variance σ_0^2. Precise asymptotics for the rough Heston case are detailed in Chapter 4, Section 4.4.2.

Theorem 8.15. *Let $H \in (0, \frac{1}{2}]$. Then for small enough $x < 0$, we have put price asymptotics, as $t \downarrow 0$,*

$$\mathrm{P}\left(xt^{\frac{1}{2}-H}, t\right) \sim \exp\left\{-\frac{\Lambda(x)}{t^{2H}}\right\} t^{\frac{1}{2}+2H} \frac{A(x)}{\Lambda'(x)^2 \sigma_x \sqrt{2\pi}}.$$

Similarly, for small enough $x > 0$ and assuming 1^+ moments for e^{X_t}, we have call price asymptotics

$$\mathrm{C}\left(xt^{\frac{1}{2}-H}, t\right) \sim \exp\left\{-\frac{\Lambda(x)}{t^{2H}}\right\} t^{\frac{1}{2}+2H} \frac{A(x)}{\Lambda'(x)^2 \sigma_x \sqrt{2\pi}}.$$

The function A, which tends to 1 as $x \to 0$, can be written as

$$A(x) = \begin{cases} \mathbb{E}\left[\exp\left\{\Lambda'(x)\Delta_2^x\right\}\right], & \text{if } H < \frac{1}{2}, \\ e^k \mathbb{E}\left[\exp\left\{\Lambda'(x)\Delta_2^x\right\}\right], & \text{if } H = \frac{1}{2}, \end{cases} \tag{8.31}$$

where Δ_2^x is a certain quadratic Wiener functional, described in [142, equation (7.5)].

Note that (8.31) can be further expanded, as detailed in [143]. We outline the proof, keeping all notation as previously used in Section 8.4.1. As therein, the short-time problem is analyzed as a small noise problem, with $t \equiv \varepsilon^2$,

$$\mathrm{C}\left(x\varepsilon^{1-2H}, \varepsilon^2\right) = \mathbb{E}\left[\left(e^{X_1^\varepsilon} - e^{x\varepsilon^{1-2H}}\right)_+\right] = \mathbb{E}\left[\left(e^{\widehat{X}_1^\varepsilon \varepsilon^{1-2H}} - e^{x\varepsilon^{1-2H}}\right)_+\right].$$

Our assumptions imply that there is an LDP for $\widehat{X}_1^\varepsilon$ with speed $\widehat{\varepsilon}^2 \equiv \varepsilon^{4H}$ and rate function given in (8.22). The subsequent analysis relies on having a unique minimizer $\mathrm{h}^x = (h^x, \overline{h}^x)$ in (8.22) which is trivially true when $x = 0$, with $(h^x, \overline{h}^x)_{k=0} \equiv (0, 0)$, further seen to be nondegenerate[47] provided $\sigma_0 > 0$, and this remains valid for small $x > 0$. Apply Girsanov's theorem, $\widehat{\varepsilon}(W, \overline{W}) \to \widehat{\varepsilon}(W, \overline{W}) + (h^x, \overline{h}^x)$,

$$\mathrm{C}\left(\frac{x}{\varepsilon^{2H-1}}, \varepsilon^2\right) = \mathbb{E}\left[\exp\left\{-\frac{1}{\widehat{\varepsilon}} \int_0^1 \left(\dot{h}^x, \dot{\overline{h}}^x\right)^\top d(W, \overline{W}) - \frac{\Lambda(x)}{\widehat{\varepsilon}^2}\right\} \left(e^{\widehat{Z}_1^\varepsilon \varepsilon^{1-2H}} - e^{x\varepsilon^{1-2H}}\right)_+\right],$$

with

$$\widehat{Z}_1^\varepsilon = \varphi_1\left(\widehat{\varepsilon}W + h^x, \widehat{\varepsilon}\overline{W} + \overline{h}^x\right) + O(\varepsilon^2) = x + \widehat{\varepsilon} g_1^x + \widehat{\varepsilon}^2 g_2^x + o\left(\widehat{\varepsilon}^2\right).$$

First-order optimality of h^x implies that

$$\int_0^1 \left(\dot{h}^x, \dot{\overline{h}}^x\right)^\top d\left(W, \overline{W}\right) = \Lambda'(x) g_1^x(\omega), \tag{8.32}$$

which in conjunction with Itô isometry gives the useful relation

$$2\Lambda(x) = \left\|\left(h^x, \overline{h}^x\right)\right\|_{\mathcal{H}}^2 = \Lambda'(x)^2 \mathbb{V}(g_1^x).$$

We then see that

$$\mathrm{C}\left(x\varepsilon^{1-2H}, \varepsilon^2\right) = \exp\left\{-\frac{\Lambda(x)}{\varepsilon^{4H}}\right\} \mathbb{E}\left[\exp\left\{-\frac{\Lambda'(x) g_1^x}{\widehat{\varepsilon}}\right\} \left(e^{\widehat{Z}_1^\varepsilon \varepsilon^{1-2H}} - e^{k\varepsilon^{1-2H}}\right)_+\right].$$

[47]The notion of nondegenerate minimizer among all k-admissible $(h, \overline{h})$, in the sense that $\varphi_1(h, \overline{h}) = k$ as in (8.22), requires some care.

Using large deviations, this expectation can be localized in path space,[48] and also inserting the expansion of $\widehat{Z}_1^\varepsilon$ leads to an expectation that allows for a conditionally Gaussian evaluation,[49] which leads to stated factor of order ε^{1+4H}. Substitution of ε by $t^{\frac{1}{2}}$ then leads to the stated short-time results for European option prices.

Corollary 8.16. *For sufficiently small x and $k_t = xt^{\frac{1}{2}-H}$,*

$$\sigma^2_{\text{imp}}(k_t, t) = \Sigma^2(x) + a(x)t^{2H} + o\big(t^{2H}\big), \qquad \text{as } t \downarrow 0, \tag{8.33}$$

where

$$\Sigma(x) := \frac{|x|}{\sqrt{2\Lambda(x)}} \quad \text{and} \quad a(x) = \begin{cases} \dfrac{x^2}{2\Lambda(x)^2}\log\left(\dfrac{2A(x)\Lambda(x)}{\Lambda'(x)}\right), & \text{if } H < \frac{1}{2}, \\[2ex] \dfrac{x^2}{2\Lambda(x)^2}\log\left(\dfrac{2A(x)\Lambda(x)}{\Lambda'(x)\exp(x/2)}\right), & \text{if } H = \frac{1}{2}. \end{cases}$$

The proof of this is an application of Gao and Lee [157] which translates option price asymptotics into implied volatility asymptotics. Note that the form of the expansion was given in [136] for the standard Heston model. We should emphasize that both Theorem 8.15 and Corollary 8.16 are not restricted to the class of rough volatility models from Section 8.1.2, including rough Bergomi, but hold in much greater generality to models that admit a robust form; this allows for the use of rough paths and regularity structures which was pivotal to the analysis.

Remark 8.4.4. *With some preliminary localization this could also be applied to rough Heston, though in this case other methods are available.*

8.4.3 ▪ Numerics

Let us see how Corollary 8.16 can be used in practice. Following [143] we consider rough Bergomi with $\sigma^2(x,t) = \sigma^2(0)e^{\eta x - \theta\eta^2 t^{2H}/2}$, where the additional parameter θ allows one to switch between the standard rough Bergomi form ($\theta = 1$) and its time-independent pendant ($\theta = 0$). The leading order term $\Sigma(x)$ is given explicitly via the rate function $\Lambda(x)$, not sensitive to θ, which can be computed with arbitrary precision by the Ritz method, which amounts to the infimum in (8.22) over a finite-dimensional approximation of the Cameron–Martin space, already suggested by Forde and Zhang [138]; see also [143]. The resulting smile approximation, referred to as the FZ expansion in Figure 8.2, is of order $O(t^{2H})$ and, in particular, can be expected to matter for small H and/or not-so-small t. Indeed, the impact of including the next order term in our expansion, $a(x)t^{2H}$ in full, with $a(x)$ computed via a Karhunen–Loeve expansion, as well as its closed-form ATM approximation $a(0)t^{2H}$, are seen to be significant. For comparison, we also include the expansion one gets from an ATM expansion of $\Lambda(x)$, i.e., at $x = 0$, the resulting expansion agrees with the one derived with CLT methods [130] and Corollary 8.14 in Section 8.3.

8.4.4 ▪ Dupire local volatility

We now consider Dupire's local volatility [124], disregarding again rates and dividends. Given a stochastic volatility model with normalized log-price process $X = \log S/S_0$ and stochastic variance process $\mathrm{d}\langle X\rangle_t/\mathrm{d}t = V_t$, we see that the local variance surface given by

$$\sigma^2_{\text{Dup}}(k,t) = \mathbb{E}[V_t | X_t = k]$$

[48] In fact, in rough path (model) space which is key to controlling the remainder of the expansion of $\widehat{Z}_1^\varepsilon$.

[49] This involves some anticipating stochastic calculus for which rough paths and regularity again come in very handy.

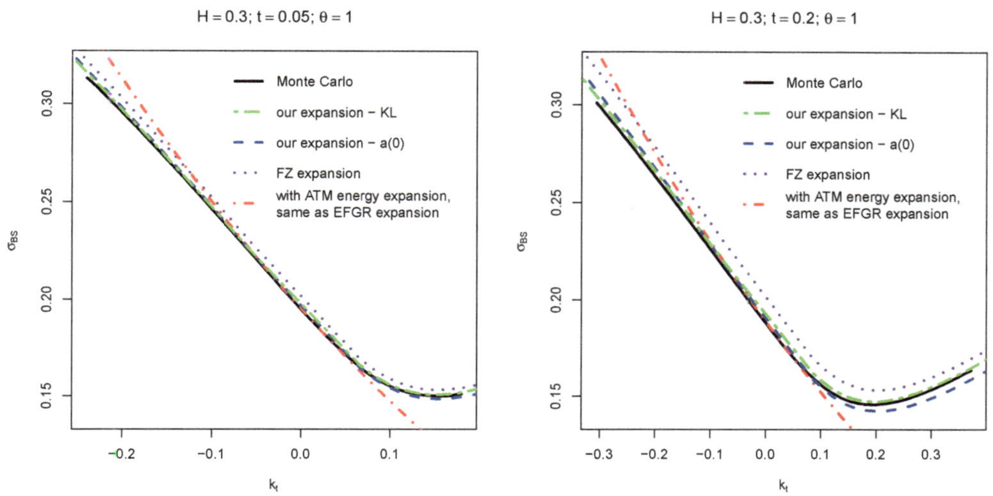

Figure 8.2: Implied volatility smile approximations in rough Bergomi ($\theta = 1$) with parameters $\sigma_0 = 0.2, \eta = 1.5, \varrho = -0.7, H = 0.3$ and expiry $t \in \{0.05, 0.2\}$. The Monte Carlo price is computed via the hybrid scheme [50] with $\kappa = 2$, with 10^9 simulations and 500 time steps of length $t/500$. Following [143], the rate function is computed using the Ritz method with $N = 8$ Haar basis functions, and the coefficient $a(\cdot)$ is computed using the Karhunen–Loeve (KL) decomposition with $N = 300$ Haar basis functions.

has the important property that $d\widetilde{X} = -\frac{1}{2}\sigma^2_{\mathrm{Dup}}(\widetilde{X}_t, t)dt + \sigma_{\mathrm{Dup}}(\widetilde{X}_t, t)dW$ mimics the time-marginals (and hence all European option prices) of X. Following [187, 73], this is valid in the generality of (possibly non-Markovian) Itô diffusions and covers our model specification (8.5), i.e., $V(t,\omega) = \sigma^2(\widehat{W}_t, t)$ for situations of interest, including rough Bergomi. The following result is taken from [69]. For simplicity we only consider the case of time-independent volatility function $\sigma = \sigma(x)$.

Theorem 8.17. *Under mild assumptions on $\sigma(\cdot)$, for $|\varrho| < 1$ and small x,*

$$\sigma^2_{\mathrm{Dup}}\left(xt^{\frac{1}{2}-H}, t\right) = \mathbb{E}\left[\sigma^2\left(\widehat{W}_t\right)\big|X_t = xt^{\frac{1}{2}-H}\right] = \sigma^2\left(\widehat{h}^x_1\right) + o(1), \qquad \text{as } t \downarrow 0,$$

with $\widehat{h}^x_1 = \left(K^H * \dot{h}^x\right)_1 = \int_0^1 K^H(1,t)\dot{h}^x_t dt$.

We outline the proof. Malliavin integration by parts, as in [141], gives

$$\mathbb{E}[V_t|X_t = x] = \frac{\mathbb{E}[V_t\,\delta(X_t - x)]}{\mathbb{E}[\delta(X_t - x)]} = \frac{\mathbb{E}\left[V_t\,\mathbf{1}_{X_t \geq x}\Pi_t\right]}{\mathbb{E}\left[\mathbf{1}_{X_t \geq x}\Pi_t\right]}$$

with $\bar{\varrho}\Pi_t = \int_0^t (1/\sigma(\widehat{W}_s))d\overline{W}_s$, provided $\bar{\varrho} \neq 0$ and $\sigma > 0$ satisfies a mild lower bound. The assumption "x small" guarantees that (8.22) admits a unique minimizer $\mathrm{h}^x = (h^x, \overline{h}^x) \in \mathcal{H}$. We proceed by Girsanov, as in the proof of Theorem 8.15, followed by a large deviation argument that allows one to localize $\mathbb{E}[\mathbf{1}_{X_t \geq x}\Pi_t]$ to a neighborhood of h^x in path space,[50] and the same is true

[50]In fact, in rough path (model space), as in the proof of Theorem 4.4.

for $\mathbb{E}[V_t \mathbb{1}_{X_t \geq x} \Pi_t] = \mathbb{E}[\sigma^2(\widehat{W}_t)\mathbb{1}_{X_t \geq x}]$ under a mild growth assumption on σ. The remainder of the proof is concerned with precise stochastic (rough) Taylor expansions, using also nondegeneracy of h^x. (As noted earlier, this can be shown for small $x > 0$.)

As an immediate consequence, similar to (8.27), we obtain the discrete local skew-formula

$$\mathcal{S}_{\text{loc}}(t; x, x') := \frac{\sigma_{\text{Dup}}\left(xt^{\frac{1}{2}-H}, t\right) - \sigma_{\text{Dup}}\left(x' t^{\frac{1}{2}-H}, t\right)}{xt^{\frac{1}{2}-H} - x' t^{\frac{1}{2}-H}} \sim \frac{\Sigma_{\text{loc}}(x) - \Sigma_{\text{loc}}(x')}{x - x'} t^{H-\frac{1}{2}}$$

as t tends to zero. Thanks to nondegeneracy, $\Sigma_{\text{loc}}(x)$ is differentiable at $x = 0$, and one computes

$$\Sigma'_{\text{loc}}(0) = (\ldots)\frac{\sqrt{2H}\varrho\sigma'_0}{(H+\frac{1}{2})(H+\frac{3}{2})\sigma_0}$$

so as to have the short-dated ATM *local volatility skew*

$$\mathcal{S}_{\text{loc}} \sim \Sigma'_{\text{loc}}(0)t^{H-\frac{1}{2}}.$$

Comparison with (8.29) and $\mathcal{S}_{\text{BS}} \sim \Sigma'(0)t^{H-\frac{1}{2}}$ shows that, as $t \downarrow 0$ and ATM,

$$\mathcal{S}_{\text{BS}}/\mathcal{S}_{\text{loc}} \to \frac{1}{H+\frac{3}{2}},$$

which generalizes the so-called $\frac{1}{2}$-rule [163] to the rough regime. (For Hurst parameter $H = \frac{1}{2}$, we are in a diffusive regime and then indeed fall back into the $\frac{1}{2}$-rule.) Numerical simulations [69] show that the $1/(H+\frac{3}{2})$-rule is remarkably robust, also for not-so-small t.

8.5 ▪ Rough volatility at criticality: $H = 0$

As has been highlighted many times, volatility is *rough*, i.e., fractional with Hurst index H close to 0, leading to ever faster skew explosion as H becomes smaller. There is, however, a certain tension between the asymptotic formulas derived so far in the extreme case $H \approx 0$, which we shall discuss using the example of short-time asymptotics of the ATM volatility skew.

Remark 8.5.1. *Going even further,* hyper-rough *stochastic volatility models with regularity $-1/2 < H$ have been suggested in the literature (see [217]), taking into account that integrated variance rather than instantaneous variance is truly fundamental for stochastic volatility models.*

As shown in (8.27), (8.12), and Corollary 8.14, the ATM skew behaves like

$$c_H T^{H-\frac{1}{2}}, \tag{8.34}$$

as T approaches zero. For instance, in Example 8.2, we have

$$c_H = \text{const}\varrho\eta\frac{\sqrt{2H}}{(H+\frac{1}{2})(H+3/2)},$$

which tends to zero as $H \downarrow 0$. Hence, we have two limits ($T \downarrow 0$, $H \downarrow 0$) with seemingly conflicting skew behaviors (skew exploding ever faster and skew flattening, respectively). We say seemingly because the relevant limit imposed by market prices of options, specifically the curvature of the smile, is $\eta\sqrt{2H}$, which converges to a constant as $T \downarrow 0$. Hence, the behavior of the ATM skew for short times when $H \approx 0$ is, a priori, unclear. Having said that, the financial interpretation of H as controlling the steepness of the power-law explosion of the ATM skew for short time clearly favors a regime with ever steepening skews as H becomes small.

8.5.1 ▪ Skew flattening in the rough Bergomi model as $H \downarrow 0$

Unfortunately, this is not the case with the standard rough stochastic volatility models of rough Bergomi type. Indeed, [133] shows that the ATM skew converges to zero as H tends to zero. More precisely, the authors of that paper show that the implied volatility smile becomes symmetric as $H \downarrow 0$, and, using the usual normalization of the kernel, the rough Bergomi model actually weakly converges to the Black–Scholes model as $H \downarrow 0$; see [133, Section 2.2]. Superficially, this means that the constant $c_H \downarrow 0$ "wins" over the steepening of the power law, but, in fact, the flattening of the skew is a consequence of a *decorrelation* phenomenon: When $H \downarrow 0$, the martingale part $\int_0^t \sqrt{V_s}\mathrm{d}W_s$ of the log-price X weakly converges to a time-change $B_{\xi([0,t])}$ of an independent Brownian motion B [133, Theorem 2.1], where ξ denotes a *Gaussian multiplicative chaos* [259].

The argument of [133] is fundamentally based on Jacod's theorem of stable convergence [209]: let M^n be a sequence of continuous local martingales with

$$\langle M^n \rangle \to Q \text{ in prob.}, \quad \langle M^n, N \rangle \to 0 \text{ in prob.}, \quad \langle M^n, N^\perp \rangle = 0$$

for a continuous process Q, a continuous local martingale N, and any bounded martingale $N^\perp$ which is orthogonal to N. Then, $M^n \to B_Q$ stably in law, where B is a Brownian motion independent of everything else.

Now consider the rough Bergomi model (8.7)

$$V_t = v_t^2 = V_0 \exp\left\{\gamma \int_0^t (t-s)^{H-\frac{1}{2}}\mathrm{d}W_s - \frac{\gamma^2}{2H}t^{2H}\right\}.$$

Note that, compared to (8.7), we updated the parameters as $\gamma := \eta\sqrt{2H}$. Otherwise, convergence would be trivial (see [133, Remark 2.2] for details). Let H go to zero, with γ fixed in $(0, \sqrt{2})$,

$$\langle \log S \rangle_t = \int_0^t V_s \mathrm{d}s \to M([0,t]) =: Q_t,$$

where M is a Gaussian multiplicative chaos, and

$$\langle \log S, W \rangle_t = \varrho \int_0^t \sqrt{V_s}\mathrm{d}s \to 0$$

because

$$\sqrt{V_t} = \sqrt{V_0}\exp\left\{Z - \frac{1}{2}\mathbb{E}[Z^2] - \frac{3\gamma^2}{16H}t^{2H}\right\}, \quad Z = \frac{\gamma}{2}\int_0^t (t-s)^{H-\frac{1}{2}}\mathrm{d}W_s.$$

By Jacod's theorem, the limit of the local martingale part of $\log S$ is a time-changed Brownian motion B_Q, where B is independent of Q. Hence, there is no leverage and no skew in the limit.

Note that the flattening of the skew as $H \downarrow 0$ is not a purely theoretical phenomenon that is unobservable in practice. Figure 8.3 shows that the skew does, indeed, flatten in the standard rough Bergomi model as $H \downarrow 0$. In contrast, a rather minor modification of the rough Bergomi model—modulating the kernel by a log-factor recovering square integrability at $H = 0$—prevents the flattening of the skew as $H \downarrow 0$ and, indeed, recovers the $T^{H-\frac{1}{2}}$ explosion of the ATM skew for all H, including $H = 0$. In the following, we shall provide an exposition of this model introduced in [42], as well as the asymptotic results obtained for this model. See, once again, Figure 8.3 for a numerical justification of the above claim for the log-modulated rough Bergomi model.

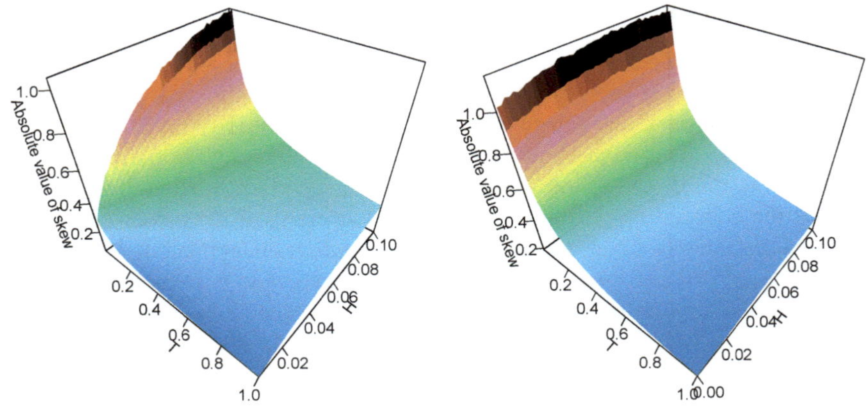

Figure 8.3: Left: Rough Bergomi model. Right: Log-modulated rough Bergomi model. ATM implied volatility skews (absolute values) in the rough Bergomi model (with and without log-modulation) plotted against expiry t and Hurst index H. Skews are computed by Monte Carlo simulation based on exact simulation of the underlying (log-moderated) fBm. Note that $H = 0$ is included in the plot in the log-modulated case. Parameter values of the rough Bergomi model are $\eta = 2.0$, $\varrho = -0.7$, $\xi(t) \equiv 0.04$. Additional parameters for the log-fBm (see below for details) are $\zeta = 0.1$, $p = 2.0$. Note how this seems to be in keeping with the findings in [133] of a vanishing skew as $H \downarrow 0$ in rough Bergomi.

Remark 8.5.2. *An alternative to modulating the kernel is to renormalize the process so as to obtain a nontrivial limit. For instance,* [247] *consider the limit of*

$$\frac{B_t^H - \frac{1}{t}\int_0^t B_u^H \mathrm{d}u}{\sqrt{H}}$$

as H tends to zero for a standard fBm B^H with Hurst index H. The limiting process is seen to be approximately a log-correlated Gaussian field. When translated into stochastic volatility models, this type of limiting model corresponds to multifractal random walks*; see* [235]. *We also refer the reader to* [189] *and the references therein for more general limiting procedures.*

8.5.2 ▪ The log-modulated rough Bergomi model

This subsection largely follows the presentation of [42]. Ultimately, the reason for the issues at $H = 0$ is that the Riemann–Liouville kernel $K(r) \simeq r^{H-\frac{1}{2}}$ fails to be square integrable for $H = 0$, which is also reflected by the normalization constant $\sqrt{2H}$ tending to zero as H becomes small. The main idea of [42] is to *modulate* the Riemann–Liouville kernel by a logarithmic factor $\log(1/r)^{-p}$ so as to retain square integrability even as $H = 0$. To otherwise stay close to the original kernel, and also to avoid the singularity of the modulation at $r = 1$, we only do so close to the origin.

Define a log-modulated kernel K for $r > 0$ by

$$K(r) := Cr^{H-\frac{1}{2}} \max\left\{\zeta \log\left(\frac{1}{r}\right), 1\right\}^{-p} = \begin{cases} Cr^{H-\frac{1}{2}}\zeta^{-p}\log\left(\frac{1}{r}\right)^{-p}, & 0 \le r \le \chi, \\ Cr^{H-\frac{1}{2}}, & r > \chi, \end{cases} \tag{8.35}$$

where we introduce the shorthand notation $\chi := \mathrm{e}^{-1/\zeta}$. The constants are chosen to satisfy

$0 \le H < \frac{1}{2}$, $\zeta > 0$, and $p > 1$. The constant $C = C_{H,p,\zeta}$ is chosen such that the *log-modulated fBm*

$$\widehat{W}_t := \int_0^t K(t-s)\mathrm{d}W_s, \quad t \ge 0, \tag{8.36}$$

satisfies $\mathbb{V}\left[\widehat{W}_1\right] = 1$. This implies that

$$C_{H,\zeta,p} := \begin{cases} \left(\frac{1}{\zeta}\mathrm{E}_{2p}\left(\frac{2H}{\zeta}\right) + \frac{1-\chi^{2H}}{2H}\right)^{-\frac{1}{2}}, & H > 0, \\ \left(\frac{2p}{(2p-1)\zeta}\right)^{-\frac{1}{2}}, & H = 0, \end{cases} \tag{8.37}$$

in terms of the *exponential integral*

$$\mathrm{E}_b(x) := \int_1^\infty \mathrm{e}^{-xt}t^{-b}\mathrm{d}t, \quad x \ge 0, \quad b > 1.$$

For any $H \ge 0$, $\widehat{W}$ is a Gaussian process with continuous sample paths. If $H > 0$, then the trajectories of $\widehat{W}$ are a.s. α-Hölder continuous for any $0 < \alpha < H$; see [42, Corollary 3.4]. For $H = 0$, the trajectories are continuous (but not Hölder continuous), and even the modulus of continuity can be determined following [183, Section 3]. In the same paper, an LDP for a general class of Gaussian Volterra processes, including the log-modulated fBm, is derived. However, short-time asymptotics are difficult to derive from such LDPs due to the lack of scale invariance. Given the log-modulated fBm $\widehat{W}$ defined in (8.36) driven by a standard Brownian motion W, we now define an asset price process

$$\frac{\mathrm{d}S_t}{S_t} = \sigma\left(\widehat{W}_t, t\right)\left(\varrho\mathrm{d}W_t + \bar{\varrho}\mathrm{d}\overline{W}_t\right). \tag{8.38}$$

Remark 8.5.3. *In comparison to the Riemann–Liouville kernel, the log-modulated kernel* (8.35) *depends on two additional constants, ζ and p. In* [42], *the authors suggest a generic choice $\zeta = 1/10$ and $p = 2$. In principle, both parameters, but especially p, could also be calibrated from market prices, but given the general nature of the log-modulated kernel as a perturbation of the Riemann–Liouville kernel, such a calibration might be delicate.*

Example 8.18. The assumptions can also be justified in the case of the rough Bergomi model based on the log-modulated fBm [42, Section 6.1]. Indeed, even for the log-modulated kernel (8.35), we still have $\mathcal{K} = \frac{\sqrt{2H}}{H+\frac{1}{2}}$, $0 \le H < \frac{1}{2}$ (as in Example 8.12). Additionally,

$$\mathfrak{s}(t)^2 = \begin{cases} \frac{C^2\zeta^{-2p}}{2p-1}\log(1/t)^{1-2p}, & H = 0, \\ \frac{C^2\zeta^{-2p}}{2H}\log(1/t)^{-2p}t^{2H}, & H > 0, \end{cases}$$

with $C = C_{H,\zeta,p}$ as defined in (8.37). In this case, we note that the rough Bergomi model is obtained by $\sigma^2(x,t) := \xi(t)\exp\left(\eta x - \frac{1}{2}\eta^2\mathfrak{s}(t)^2\right)$. Note that $\mathcal{K} = 0$ for $H = 0$, which implies that the expansion of Theorem 8.11 is not precise enough to identify the skew, but a higher-order expansion would be needed. So, for $H = 0$, we can only state that

$$\sqrt{t}\log(1/t)^{p-\frac{1}{2}}\frac{\sigma_{\mathrm{imp}}(z\sqrt{t},\sqrt{t}) - \sigma_{\mathrm{imp}}(z'\sqrt{t},\sqrt{t})}{\sqrt{t}z - \sqrt{t}z'} \xrightarrow{t\downarrow 0} 0.$$

In the case $H > 0$, we obtain

$$\frac{\sigma_{\text{imp}}(z\sqrt{t},\sqrt{t}) - \sigma_{\text{imp}}(z'\sqrt{t},\sqrt{t})}{\sqrt{t}z - \sqrt{t}z'} \sim \frac{\varrho\eta C\zeta^{-p}}{(2H+3)(H+\frac{1}{2})} t^{H-\frac{1}{2}} \log(1/t)^{-p}$$

as $t \downarrow 0$ with $C = \left[\frac{1}{\zeta}\mathrm{E}_{2p}\left(\frac{2H}{\zeta}\right) + \frac{1-\chi^{2H}}{2H}\right]^{-\frac{1}{2}}$.

Remark 8.5.4. *Example* 8.18 *can be complemented by a small volatility of volatility analysis provided in* [42, Corollary 5.6]. *In fact, assume that the variance satisfies*

$$V_t^n = \xi(t)\mathcal{E}\left(\varepsilon_n \widehat{W}_t\right)$$

with a sequence $\varepsilon_n \downarrow 0$ of vol-of-vol parameters. Then, asymptotically as $\varepsilon_n \downarrow 0$, [42, Theorem 5.5] *shows that for any $H \geq 0$,*

$$\frac{\sigma_{\text{imp}}(k,T) - \sigma_{\text{imp}}(k',T)}{k-k'} \sim \frac{\varepsilon_n T^{-\frac{1}{2}}}{2}\left(\int_0^T \xi(s)\mathrm{d}s\right)^{-\frac{3}{2}} \int_0^T \xi(s) \int_0^s K(s-r)\sqrt{\xi(r)}\mathrm{d}r\mathrm{d}s.$$

If we now formally consider the short-time behavior of the above skew asymptotic, then we obtain

$$\frac{\sigma_{\text{imp}}(k,T) - \sigma_{\text{imp}}(k',T)}{k-k'} \sim -a\varrho \log\left(\frac{1}{T}\right)^{-p} T^{H-\frac{1}{2}}\varepsilon_n, \qquad \text{as} \quad T \downarrow 0,\ \varepsilon_n \downarrow 0,$$

with $a = \frac{C_{H,\zeta,p}\zeta^{-p}}{(H+\frac{1}{2})(2H+3)}$ and $C_{H,\zeta,p}$ given as in (8.37). *This formula is formally valid for any H, including $H = 0$, and coincides with the formula established in Example* 8.18 *for $H > 0$.*

Chapter 9

The forest expansion of forward variance models

Peter K. Friz, Jim Gatheral

In this nontechnical introduction to diamond trees and forests, we focus on their application to computation in stochastic volatility models (rough volatility models in particular) written in forward variance form.

9.1 ▪ Diamond trees and forests

From [9, 144], we have the following definition of the diamond product.

Definition 9.1. *Given two continuous semimartingales A, B with integrable covariation process $\langle A, B\rangle$, the diamond product is defined by*

$$(A \diamond B)_t(T) := \mathbb{E}\left[\langle A, B\rangle_{t,T} \middle| \mathcal{F}_t\right] = \mathbb{E}[\langle A, B\rangle_T|\mathcal{F}_t] - \langle A, B\rangle_t\ .$$

We shall assume that all martingales admit a continuous version. It is then clear that $A \diamond B$ is another continuous semimartingale (but in general, due to the covariation term $\langle A, B\rangle$, not a martingale). The diamond product has the following properties.

- It is commutative: $A \diamond B = B \diamond A$.
- It is nonassociative: $(A \diamond B) \diamond C \neq A \diamond (B \diamond C)$.
- $A \diamond B$ depends only on the respective local martingale parts of A and B.

Diamond products of (sufficiently integrable, continuous) semimartingales naturally lead to binary "diamond" trees such as $(A \diamond B) \diamond C$. Diagrammatically, the diamond product of two trees $\mathbb{T}_1$ and $\mathbb{T}_2$ is represented by *root joining*,

$$\mathbb{T}_1 \diamond \mathbb{T}_2 =$$,

where the two binary trees $\mathbb{T}_1$ and $\mathbb{T}_2$ are represented as the single leaves • and ◦. We regard linear combinations of diamond trees, as seen in (9.4), as *forests*. This tree formalism is extremely convenient when it comes to doing explicit computations as we will see later on.

The following theorem, adapted from [144], expresses the cumulant generating function (CGF) of any continuous semimartingale and its quadratic variation as a sum of forests of diamond trees.

Theorem 9.2. *Let Y_T be a real-valued, $\mathcal{F}_T$-measurable random variable with associated martingale $Y_t = \mathbb{E}_t[Y_T]$. Under natural integrability conditions, with a, b small enough, there is a.s. convergence of*

$$\log \mathbb{E}\left[\mathrm{e}^{aY_T + b\langle Y\rangle_T} \middle| \mathcal{F}_t\right] = aY_t + b\langle Y\rangle_t + \sum_{k\geq 2} \mathbb{G}_t^k(T), \tag{9.1}$$

where

$$\mathbb{G}^2 = \left(\frac{1}{2}a^2 + b\right)(Y \diamond Y)_t(T),$$

$$\mathbb{G}^k = \frac{1}{2}\sum_{j=2}^{k-2} \mathbb{G}^{k-j} \diamond \mathbb{G}^j + (a\,Y \diamond \mathbb{G}^{k-1}) \text{ for } k > 2. \tag{9.2}$$

For the multivariate case, $Y = (Y^1, \ldots, Y^d)$ and $a \in \mathbf{R}^d, b \in \mathbf{R}^{d\times d}$, we replace aY and $b\langle Y\rangle$ by $\sum_i a_i Y^i$ and $\sum_{i,j} b_{ij}\langle Y^i, Y^j\rangle$ and further have $\mathbb{G}^2 = \sum_{i,j}\left(\frac{1}{2}a_i a_j + b_{ij}\right)(Y^i \diamond Y^j)_t(T)$.

Proof of Theorem 9.2 (sketch). For a generic (continuous) semimartingale Z, sufficiently integrable, let

$$\Lambda_t^T = \log \mathbb{E}_t\left[\mathrm{e}^{Z_{t,T}}\right].$$

Then, noting that $\Lambda_T^T = 0$,

$$\mathbb{E}_t\left[\mathrm{e}^{Z_T}\right] = \mathbb{E}_t\left[\mathrm{e}^{Z_T + \Lambda_T^T}\right] = \mathrm{e}^{Z_t + \Lambda_t^T}.$$

The stochastic logarithm $\mathcal{L}(\mathbb{E}_\bullet(Z_T)) = Z + \Lambda^T + \frac{1}{2}\langle Z + \Lambda^T\rangle$ is a martingale. Thus,

$$\begin{aligned}\Lambda_t^T &= \mathbb{E}_t\left[Z_{t,T} + \tfrac{1}{2}\langle Z + \Lambda^T\rangle_{t,T}\right] \\ &= \mathbb{E}_t\left[Z_{t,T}\right] + \tfrac{1}{2}((Z + \Lambda^T) \diamond (Z + \Lambda^T))_t(T).\end{aligned} \tag{9.3}$$

Now with[51] $Z = \epsilon a Y + \epsilon^2\, b\langle Y\rangle$ we get

$$\Lambda_t^T(\epsilon) = \epsilon a\, \mathbb{E}_t\left[Y_{t,T}\right] + \epsilon^2\, b\,(Y \diamond Y)_t(T) + \tfrac{1}{2}\left(\epsilon a Y + \Lambda_t^T(\epsilon)\right)_t^{\diamond 2}(T)\,.$$

Put $\Lambda_t^T(\epsilon) = \epsilon^2 \mathbb{G}_t^2 + \epsilon^3 \mathbb{G}_t^3 + \ldots$, and match coefficients of ϵ^n to obtain the result. □

The process Λ^T is the correction required for $\mathrm{e}^{Z+\Lambda^T}$ to be a martingale, with $\Lambda_T^T = 0$. In particular, if e^Z is already a martingale, then $\Lambda^T \equiv 0$ on $[0, T]$.

Remark 9.1.1. *Note that the convergent $\mathbb{G}$-sum is exactly equal to Λ^T, which satisfies the "abstract Riccati" equation* (9.3),

$$\Lambda_t^T = \mathbb{E}_t Z_{t,T} + \tfrac{1}{2}\left((Z + \Lambda^T) \diamond (Z + \Lambda^T)\right)_t(T).$$

To make the recursion (9.2) more concrete, consider the first few forests written diagrammatically with ∘ as shorthand for Y, interpreted as single leaf:

$\mathbb{G}^2 = (\frac{1}{2}a^2 + b)$ 

$\mathbb{G}^3 = a\,(\frac{1}{2}a^2 + b)$ 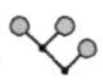

$\mathbb{G}^4 = \frac{1}{2}(\frac{1}{2}a^2 + b)^2$  $+ a^2\,(\frac{1}{2}a^2 + b)$ 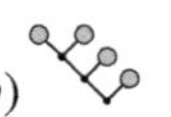

$\mathbb{G}^5 = a\,(\frac{1}{2}a^2 + b)^2$  $+ \frac{1}{2}a\,(\frac{1}{2}a^2 + b)^2$ 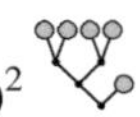 $+ a^3\,(\frac{1}{2}a^2 + b)$ 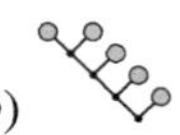 (9.4)

[51] Recall that terms of bounded variation such as $\langle Y\rangle$ do not contribute to diamond products.

We note that when $b = -\frac{1}{2}a^2$, all of the corrector terms $\mathbb{G}^k$ vanish, as they must, because in this case $e^Z = \exp\left\{aY - \frac{1}{2}a^2\langle Y\rangle\right\}$ is the exponential martingale and $\Lambda = 0$. With $a = -b/2$, the $\mathbb{G}$-recursion (9.2) becomes precisely the $\mathbb{F}$-recursion (equation (3.1) of [9] derived in the context of stochastic volatility modeling), whereas the case $b = 0$ yields the $\mathbb{K}$ (cumulant) recursion (equations (3.4) and (3.9) of [226]), derived in the context of renormalization of the sine-Gordon model in quantum physics.

In [144], a number of applications of the $\mathbb{K}$ expansion are given, including a neat derivation of the moment generating function (mgf) of the Lévy area. Other applications include the computation of likelihood functions in statistics, the computation of correlation functions in statistical physics, and the computation of amplitudes in quantum field theory.

However, in the context of (rough) stochastic volatility modeling that originally gave rise to diamond expansions, the $\mathbb{G}$-expansion and its special case, the $\mathbb{F}$-expansion (with $b = 0$), are the relevant ones.

9.2 ▪ Model-free results under stochastic volatility

Consider a stochastic volatility model written in forward variance form. Specifically, let S be a strictly positive continuous martingale (the stock price). Then $X := \log S/S_0$ is a continuous semimartingale with quadratic variation process $\langle X\rangle$, which is assumed continuously differentiable so as to have a well-defined instantaneous variance, defined via

$$V_t\,\mathrm{d}t := \mathrm{d}\langle X\rangle_t\,,$$

and then forward variance as the conditional expectation of future instantaneous variance

$$\xi_t(T) = \mathbb{E}[\,V_T|\mathcal{F}_t].$$

Forward variances are tradable assets (unlike spot variance), constituting a family of martingales indexed by their individual time horizons T.

The fair strike of a (total) variance swap maturing at time T is given by

$$M_t(T) = \int_t^T \xi_t(u)\,\mathrm{d}u$$

and the fair strike of a forward-starting variance swap, starting at time T and maturing at time $T + \Delta$ by

$$\zeta_T(T) = \int_T^{T+\Delta} \xi_T(u)\,\mathrm{d}u = \mathbb{E}_T \int_T^{T+\Delta} V_u\,\mathrm{d}u = \mathbb{E}_T\langle X\rangle_{T,T+\Delta}.$$

Its price process $\zeta_t(T) = \mathbb{E}_t\langle X\rangle_{T,T+\Delta}$ for $t \in [0,T]$ defines a martingale $\zeta = \zeta(T)$.

We now show how a straightforward application of Theorem 9.2 gives a *model-free* expression for the mgf of $\log S$, the variance swap, and the forward-starting variance swap. As for the practical interest, if $S = S_0 e^X$ represents the SPX index, and Δ is 30 days, we get the joint mgf of SPX, the variance swap, and VIX^2.

Corollary 9.3. *Under natural integrability conditions, for $a, b, c \in \mathbb{R}$ sufficiently small,*

$$\mathbb{E}\left[e^{a X_T + b\langle X\rangle_T + c\,\zeta_T(T)}\middle|\mathcal{F}_t\right] = \exp\left\{a\,X_t + b\,\langle X\rangle_t + c\,\zeta_t(T) + \sum_{k=2}^{\infty} \mathbb{G}_t^k(T; a, b, c)\right\}, \tag{9.5}$$

with $0 \le t \le T$, where the $\mathbb{G}^k$'s are given recursively by (9.2), *starting with*

$$\mathbb{G}^2 = \left(\tfrac{1}{2}a(a-1) + b\right)X \diamond X + ac\,X \diamond \zeta + \tfrac{1}{2}c^2\,\zeta \diamond \zeta\,. \tag{9.6}$$

Proof. Re-express the exponent at terminal time T in terms of the martingale $Y = X + \frac{1}{2}\langle X\rangle$,

$$a\,X_T + b\,\langle X\rangle_T + c\,\zeta_T(T) = a\,Y_T + c\,\zeta_T(T) + \left(b - \tfrac{1}{2}a\right)\langle Y\rangle_T\,.$$

This is the inner product of $(a,c) \in \mathbf{R}^2$ with the 2-dimensional martingale $(Y,\zeta) = (Y,\zeta(T)$, plus diag$(b - a/2, 0)$ contracted against its (2×2)-covariation matrix, evaluated at time T. The bivariate formulation of Theorem 9.2 now gives the right-hand side exponent

$$a\,Y_t + c\,\zeta_t(T) + \left(b - \tfrac{1}{2}a\right)\langle Y\rangle_t = a\,X_t + b\,\langle X\rangle_t + c\,\zeta_t(T)$$

plus the $\mathbb{G}$-series, started with

$$\mathbb{G}^2 = \tfrac{1}{2}a^2 Y \diamond Y + acY \diamond \zeta + \tfrac{1}{2}c^2\zeta \diamond \zeta + (b - a/2)Y \diamond Y.$$

It remains to collect terms, noting that Y can be safely replaced by X, since they only differ by a bounded variation term invisible to the diamond product. □

Remark 9.2.1. *Martingality of $S = S_0\mathrm{e}^X$ is seen in* (9.5) *upon setting $b = c = 0$ and $a = 1$. This is perfectly reflected in our $\mathbb{G}$ expansion, which in this case, and also when $a = 0$, vanishes altogether, as seen directly from* (9.6). *In this sense, martingality and total probability constraints are preserved at arbitrary truncation of the $\mathbb{G}$ expansion of* [144] *and its $\mathbb{F}$-predecessor in* [9]. *We note that this is not achieved by the (cumulant) recursion of* [226]*: applied in its bivariate form, with $X = (1, -1/2)\cdot(Y, \langle Y\rangle)$, the martingality constraint is only seen upon summing a nontrivial infinite sum, whose value turns out to be zero. (In* [144] *this is explained by forest reordering.)*

Let $X \equiv \bullet$ and $\zeta \equiv \bullet$ so that we can write the first term of the $\mathbb{G}$-series in Corollary 9.3 as

$$\mathbb{G}^2 = \left(\tfrac{1}{2}a\,(a-1) + b\right)\bullet\!\vee\!\bullet + ac\,\bullet\!\vee\!\bullet + \tfrac{1}{2}c^2\,\bullet\!\vee\!\bullet.$$

We could define $\bullet\!\vee\!\bullet =: \circ$, meaning $(X \diamond X) = M = \circ$, resulting in trees with leaves of three different colors, in which case X_t would represents the log-stock price and $M_t(T)$ the variance swap. Then

$$\mathbb{G}^2 = \left(\tfrac{1}{2}a\,(a-1) + b\right)\circ + ac\,\bullet\!\vee\!\bullet + \tfrac{1}{2}c^2\,\bullet\!\vee\!\bullet.$$

In general, we can always identify subtrees in this way and assign them a new variable name (and leaf color).

Upon setting $b = c = 0$ in Corollary 9.3, we get the $\mathbb{F}$-expansion of [144]. Upon Wick rotation, replacing a by ia, and shifting the integer index, we recover the $\tilde{\mathbb{F}}$-expansion of [9].

Corollary 9.4. *The conditional CGF is given by*

$$\psi_t(T;a) = \log \mathbb{E}_t\left[\mathrm{e}^{\mathrm{i}\,a\,X_{t,T}}\right] = -\frac{1}{2}a\,(a+\mathrm{i})\,M_t(T) + \sum_{k=1}^{\infty}\tilde{\mathbb{F}}_k(a), \tag{9.7}$$

where the $\tilde{\mathbb{F}}_k$ satisfy the recursion $\tilde{\mathbb{F}}_0 = -\frac{1}{2}a(a+\mathrm{i})\,M_t = -\frac{1}{2}a(a+\mathrm{i})\circ$, and for $k > 0$,

$$\tilde{\mathbb{F}}_k = \frac{1}{2}\sum_{j=0}^{k-2}\left(\tilde{\mathbb{F}}_{k-2-j} \diamond \tilde{\mathbb{F}}_j\right) + \mathrm{i}\,a\left(X \diamond \tilde{\mathbb{F}}_{k-1}\right). \tag{9.8}$$

Having generated a model-free expression for the CGF for any stochastic volatility model written in forward variance form, we can compute model-free expressions for attainable claims, such as variance and gamma swaps. (See, e.g., [227].)

As is well known, the fair value of the variance swap is given by the fair value of the log-strip,

$$\mathbb{E}\left[X_{t,T}\middle|\mathcal{F}_t\right] = (-\mathrm{i})\,\psi_t'(T;0) = -\tfrac{1}{2}\,M_t(T),$$

and the fair value of the gamma swap is given by the fair value of the entropy contract,

$$\mathbb{E}\left[X_{t,T}\,\mathrm{e}^{X_{t,T}}\middle|\mathcal{F}_t\right] = -\mathrm{i}\,\psi_t'(T;-\mathrm{i}).$$

From (9.7) and the recursion (9.8), it is easy to see that only trees containing a single $\circ$ leaf will survive in the sum after differentiation when $a = -\mathrm{i}$ so that

$$\sum_{k=1}^{\infty} \tilde{\mathbb{F}}_k'(-\mathrm{i}) = \tfrac{\mathrm{i}}{2} \sum_{k=1}^{\infty} X^{\diamond k} M$$

$= \frac{\mathrm{i}}{2}\Big\{$ 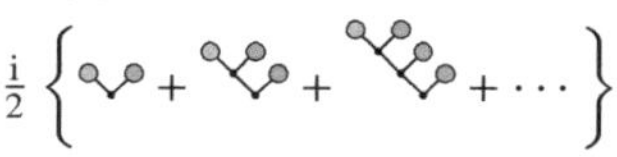 $\Big\}.$

Then the fair value of a gamma swap is given by

$$\mathcal{G}_t(T) = 2\,\mathbb{E}\left[X_{t,T}\,\mathrm{e}^{X_{t,T}}\middle|\mathcal{F}_t\right] =$$ 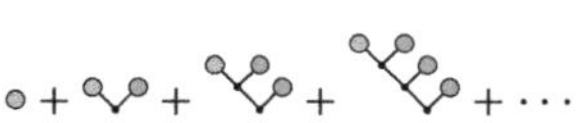 (9.9)

We deduce that the fair value of a leverage swap is given by

$$\mathcal{L}_t(T) := \mathcal{G}_t(T) - M_t(T) = \sum_{k=1}^{\infty} X^{\diamond k} M \tag{9.10}$$

$=$ 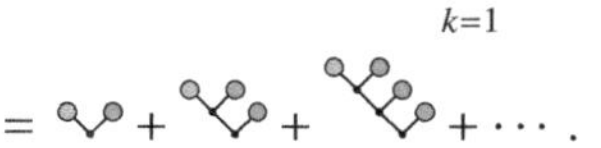

The completely model-free and explicit expression (9.10) for the leverage swap is in terms of diamond products of products of the spot and volatility processes. In particular, if spot and volatility processes are uncorrelated, 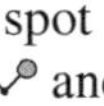 and higher-order terms vanish, and we see that the fair value of the leverage swap is zero.

In [152], Fukasawa shows that

$$\mathbb{E}_t\left[\int_t^T \mathrm{d}\langle S, M(T)\rangle_r\right] = S_t\{\mathcal{G}_t(T) - M_t(T)\}.$$

Thus, by definition of the diamond product and by the definition (9.10) of the leverage swap,

$$(S \diamond M)_t(T) = S_t\,\mathcal{L}_t(T). \tag{9.11}$$

In other words, the fair value of a leverage swap is given by the quadratic covariation of the underlying and the variance swap, for any admissible stochastic volatility model.

Comparing equation (9.10) with equation (9.11), we are led to the following identity, for which we offer a direct diamond proof.

Lemma 9.5.

$$(S \diamond M)_t(T) = \left(\mathrm{e}^X \diamond M\right)_t(T) = S_t \sum_{k=1}^{\infty} \left(X^{k\diamond} M\right)_t(T).$$

Proof. Note that

$$(S \diamond M)_t(T) = \mathbb{E}\left[\int_t^T \mathrm{d}\langle S, M\rangle_u \middle| \mathcal{F}_t\right] = \mathbb{E}\left[\int_t^T S_u \, \mathrm{d}\langle X, M\rangle_u \middle| \mathcal{F}_t\right].$$

Now define $U_t := \mathbb{E}\left[\int_t^T \mathrm{d}\langle X, M\rangle_u \middle| \mathcal{F}_t\right] = (X \diamond M)_t(T)$. Applying Itô's formula to the process $S\,U$ we get

$$\begin{aligned}
&\mathbb{E}\left[\int_t^T S_u \, \mathrm{d}\langle X, M\rangle_u \middle| \mathcal{F}_t\right] \\
&= S_t \, \mathbb{E}\left[\int_t^T \mathrm{d}\langle X, M\rangle_u \middle| \mathcal{F}_t\right] + \mathbb{E}\left[\int_t^T S_u \, \mathrm{d}\langle X, U\rangle_u \middle| \mathcal{F}_t\right] \\
&= S_t \, (X \diamond M)_t(T) + \mathbb{E}\left[\int_t^T S_u \, \mathrm{d}\langle X, (X \diamond M)\rangle_u \middle| \mathcal{F}_t\right].
\end{aligned}$$

Then a recursive argument gives us that

$$\mathbb{E}\left[\int_t^T \mathrm{d}\langle S, M\rangle_u \middle| \mathcal{F}_t\right] = S_t \sum_{k=1}^{\infty} \left(X^{k\diamond} M\right)_t(T). \qquad \square$$

What is this good for? Since the leverage swap is an attainable claim, it can be expressed as a portfolio of European options whose value is given in principle by the market—only in principle, because in practice interpolation and extrapolation are required, market prices being available only for certain discrete strikes and expirations. Since (9.10) gives the model value of the leverage swap for any given maturity T, and we can estimate the value of the leverage swap for each T from the market, models can be efficiently calibrated.

However, there is a catch. This model calibration procedure works only when we know how to compute or easily approximate diamond trees.

9.2.1 ▪ The Bergomi–Guyon smile expansion

Following [58], consider a forward variance model of the form

$$\begin{aligned}
\frac{\mathrm{d}S_t}{S_t} &= \sqrt{V_t}\, \mathrm{d}Z_t \\
\mathrm{d}\xi_t(u) &= \lambda(t, u, \xi_t)\, \mathrm{d}W_t.
\end{aligned} \tag{9.12}$$

To expand such a model for small volatility of volatility, Bergomi and Guyon scale the volatility of volatility function $\lambda(\cdot)$ by a dimensionless parameter ε so that $\lambda \mapsto \varepsilon\,\lambda$. Setting $\varepsilon = 1$ at the end then gives the required expansion.

Let $k = \log K/S$ be the log-strike and T be the option expiration. For ease of notation and without loss of generality, set $t = 0$ and drop references to the initial time. The Bergomi–Guyon smile expansion (equation (14) of [58]) then reads

$$\sigma_{\mathrm{BS}}(k, T) = \hat{\sigma}_T + \mathcal{S}_T\, k + C_T\, k^2 + O(\varepsilon^3) \tag{9.13}$$

where the coefficients $\hat{\sigma}_T$, $\mathcal{S}_T$, and C_T are expressed in terms of what Bergomi and Guyon call autocorrelation functionals. The beauty of the Bergomi–Guyon smile expansion is that it relates observed properties of the implied volatility surface such as at-the-money volatility $\hat{\sigma}_T$ and the at-the-money volatility skew $\mathcal{S}_T$ directly to quantities that are computable from the formulation of the model, for any model written in forward variance form.

On the other hand, the $\tilde{\mathbb{F}}$-expansion (9.7) for the cgf can be expressed as a formal power series in (our own parameter) ϵ whose power counts the forest index ℓ. That is, from (9.7),

$$\psi(T;a) = \log \mathbb{E}\left[e^{\mathrm{i}\,a X_T}\right] = -\frac{1}{2} a\,(a+\mathrm{i})\,M(T) + \sum_{\ell=1}^{\infty} \epsilon^\ell\, \tilde{\mathbb{F}}_\ell(a). \tag{9.14}$$

It is well known [85, 229] that the price of a European option may be computed by inverting the characteristic function of a given stochastic process. Equivalently, equation (5.7) of [163] expresses the implied total variance smile $\Sigma(k,T) := \sigma^2_{\rm BS}(k,T)\,T$ in terms of the characteristic function,

$$\int_0^\infty \frac{\mathrm{d}u}{u^2+\frac{1}{4}}\,\mathrm{Re}\left[e^{-iuk}\left(e^{\psi(T;u-i/2)} - e^{-\frac{1}{2}\left(u^2+\frac{1}{4}\right)\Sigma(k,T)}\right)\right] = 0. \tag{9.15}$$

Dropping explicit references to T, let

$$\Sigma(k) = \sum_{\ell=0}^{\infty} \epsilon^\ell\, a_\ell(k).$$

Then, substituting from (9.14), (9.15) becomes

$$\int_0^\infty \frac{\mathrm{d}u}{u^2+\frac{1}{4}}\,\mathrm{Re}\left[e^{-iuk}\left\{e^{-\frac{1}{2}\left(u^2+\frac{1}{4}\right)M(T)+\sum_{\ell=1}^{\infty}\epsilon^\ell\,\tilde{\mathbb{F}}_\ell(u-i/2)} - e^{-\frac{1}{2}\left(u^2+\frac{1}{4}\right)\sum_{\ell=0}^{\infty}\epsilon^\ell\,a_\ell(k)}\right\}\right] = 0.$$

Setting $\epsilon = 0$ gives $a_0(k) = M(T)$, as expected. Then matching powers of ϵ to first order gives

$$a_1(k) = \left(\frac{k}{M} + \frac{1}{2}\right)(X \diamond M), \tag{9.16}$$

where we have put $M = M(T)$ for short. In Appendix A of [9], an explicit algorithm using Hermite and Bell polynomials that matches coefficients of ϵ to any desired order is given. This algorithm gives the second-order coefficient as

$$\begin{aligned} a_2(k) = {} & \frac{1}{4}(X \diamond M)^2 \left\{-\frac{5k^2}{M^3} - \frac{2k}{M^2} + \frac{3}{M^2} + \frac{1}{4M}\right\} \\ & + \frac{1}{4}(M \diamond M)\left\{\frac{k^2}{M^2} - \frac{1}{M} - \frac{1}{4}\right\} + (X \diamond (X \diamond M))\left\{\frac{k^2}{M^2} + \frac{k}{M} - \frac{1}{M} + \frac{1}{4}\right\}. \end{aligned}$$

It is straightforward to verify that the resulting expansion $\Sigma(k) = M + \epsilon\,a_1(k) + \epsilon^2\,a_2(k) + O(\epsilon^3)$ agrees with the Bergomi–Guyon smile expansion to second order in ϵ, up to the identification $\epsilon \rightsquigarrow \varepsilon$. In this sense, the $\tilde{\mathbb{F}}$-expansion allows us to extend the Bergomi–Guyon expansion to all orders.

9.3 ▪ Affine forward variance models

Diamond trees turn out to be particularly easy to compute in affine forward variance (AFV) models. As shown in Chapter 5, affine forward variance (AFV) models must take the form

$$\begin{aligned} \frac{\mathrm{d}S_t}{S_t} &= \sqrt{V_t}\,\mathrm{d}Z_t, \\ \mathrm{d}\xi_t(u) &= \kappa(u-t)\,\sqrt{V_t}\,\mathrm{d}W_t, \qquad t \le u, \end{aligned} \tag{9.17}$$

with $\mathrm{d}\langle W, Z\rangle_t = \varrho\,\mathrm{d}t$. AFV models are therefore essentially all Heston models with different choices of the kernel function κ. For example, the classical Heston model corresponds to the choice

$$\kappa(\tau) = \nu\, e^{-\lambda\tau},$$

and the rough Heston model of [129] to the choice

$$\kappa(\tau) = \nu\, \tau^{\alpha-1}\, E_{\alpha,\alpha}(-\lambda\,\tau^{\alpha}),$$

with $\nu, \lambda > 0$ and $\alpha = H + \frac{1}{2} \in \left(\frac{1}{2}, 1\right]$. Integrating the forward variance equation gives

$$\mathrm{d}\zeta_t(T) = \left(\int_T^{T+\Delta} \kappa(u-t)\,\mathrm{d}u\right) \sqrt{V_t}\,\mathrm{d}W_t =: \bar{\kappa}(T-t)\,\sqrt{V_t}\,\mathrm{d}W_t \tag{9.18}$$

which has the same form as (9.17).

Lemma 9.6. *In the AFV model (9.17) all diamond trees (with leaves of two types $X = \bullet$ and $\zeta = \bullet$, respectively), and hence all forests terms $\mathbb{G}_t^k$ in (9.5), are of the form*

$$\int_t^T \xi_t(u)\, h(T-u)\,\mathrm{d}u \tag{9.19}$$

for some integrable function h.

Proof. As we computed earlier, $\bullet$ =  = $\int_t^T \xi_t(s)\,\mathrm{d}s$, and from (9.18),

 $= \varrho \int_t^T \xi_t(u)\,\bar{\kappa}(T-u)\,\mathrm{d}u\,,$  $= \int_t^T \xi_t(u)\,\bar{\kappa}(T-u)^2\,\mathrm{d}u.$ (9.20)

We thus see that the claim holds for all diamond trees with two leaves and proceed by induction. Consider two trees

$$\mathbb{T}_t^i = \int_t^T \xi_t(u)\, h^i(T-u)\,\mathrm{d}u, \qquad i = 1, 2,$$

of the supposed form. Then

$$\begin{aligned}
\left(\mathbb{T}^1 \diamond \mathbb{T}^2\right)_t(T) &= \mathbb{E}\left[\int_t^T \mathrm{d}\langle \mathbb{T}^1, \mathbb{T}^2\rangle_u \middle| \mathcal{F}_t\right] \\
&= \mathbb{E}\left[\int_t^T \int_u^T \int_u^T h^1(T-s)\, h^2(T-r)\,\mathrm{d}s\,\mathrm{d}r\,\mathrm{d}\langle \xi(s), \xi(r)\rangle_u \middle| \mathcal{F}_t\right] \\
&= \mathbb{E}\left[\int_t^T V_u\, \kappa(s-u)\,\kappa(r-u)\,\mathrm{d}u \int_u^T h^1(T-s)\,\mathrm{d}s \int_u^T h^2(T-r)\,\mathrm{d}r \middle| \mathcal{F}_t\right] \\
&= \int_t^T \xi_t(u)\, h^{12}(T-u)\,\mathrm{d}u,
\end{aligned}$$

and the induction step is completed upon setting

$$h^{12}(T-u) = \int_u^T h^1(T-s)\,\kappa(T-s)\,\mathrm{d}s \int_u^T h^2(T-r)\,\kappa(T-r)\,\mathrm{d}r. \qquad \square$$

At this stage it is tempting to combine Lemma 9.6 with Theorem 9.3 to compute the triple-joint mgf of X_T, $\langle X\rangle_{t,T}$, and $\zeta_T(T)$ by summing the full $\mathbb{G}$-expansion for an AFV model. Since each tree in the $\mathbb{G}$-expansion has the form $\int_t^T \xi_t(u)\, h(T-u)\, du$, it follows that the mgf is necessarily of the convolutional form

$$\log \mathbb{E}\left[\mathrm{e}^{a X_T + b\langle X\rangle_T + c\,\zeta_T(T)} \middle| \mathcal{F}_t\right] = a\, X_t + b\,\langle X\rangle_t + c\,\zeta_t(T) + \int_t^T \xi_t(u)\, g(T-u; a,b,c,\Delta)\, du,$$

which amounts to an infinite-dimensional version of the classical affine ansatz. Inserting $\Lambda_t(T) = \int_t^T \xi_t(u)\, g(T-u; a,b,c,\Delta)\, du$ directly into the "abstract Riccati" equation (9.3), we readily obtain that the triple-joint MGF satisfies a convolution Riccati equation of the type considered in [4, 168]. We summarize this in the following theorem.

Theorem 9.7. *Let*

$$\begin{aligned} \mathrm{d}X_t &= -\tfrac{1}{2}\, V_t\, \mathrm{d}t + \sqrt{V_t}\, \mathrm{d}Z_t, \\ \mathrm{d}\xi_t(T) &= \kappa(T-t)\, \sqrt{V_t}\, \mathrm{d}W_t, \end{aligned}$$

with $\mathrm{d}\langle W, Z\rangle_t = \varrho\, \mathrm{d}t$, *and let* $\langle X\rangle_{t,T} = \langle X\rangle_T - \langle X\rangle_t$. *Further let* $\tau = T - t$, $\bar\kappa(\tau) = \int_\tau^{\tau+\Delta} \kappa(u)\, \mathrm{d}u$, *and define the convolution integral*

$$(\kappa \star g)(\tau) = \int_0^\tau \kappa(\tau - s)\, g(s)\, \mathrm{d}s.$$

Then

$$\mathbb{E}\left[\mathrm{e}^{a X_T + b\langle X\rangle_{t,T} + c\,\zeta_T(T)} \middle| \mathcal{F}_t\right] = \exp\left\{a\, X_t + c\,\zeta_t(T) + (\xi \star g)(T-t; a,b,c,\Delta)\right\}$$

where $g(\tau; a,b,c,\Delta)$ *satisfies the convolution Riccati integral equation*

$$g(\tau; a,b,c,\Delta) = b - \tfrac{1}{2}a + \tfrac{1}{2}(1-\varrho^2)\, a^2 + \tfrac{1}{2}\left[\varrho a + c\,\bar\kappa(\tau) + (\kappa \star g)(\tau; a,b,c,\Delta)\right]^2, \tag{9.21}$$

with the boundary condition $g(0; a,b,c,\Delta) = b + \frac{1}{2}a(a-1) + \varrho a c\, \bar\kappa(0) + \frac{1}{2}c^2\, \bar\kappa(0)^2$.

Proof. From (9.18), $d\zeta_t(T) = \sqrt{V_t}\, \mathrm{d}W_t\, \bar\kappa(T-t)$. As before, let $\Lambda_t = \int_t^T \xi_t(u) g(T-u; a,b,c,\Delta) du$. Then dropping the arguments a, b, c, Δ for ease of notation,

$$\begin{aligned} d\Lambda_t &= -\xi_t(t)\, g(T-t)\, \mathrm{d}t + \int_t^T \mathrm{d}\xi_t(s)\, g(T-s)\, \mathrm{d}s \\ &= -V_t\, g(T-t)\, \mathrm{d}t + \sqrt{V_t}\, \mathrm{d}W_t \int_t^T \kappa(s-t)\, g(T-s)\, \mathrm{d}s \\ &= -V_t\, g(T-t)\, \mathrm{d}t + \sqrt{V_t}\, \mathrm{d}W_t\, (\kappa \star g)(T-t). \end{aligned}$$

We compute

$$\begin{aligned} \mathrm{d}\langle X\rangle_t &= V_t\, \mathrm{d}t, \\ \mathrm{d}\langle X, \zeta\rangle_t &= \varrho\, V_t\, \bar\kappa(T-t)\, \mathrm{d}t, \\ \mathrm{d}\langle \zeta\rangle_t &= V_t\, \bar\kappa(T-t)^2\, \mathrm{d}t, \\ \mathrm{d}\langle X, \Lambda\rangle_t &= \varrho\, V_t\, (\kappa \star g)(T-t)\, dt, \\ \mathrm{d}\langle \Lambda\rangle_t &= V_t\, \left[(\kappa \star g)(T-t)\right]^2 \mathrm{d}t, \\ \mathrm{d}\langle \zeta, \Lambda\rangle_t &= V_t\, \bar\kappa(T-t)\, (\kappa \star g)(T-t)\, \mathrm{d}t. \end{aligned}$$

Integrating these terms from t to T and then taking a time t conditional expectation allows us to compute all diamond products in the "abstract Riccati" equation (9.3)

$$\Lambda = \left(\tfrac{1}{2}a(a-1) + b\right) {}^{\circ}\!\!\vee\!\!{}^{\circ} + ac\, {}^{\circ}\!\!\vee\!\!{}^{\bullet} + a \circ \diamond \Lambda + \tfrac{1}{2}\left[c\, \bullet + \Lambda\right]^{\diamond 2}$$

to yield

$$g(\tau) = \left(\tfrac{1}{2}a(a-1)+b\right) + \varrho a c\,\bar{\kappa}(\tau) + a\,\varrho\,(\kappa \star g)(\tau) + \tfrac{1}{2}\left[c\,\bar{\kappa}(\tau) + (\kappa \star g)(\tau)\right]^2,$$

which upon rearrangement gives (9.21). Finally, $(\kappa \star g)(0) = 0$ gives the boundary condition

$$g(0) = b + \tfrac{1}{2}\,a(a-1) + \varrho a c\,\bar{\kappa}(0) + \tfrac{1}{2}\,c^2\,\bar{\kappa}(0)^2. \qquad \square$$

9.4 ▪ Explicit computations under rough Heston

Explicit computations are easiest in the rough Heston model with $\lambda = 0$. In this case, with $\alpha = H + 1/2 \in (1/2, 1]$,

$$\begin{aligned} \mathrm{d}X_t &= -\tfrac{1}{2}V_t\,\mathrm{d}t + \sqrt{V_t}\,\mathrm{d}Z_t,\\ \mathrm{d}\xi_t(u) &= \frac{\nu}{\Gamma(\alpha)}\,(u-t)^{\alpha-1}\,\sqrt{V_t}\,\mathrm{d}W_t, \end{aligned}$$

with $\mathrm{d}\langle W, Z\rangle = \varrho\,\mathrm{d}t$. Starting with the simplest tree, the fair strike of the (total) variance swap is given by

$$M_t(T) = \circ = \text{[tree]} = (X \diamond X)_t(T) = \int_t^T \xi_t(u)\,\mathrm{d}u,$$

as we noted earlier. Thus, abbreviating bounded variation terms as "BV," we have

$$\begin{aligned} \mathrm{d}X_t &= \sqrt{V_t}\,\mathrm{d}Z_t + \mathrm{BV},\\ \mathrm{d}M_t &= \int_t^T \mathrm{d}\xi_t(u)\,\mathrm{d}u + \mathrm{BV}\\ &= \frac{\nu}{\Gamma(\alpha)}\,\sqrt{V_t}\left(\int_t^T \frac{\mathrm{d}u}{(u-t)^\gamma}\right)\mathrm{d}W_t + \mathrm{BV}\\ &= \frac{\nu\,(T-t)^\alpha}{\Gamma(1+\alpha)}\,\sqrt{V_t}\,\mathrm{d}W_t + \mathrm{BV}\,. \end{aligned}$$

We now proceed to compute the first few $\tilde{\mathbb{F}}$ forests explicitly. There is only one tree in the forest $\tilde{\mathbb{F}}_1$.

$$\begin{aligned} \text{[tree]} = (X \diamond M)_t(T) &= \mathbb{E}\left[\int_t^T \mathrm{d}\langle X, M\rangle_s \,\middle|\, \mathcal{F}_t\right]\\ &= \frac{\varrho\,\nu}{\Gamma(1+\alpha)}\,\mathbb{E}\left[\int_t^T v_s\,(T-s)^\alpha\,\mathrm{d}s \,\middle|\, \mathcal{F}_t\right]\\ &= \frac{\varrho\,\nu}{\Gamma(1+\alpha)}\int_t^T \xi_t(s)\,(T-s)^\alpha\,\mathrm{d}s. \end{aligned}$$

In order to continue to higher orders, it makes sense to define for $j \geq 0$

$$I_t^{(j)}(T) := \int_t^T \mathrm{d}s\,\xi_t(s)\,(T-s)^{j\alpha}.$$

Then

$$\begin{aligned} dI_s^{(j)}(T) &= \int_s^T \mathrm{d}u\,\mathrm{d}\xi_s(u)\,(T-u)^{j\alpha} + \mathrm{BV}\\ &= \frac{\nu\,\sqrt{v_s}}{\Gamma(\alpha)}\,\mathrm{d}W_s \int_s^T \frac{(T-u)^{j\alpha}}{(u-s)^\gamma}\,\mathrm{d}u + \mathrm{BV}\\ &= \frac{\Gamma(1+j\alpha)}{\Gamma(1+(j+1)\,\alpha)}\,\nu\,\sqrt{v_s}\,(T-s)^{(j+1)\alpha}\,\mathrm{d}W_s + \mathrm{BV}. \end{aligned}$$

With this notation,

$$(X \diamond M)_t(T) = \frac{\varrho\,\nu}{\Gamma(1+\alpha)}\, I_t^{(1)}(T).$$

There are two trees in $\tilde{\mathbb{F}}_2$:

$$= (M \diamond M)_t(T) = \mathbb{E}\left[\int_t^T \mathrm{d}\langle M, M\rangle_s \middle| \mathcal{F}_t\right] = \frac{\nu^2}{\Gamma(1+\alpha)^2}\int_t^T \xi_t(s)\,(T-s)^{2\alpha}\,\mathrm{d}s = \frac{\nu^2}{\Gamma(1+\alpha)^2}\, I_t^{(2)}(T)$$

and

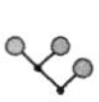

$$= (X \diamond (X \diamond M))_t(T) = \frac{\varrho\,\nu}{\Gamma(1+\alpha)}\,\mathbb{E}\left[\int_t^T \mathrm{d}\langle X, I^{(1)}\rangle_s \middle| \mathcal{F}_t\right] = \frac{\varrho^2\,\nu^2}{\Gamma(1+2\,\alpha)}\, I_t^{(2)}(T).$$

Note in passing that  gives the variance of the variance swap. Continuing to the forest $\tilde{\mathbb{F}}_3$, we have the following:

$$= (M \diamond (X \diamond M))_t(T) = \frac{\varrho\,\nu^3}{\Gamma(1+\alpha)\,\Gamma(1+2\,\alpha)}\, I_t^{(3)}(T),$$

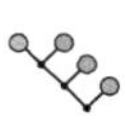

$$= (X \diamond (X \diamond (X \diamond M)))_t(T) = \frac{\varrho^3\,\nu^3}{\Gamma(1+3\,\alpha)}\, I_t^{(3)}(T),$$

$$= (X \diamond (M \diamond M))_t(T) = \frac{\varrho\,\nu^3\,\Gamma(1+2\,\alpha)}{\Gamma(1+\alpha)^2\,\Gamma(1+3\,\alpha)}\, I_t^{(3)}(T).$$

In particular, we readily identify the pattern

$$\left(X^{\diamond k} M\right)_t(T) = \frac{(\varrho\,\nu)^k}{\Gamma(1+k\,\alpha)}\, I_t^{(k)}(T), \tag{9.22}$$

which gives us a simple closed-form expression for a tree with k ● leaves and only one ● leaf.

9.4.1 ▪ Time scaling of the $\tilde{\mathbb{F}}$-expansion

By inspection, fixing model parameters, we see that $\tilde{\mathbb{F}}_k$ scales as

$$I_t^{(k)}(T) = \int_t^T \mathrm{d}s\,\xi_t(s)\,(T-s)^{k\alpha} \sim (T-t)^{k\alpha+1} \quad \text{as } T \downarrow t.$$

Thus, the $\tilde{\mathbb{F}}$-expansion of Corollary 9.4 has the interpretation of a short-time expansion, the concrete powers of which depend on the roughness parameter $\alpha = H + 1/2 \in (1/2, 1)$. The resulting diamond expansions (which can obtained by alternative methods in the rough Heston case) have been applied to construct efficient numerical schemes for the computation of the rough Heston characteristic function [82, 170].

9.4.2 ▪ The leverage swap

Recall that equation (9.10) gives a model-free expression for the fair value of the leverage swap in terms of trees with only one $\circ$ leaf. Substituting from (9.22), we obtain

$$\begin{aligned}\mathcal{L}_t(T) &= \sum_{k=1}^{\infty} \left(X^{\diamond k} M\right)_t(T) \\ &= \sum_{k=1}^{\infty} \frac{(\varrho\,\nu)^k}{\Gamma(1+k\,\alpha)} \int_t^T \mathrm{d}u\, \xi_t(u)\,(T-u)^{k\,\alpha} \\ &= \int_t^T \mathrm{d}u\, \xi_t(u)\,\{E_\alpha(\varrho\,\nu\,(T-u)^\alpha)-1\}\end{aligned}$$

where $E_\alpha(\cdot)$ denotes the Mittag–Leffler function. We have thus obtained an explicit expression for the leverage swap under rough Heston with $\lambda = 0$. As discussed earlier in Section 9.2, since we can impute the leverage swap $\mathcal{L}_t(t)$ from the smile for each expiration T, fast calibration of the rough Heston model is then possible.

9.4.3 ▪ The rough Heston implied volatility skew

As in Section 9.2.1, fix $t = 0$ and define $\Sigma(k,T) = \sigma_{\mathrm{BS}}^2(k,T)\,T$. Then, from (9.16), in a general stochastic volatility model, the short-dated ($T \downarrow 0$) total variance skew is given by

$$\partial_k \Sigma(k,T)|_{k=0} = \frac{X \diamond M}{M}.$$

Recall that under rough Heston, $M_t(T) = \circ =$  $= (X \diamond X)_t(T) = \int_t^T \xi_t(u)\,\mathrm{d}u$, and

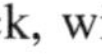 $$= (X \diamond M)_t(T) = \frac{\varrho\,\nu}{\Gamma(H+3/2)} \int_t^T \xi_t(s)\,(T-s)^{H+1/2}\,\mathrm{d}s.$$

Setting $t = 0$ in these expressions, with $\xi_0(s) \to \xi_0(0) = V_0$, time-0 spot variance as $s \downarrow 0$, an easy computation gives

$$\left.\frac{\partial}{\partial k}\right|_{k=0} \sigma_{\mathrm{BS}}^2(k,T) \sim \frac{\varrho\,\nu}{(H+3/2)\,\Gamma(H+3/2)} T^{H-1/2} = \frac{\varrho\,\nu}{\Gamma(H+5/2)} T^{H-1/2}.$$

As a sanity check, with $H = 1/2$, we recover the classical Heston model with no mean reversion, and noting that $\Gamma(3) = 2! = 2$, the well-known short-dated Heston implied variance skew formula; cf. [163, page 35]. For $H < 1/2$ we see skew explosion of order $T^{H-1/2}$, in complete agreement with the results of Chapter 4 and Chapter 8.

9.5 ▪ Explicit computations under rough Bergomi

The skeptical reader might wonder whether explicit diamond tree computations are only possible in AFV models. We now show that explicit tree computations are also possible in the rough Bergomi model.

From [39] (reprinted as Chapter 2 in this book), the rough Bergomi model may be written as

$$\begin{aligned}\frac{\mathrm{d}S_t}{S_t} &= \sqrt{V_t}\left\{\varrho\,\mathrm{d}W_t + \sqrt{1-\varrho^2}\,\mathrm{d}W_t^{\perp}\right\}, \\ V_u &= \xi_t(u)\,\mathcal{E}\left(\eta\,\sqrt{2H}\int_t^u \frac{\mathrm{d}W_s}{(u-s)^\gamma}\right)\end{aligned}$$

with $\gamma = \frac{1}{2} - H$. In forward variance form,

$$\frac{\mathrm{d}_t \xi_t(u)}{\xi_t(u)} = \tilde{\eta} \, \frac{\mathrm{d}W_t}{(u-t)^\gamma}$$

with $\tilde{\eta} := \eta \, \sqrt{2H}$. Then, again abbreviating bounded variation terms as "BV," we have

$$\begin{aligned} \mathrm{d}X_t &= \sqrt{V_t} \, \mathrm{d}Z_t + \mathrm{BV}, \\ \mathrm{d}M_t &= \tilde{\eta} \left(\int_t^T \xi_t(u) \, \frac{\mathrm{d}u}{(u-t)^\gamma} \right) \mathrm{d}W_t + \mathrm{BV}\,. \end{aligned}$$

As in the rough Heston case, we now proceed to compute the first few $\tilde{\mathbb{F}}$ forests explicitly.

9.5.1 ▪ The first-order forest $\tilde{\mathbb{F}}_1$

There is only one tree in the forest $\tilde{\mathbb{F}}_1$.

$$\begin{aligned} &= (X \diamond M)_t(T) = \mathbb{E}\left[\int_t^T \mathrm{d}\langle X, M\rangle_s \middle| \mathcal{F}_t \right] \\ &= \varrho \, \tilde{\eta} \, \mathbb{E}\left[\int_t^T \mathrm{d}s \, \sqrt{v_s} \int_s^T \xi_s(u) \, \frac{\mathrm{d}u}{(u-s)^\gamma} \middle| \mathcal{F}_t \right] \\ &= \varrho \, \tilde{\eta} \int_t^T \mathrm{d}s \int_s^T \mathbb{E}\left[\sqrt{\xi_s(s)} \, \xi_s(u) \middle| \mathcal{F}_t \right] \frac{\mathrm{d}u}{(u-s)^\gamma}. \end{aligned}$$

Applying Lemma 9.9, we have

$$\begin{aligned} &\mathbb{E}\left[\sqrt{\xi_s(s)} \, \xi_s(u) \middle| \mathcal{F}_t \right] \\ &\quad = \sqrt{\xi_t(s)} \, \xi_t(u) \, \exp\left\{ \frac{\eta^2}{2} (s-t)^{2H} \, G_\gamma\left(\frac{u-t}{s-t} \right) \right\} \exp\left\{ -\frac{\eta^2}{8} (s-t)^{2H} \right\}, \end{aligned}$$

with $G_\gamma(\cdot)$ as defined in Appendix 9.6.1. Thus

$$\begin{aligned} &= (X \diamond M)_t(T) \\ &= \varrho \, \tilde{\eta} \int_t^T \mathrm{d}s \, \sqrt{\xi_t(s)} \int_s^T \frac{\mathrm{d}u}{(u-s)^\gamma} \, \xi_t(u) \, \exp\left\{ \frac{\eta^2}{2} (s-t)^{2H} \left[G_\gamma\left(\frac{u-t}{s-t} \right) - \frac{1}{4} \right] \right\}. \end{aligned} \tag{9.23}$$

9.5.2 ▪ The second-order forest $\tilde{\mathbb{F}}_2$

There are two trees in $\tilde{\mathbb{F}}_2$, $= (M \diamond M)$ and 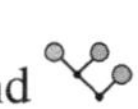 $= (X \diamond (X \diamond M))$. First, we compute

$$\begin{aligned} &= (M \diamond M)_t(T) = \mathbb{E}\left[\int_t^T \mathrm{d}\langle M, M\rangle_s \middle| \mathcal{F}_t \right] \\ &= \tilde{\eta}^2 \, \mathbb{E}\left[\int_t^T \mathrm{d}s \int_s^T \xi_s(r) \, \frac{\mathrm{d}r}{(r-s)^\gamma} \int_s^T \xi_s(u) \, \frac{\mathrm{d}u}{(u-s)^\gamma} \middle| \mathcal{F}_t \right] \\ &= 2 \, \tilde{\eta}^2 \, \mathbb{E}\left[\int_t^T \mathrm{d}s \int_s^T \xi_s(r) \, \frac{\mathrm{d}r}{(r-s)^\gamma} \int_r^T \xi_s(u) \, \frac{\mathrm{d}u}{(u-s)^\gamma} \middle| \mathcal{F}_t \right]. \end{aligned}$$

Another application of Lemma 9.9 gives, for $u \geq r$,

$$\begin{aligned}
&\mathbb{E}\left[\xi_s(r)\,\xi_s(u)|\mathcal{F}_t\right]\\
&\quad = \xi_t(r)\,\xi_t(u)\,\exp\left\{\eta^2\,(s-t)^{2H}\,G_\gamma\left(\frac{u-t}{s-t},\frac{r-t}{s-t}\right)\right\}.
\end{aligned}$$

Thus

$$\begin{aligned}
&= (M \diamond M)_t(T)\\
&= 2\,\tilde\eta^2 \int_t^T \mathrm{d}s \int_s^T \xi_t(r)\,\frac{\mathrm{d}r}{(r-s)^\gamma} \int_r^T \xi_t(u)\,\frac{\mathrm{d}u}{(u-s)^\gamma}\,\exp\left\{\eta^2\,(s-t)^{2H}\,G_\gamma\left(\frac{u-t}{s-t},\frac{r-t}{s-t}\right)\right\}.
\end{aligned} \tag{9.24}$$

Reflecting the fact that  is the variance of $\circ$, this integral may be simplified as

$$\begin{aligned}
&= (M \diamond M)_t(T)\\
&= 2\int_t^T \xi_t(u)\,\mathrm{d}u \int_t^u \xi_t(r)\,\mathrm{d}r\left[\exp\left\{\eta^2\,(r-t)^{2H}\,G_\gamma\left(\frac{u-t}{r-t}\right)\right\}-1\right].
\end{aligned} \tag{9.25}$$

Next, we compute

$$= (X \diamond (X \diamond M))_t(T) = \mathbb{E}\left[\int_t^T \mathrm{d}\langle X, (X \diamond M)\rangle_s \middle| \mathcal{F}_t\right].$$

To compute this, we need that

$$\mathrm{d}(X \diamond M)_s = \varrho\,\tilde\eta\,\{\mathrm{d}I_s + \mathrm{d}J_s\}$$

where

$$\begin{aligned}
\mathrm{d}I_s &= \frac{1}{2}\int_s^T \mathrm{d}r\,\frac{d\xi_s(r)}{\sqrt{\xi_s(r)}}\int_r^T \frac{\mathrm{d}u}{(u-r)^\gamma}\,\xi_s(u)\,\exp\left\{\frac{\eta^2}{2}\,(r-s)^{2H}\left[G_\gamma\left(\frac{u-s}{r-s}\right)-\frac{1}{4}\right]\right\}\\
&= \mathrm{d}W_s\,\frac{\tilde\eta}{2}\int_s^T \frac{\mathrm{d}r}{(r-s)^\gamma}\,\sqrt{\xi_s(r)}\int_r^T \frac{\mathrm{d}u}{(u-r)^\gamma}\,\xi_s(u)\,\exp\left\{\frac{\eta^2}{2}\,(r-s)^{2H}\left[G_\gamma\left(\frac{u-s}{r-s}\right)-\frac{1}{4}\right]\right\}
\end{aligned}$$

and

$$\begin{aligned}
\mathrm{d}J_s &= \int_s^T \mathrm{d}r\,\sqrt{\xi_s(r)}\int_r^T \frac{\mathrm{d}u}{(u-r)^\gamma}\,\mathrm{d}\xi_s(u)\,\exp\left\{\frac{\eta^2}{2}\,(r-s)^{2H}\left[G_\gamma\left(\frac{u-s}{r-s}\right)-\frac{1}{4}\right]\right\}\\
&= \mathrm{d}W_s\,\tilde\eta\int_s^T \mathrm{d}r\,\sqrt{\xi_s(r)}\int_r^T \xi_s(u)\,\frac{\mathrm{d}u}{(u-s)^\gamma\,(u-r)^\gamma}\,\exp\left\{\frac{\eta^2}{2}\,(r-s)^{2H}\left[G_\gamma\left(\frac{u-s}{r-s}\right)-\frac{1}{4}\right]\right\}.
\end{aligned}$$

Also from Lemma 9.9 again,

$$\begin{aligned}
&\mathbb{E}\left[\sqrt{\xi_s(s)}\,\sqrt{\xi_s(r)}\,\xi_s(u)\middle|\mathcal{F}_t\right]\\
&\quad = \sqrt{\xi_t(s)}\,\sqrt{\xi_t(r)}\,\xi_t(u)\\
&\qquad\quad \times\exp\left\{\eta^2\,(s-t)^{2H}\left[\frac{1}{4}G_\gamma\left(\frac{r-t}{s-t}\right)+\frac{1}{2}G_\gamma\left(\frac{u-t}{s-t}\right)+\frac{1}{2}G_\gamma\left(\frac{u-t}{s-t},\frac{r-t}{s-t}\right)\right]\right\}\\
&\qquad\quad \times\exp\left\{-\frac{1}{8}\,\eta^2\left[(s-t)^{2H}+(r-t)^{2H}-(r-s)^{2H}\right]\right\}.
\end{aligned}$$

Define

$$\begin{aligned}
&\eta^2 F(s-t,r-t,u-t)\\
&\quad= \eta^2\,(s-t)^{2H}\left[\frac{1}{4}G_\gamma\Big(\frac{r-t}{s-t}\Big)+\frac{1}{2}G_\gamma\Big(\frac{u-t}{s-t}\Big)+\frac{1}{2}G_\gamma\Big(\frac{u-t}{s-t},\frac{r-t}{s-t}\Big)\right]\\
&\qquad\qquad+\frac{\eta^2}{2}\,(r-s)^{2H}\,G_\gamma\Big(\frac{u-s}{r-s}\Big)-\frac{1}{8}\,\eta^2\left[(s-t)^{2H}+(r-t)^{2H}\right].
\end{aligned}$$

Then

$$(X\diamond(X\diamond M))_t(T)=(\varrho\,\tilde\eta)^2\left\{\frac{1}{2}\,I_t+J_t\right\}$$

with

$$I_t=\int_t^T \mathrm{d}s\,\sqrt{\xi_t(s)}\int_s^T\frac{\mathrm{d}r}{(r-s)^\gamma}\,\sqrt{\xi_t(r)}\int_r^T\frac{\mathrm{d}u}{(u-r)^\gamma}\,\xi_t(u)\,\exp\left\{\eta^2 F(s-t,r-t,u-t)\right\}\tag{9.26}$$

and

$$J_t=\int_t^T \mathrm{d}s\,\sqrt{\xi_t(s)}\int_s^T \mathrm{d}r\,\sqrt{\xi_t(r)}\int_r^T\frac{\mathrm{d}u}{(u-s)^\gamma\,(u-r)^\gamma}\,\xi_t(u)\,\exp\left\{\eta^2 F(s-t,r-t,u-t)\right\}.\tag{9.27}$$

One might ask whether it is possible to simplify these expressions using integration by parts as we did for $M\diamond M$. Unfortunately, it seems that we cannot.

In summary, though computations of diamond trees under rough Bergomi are possible in closed form as shown above, computations get more and more complicated for higher-order forests, and numerical implementation is far from straightforward. To make further progress, smart approximations will doubtless be required.

9.6 ▪ Appendix

9.6.1 ▪ Conditional expectations of products of forward variances under rough Bergomi

We will need the following lemma which is obtained by explicit integration.

Lemma 9.8. *Let $u_j > u_i \geq s > t$ and $\gamma = \frac{1}{2} - H$. Then*

$$\int_t^s \frac{\mathrm{d}r}{(u_j-r)^\gamma\,(u_i-r)^\gamma}=\frac{1}{2H}\,(s-t)^{2H}\,G_\gamma\left(\frac{u_j-t}{s-t},\frac{u_i-t}{s-t}\right)$$

where, for $y\geq x\geq 1$,

$$\begin{aligned}
G_\gamma(y,x)&=2H\int_0^1\frac{\mathrm{d}r}{(y-r)^\gamma\,(x-r)^\gamma}\\
&=\frac{1-2\gamma}{(1-\gamma)\,(y-x)}\left\{x^{1-\gamma}y^{1-\gamma}\,{}_2F_1\Big(1,2-2\gamma;2-\gamma;\frac{y}{y-x}\Big)\right.\\
&\qquad\left.-(x-1)^{1-\gamma}(y-1)^{1-\gamma}\,{}_2F_1\Big(1,2-2\gamma;2-\gamma;\frac{y-1}{y-x}\Big)\right\}.
\end{aligned}\tag{9.28}$$

Equation (9.28) *has the special cases*

$$G_\gamma(y) := G_\gamma(y,1) = \frac{1-2\gamma}{(1-\gamma)(y-1)}\, y^{1-\gamma}\, {}_2F_1\left(1, 2-2\gamma; 2-\gamma; \frac{y}{y-1}\right)$$

with $y > 1$. *In particular,* $G_\gamma(1) = G_\gamma(1,1) = 1$.

Lemma 9.9. *Consider the rough Bergomi model*

$$V_u = \xi_u(u) = \xi_t(u)\, \mathcal{E}\left(\tilde\eta \int_t^u \frac{dW_r}{(u-r)^\gamma}\right),$$

and let $u_n > \cdots > u_i > u_{i-1} > \cdots \geq s > t$, $i = 1, \ldots, n$. *Then,*

$$\begin{aligned}
\mathbb{E}&\left[\prod_{i=1}^n \xi_s(u_i)^{\alpha_i} \middle| \mathcal{F}_t\right] \\
&= \left(\prod_{i=1}^n \xi_t(u_i)^{\alpha_i}\right) \exp\left\{\eta^2\,(s-t)^{2H} \sum_{j>i} \alpha_i\,\alpha_j\, G_\gamma\left(\frac{u_j-t}{s-t}, \frac{u_i-t}{s-t}\right)\right\} \\
&\qquad \times \exp\left\{\frac{1}{2}\eta^2 \sum_{i=1}^n \alpha_i\,(\alpha_i-1)\left[(u_i-t)^{2H} - (u_i-s)^{2H}\right]\right\}.
\end{aligned}$$

Proof. Let X be a Gaussian random variable, and let $\alpha \in \mathbb{R}$. Then

$$\mathcal{E}(X)^\alpha = \mathcal{E}(\alpha\, X)\, \exp\left\{\frac{\alpha\,(\alpha-1)}{2}\, V[X]\right\}.$$

Also, with $\gamma = \frac{1}{2} - H$, $\tilde\eta = \eta\,\sqrt{2H}$, and $u > s > t$,

$$\xi_s(u) = \xi_t(u)\, \mathcal{E}\left(\tilde\eta \int_t^s \frac{dW_r}{(u-r)^\gamma}\right),$$

so

$$\xi_s(u)^\alpha = \xi_t(u)^\alpha\, \mathcal{E}\left(\alpha\,\tilde\eta \int_t^s \frac{dW_r}{(u-r)^\gamma}\right) \exp\left\{\frac{1}{2}\eta^2\, \alpha\,(\alpha-1)\left[(u-t)^{2H} - (u-s)^{2H}\right]\right\}.$$

Also, if X_i, $i = 1, \ldots, n$, are zero mean Gaussian random variables, then

$$\prod_{i=1}^n \mathcal{E}(X_i) = \mathcal{E}\left(\sum_{i=1}^n X_i\right) \exp\left\{\sum_{j>i}^n \mathrm{Cov}[X_i X_j]\right\}.$$

Finally, assuming without loss of generality that $u_j > u_i$, we have from Lemma 9.8 that

$$\begin{aligned}
\mathbb{E}\left[\alpha_j\,\tilde\eta \int_t^s \frac{dW_r}{(u_j-r)^\gamma}\, \alpha_i\,\tilde\eta \int_t^s \frac{dW_{r'}}{(u_i-r')^\gamma} \middle| \mathcal{F}_t\right] &= \alpha_i\,\alpha_j\,\tilde\eta^2 \int_t^s \frac{dr}{(u_j-r)^\gamma\,(u_i-r)^\gamma} \\
&= \alpha_i\,\alpha_j\,\eta^2\,(s-t)^{2H}\, G_\gamma\left(\frac{u_j-t}{s-t}, \frac{u_i-t}{s-t}\right).
\end{aligned}$$

□

Chapter 10

Regularity structure for rough volatility[52]

Christian Bayer, Peter K. Friz, Paul Gassiat, Jörg Martin, Benjamin Stemper

Introduction

We are interested in stochastic volatility (SV) models given in Itô differential form

$$dS_t/S_t = \sigma_t dB_t \equiv \sqrt{V_t} dB_t. \tag{10.1}$$

Here, B is a standard Brownian motion and σ (resp., V) is the *stochastic volatility* (resp., *variance*) process. Many classical Markovian asset price models fall into this framework, including Dupire's local volatility model, and the SABR, Stein–Stein, and Heston models. In all named SV models, one has Markovian dynamics for the variance process

$$dv_t = g(V_t) dW_t + h(V_t) dt. \tag{10.2}$$

Constant correlation $\varrho := d\langle B, W\rangle_t / dt$ is incorporated by working with a 2D standard Brownian motion $(W, \overline{W})$,

$$B := \varrho W + \bar{\varrho}\overline{W}, \qquad \bar{\varrho} := \sqrt{1-\varrho^2}.$$

This chapter is concerned with an important class of non-Markovian (fractional) SV models, dubbed **rough volatility (RV) models**, in which case σ_t (equivalently, $V_t \equiv \sigma_t^2$) is modeled via a fractional Brownian motion (fBm) in the regime $H \in \left(0, \frac{1}{2}\right)$. The term "rough" stems from the fact that in such models stochastic volatility (variance) sample paths are $(H - \kappa)$-Hölder continuous, for any $\kappa > 0$, and hence "rougher" than Brownian sample paths. Note the stark contrast to the idea of "trending" fractional volatility, which amounts to taking $H > \frac{1}{2}$. The evidence for the rough regime (recent calibration suggests H as low as 0.05) is now overwhelming—under both the physical and pricing measures; see [11, 151, 153, 166, 39, 138, 241]. It should be noted, however, that these different regimes can be easily mixed, so that rough volatility governs the

[52]This chapter is derived from "A regularity structure for rough volatility," *Mathematical Finance*, 30 (2019), pp. 782–832, https://doi.org/10.1111/mafi.12233.

short-time behavior, while trending volatility affects the long-time behavior; we refer the reader to [99, 97, 13, 51] for more information on this.

In the above references on rough volatility models much attention has in fact been given to "simple" rough volatility models of the form

$$\sigma_t = f(\widehat{W}_t), \tag{10.3}$$

$$\widehat{W}_t := \int_0^t K(s,t)\mathrm{d}W_s, \tag{10.4}$$

$$K(s,t) := \sqrt{2H}|t-s|^{H-\frac{1}{2}}\mathbf{1}_{t>s}, \quad H \in \left(0, \frac{1}{2}\right). \tag{10.5}$$

(Later on, we will allow for explicit time dependence of f in order to cover the rough Bergomi model [39].) In other words, volatility is an explicit function of an fBm, with fixed Hurst parameter. More specifically, following [39], we work with the Volterra fBm, a.k.a. the Riemann–Liouville fBm, but other choices such as the Mandelbrot van Ness fBm, with suitably modified kernel K, are possible. Note that, in contrast to many classical SV models (such as Heston), the stochastic volatility is explicitly given, and no rough or stochastic differential equation needs to be solved (hence the term "simple"). Rough volatility not only provides remarkable fits to both time series and option price data but also has a market microstructure justification: starting with a Hawkes process model, Rosenbaum and coworkers [126, 129, 131] find, in a suitable scaling limit, functions $f(\cdot), v(\cdot), u(\cdot)$ such that

$$\sigma_t := f(\widehat{Z}_t) \quad \ldots \quad \text{“nonsimple rough volatility (RV)”} \tag{10.6}$$

$$Z_t = z + \int_0^t K(s,t)v(Z_s)\mathrm{d}s + \int_0^t K(s,t)u(Z_s)\mathrm{d}W_s. \tag{10.7}$$

Such stochastic Volterra dynamics provide a natural generalization of simple rough volatility. We refer to this class of models as "nonsimple": in contrast to the aforementioned simple model, (10.7) generally does not admit a closed-form solution.

10.0.1 ▪ Markovian stochastic volatility models

For comparison with rough volatility, which will be discussed in more detail, we first mention a selection of tools and methods well known for *Markovian* SV models:

- PDE methods are ubiquitous in (low-dimensional) pricing problems;
- Monte Carlo methods, noting that knowledge of strong and weak rates of convergence of time discretizations of stochastic differential equations (typically with rates $\frac{1}{2}$ and 1, respectively) is the starting point of modern multilevel Monte Carlo (MLMC) methods;
- Quasi Monte Carlo (QMC) methods are widely used; related in spirit we have the Kusuoka–Lyons–Victoir cubature approach, popularized in the form of the Ninomiya–Victoir splitting scheme, available in standard software packages;
- The Freidlin–Wentzell theory of small noise large deviations is essentially immediately applicable, as are various "strong" large deviation (a.k.a. exact asymptotic) results, used, for example, to derive the famous SABR formula.

For several reasons it can be useful to write model dynamics in *Stratonovich form*: From a PDE perspective, the operators then take a sum-of-squares form which can be exploited in many ways (think Hörmander theory, Malliavin calculus, etc.). From a numerical perspective, we note that the Kusuoka–Lyons–Victoir scheme [225, 233] also requires the full dynamics to be rewritten in Stratonovich form. In fact, viewing the Ninomiya–Victoir scheme [248] as level-5 cubature, in the sense of [233], we see that its level-3 variant is nothing but the familiar Wong–Zakai approximation for diffusions. Another financial example that requires a Stratonovich formulation comes from interest rate model validation [110], based on the Stroock–Varadhan support theorem. We further note that QMC (based on Sobol' numbers, say) works particularly well if the noise has a multiscale decomposition, as obtained by interpreting a (piecewise) linear Wong–Zakai approximation as a Haar wavelet expansion of the driving white noise. Indeed, the naturally induced order of random coefficients, in terms of their importance, leads to a lower "effective dimension" of the integration problem; see, for instance, [5].

10.0.2 ▪ Complications with rough volatility

Due to loss of Markovianity, PDE methods are not applicable and neither are (off-the-shelf) Freidlin–Wentzell large deviation estimates (but see [138]). Moreover, the variance process in rough volatility models is not a semimartingale, which complicates, the use of several established stochastic analysis tools. In particular, *rough volatility admits no Stratonovich formulation.* Closely related is the fact that, one lacks a (Wong–Zakai type of) approximation theory for rough volatility. To see this, focus on the "simple" situation, that is (10.1), (10.3), so that

$$S_t/S_0 = \mathcal{E}\Big(\int_0^{\cdot} f\big(\widehat{W}_s\big)\mathrm{d}B_s\Big)(t)\,. \tag{10.8}$$

Inside the (classical) stochastic exponential $\mathcal{E}(M)(t) = \exp(M_t - \frac{1}{2}[M]_t)$ we have the martingale term

$$\int_0^t f(\widehat{W}_s)\mathrm{d}B_s = \varrho \underbrace{\int_0^t f(\widehat{W}_s)\mathrm{d}W_s}\, + \bar{\varrho}\int_0^t f(\widehat{W}_s)\mathrm{d}\overline{W}_s. \tag{10.9}$$

In essence, the trouble is due to the underbraced, innocent looking Itô integral. Indeed, any naive attempt to put it in Stratonovich form,

$$\text{“}\int_0^t f(\widehat{W}_s)\circ \mathrm{d}W_s := \int_0^t f(\widehat{W}_s)\mathrm{d}W_s + (\text{Itô–Stratonovich correction}),\text{”} \tag{10.10}$$

or, in the spirit of Wong–Zakai approximations,

$$\text{“}\int_0^t f(\widehat{W}_s)\circ \mathrm{d}W_s := \lim_{\varepsilon\to 0}\int_0^t f(\widehat{W}^\varepsilon_s)\mathrm{d}W^\varepsilon_s\text{,”} \tag{10.11}$$

must fail for $H < \frac{1}{2}$. The Itô–Stratonovich correction is given by the quadratic covariation, defined (whenever possible) as the limit, in probability, of

$$\sum_{[u,v]\in\pi} (f(\widehat{W}_v) - f(\widehat{W}_u))(W_v - W_u), \tag{10.12}$$

along any sequence (π^n) of partitions with mesh-size tending to zero. Disregarding trivial situations we see that this limit does not exist. For instance, when $f(x) = x$, fractional scaling immediately gives divergence (at rate $H - \frac{1}{2}$) of the expression (10.12). This issue also arises in the context of option pricing (Theorem 10.2 and Section 10.5). All theses problems remain present, of course, for the more complicated situation of "nonsimple" rough volatility, as discussed in Section 10.4.

10.0.3 ▪ Description of main results

Motivated by singular stochastic PDE (SPDE) theory, such as Hairer's work on the Kardar–Parisi–Zhang (KPZ) equation [190] and the Hairer–Pardoux "renormalized" Wong–Zakai theorem [194], we provide a (necessarily renormalized) strong approximation theory for rough volatility. Rough path theory, despite its very purpose of dealing with low regularity paths, is not applicable to the problem at hand. (We shall elaborate on this at the beginning of Section 10.1.) In essence, what one needs is a more flexible type of rough path theory, which is exactly what Hairer's theory of regularity structures [191] supplies. As consequence of fundamental continuity statements in "model" (think "rough path") metrics, we will discuss short-time large deviations for rough volatility models. Following, e.g., [147, Section 9.3], we also envision support results in "rough" interest rate models, in the spirit of [110].

To state our basic approximation results, we write $\dot{W}^{\varepsilon} := \partial_t W^{\varepsilon}$ for a suitable approximation at scale ε to white noise, with the induced approximation to fBm denoted by $\widehat{W}^{\varepsilon}$. Throughout, the Hurst parameter $H \in (0, \frac{1}{2}]$ is fixed, and f is a smooth function, such that (10.8) is a (local) martingale, as required by standard financial theory. More precisely, let W^{ε} denote the Haar wavelet construction of the Brownian motion W truncated at level $N = -\log_2(\epsilon)$; see Section 10.2.4 for details. Then, $\dot{W}^{\epsilon}$ is simply defined as the time derivative of the (piecewise-linear) process W^{ϵ}, and $\widehat{W}^{\epsilon}$ is obtained by integrating (in a pathwise fashion) $\dot{W}^{\epsilon}$ against the Volterra kernel K, i.e., $\widehat{W}^{\epsilon}_t = \int_0^t K(t,s)\dot{W}^{\epsilon}_s ds$. B^{ϵ} denotes the analogous construction for a correlated Brownian motion B, where $B_t = \varrho W_t + (1-\varrho^2)^{\frac{1}{2}}\overline{W}_t$, with some independent Brownian motion $\overline{W}$.

Theorem 10.1. *Consider a simple rough volatility model with dynamics* $\mathrm{d}S_t/S_t = f(\widehat{W}_t)\mathrm{d}B_t$, *i.e., driven by Brownian motions B and W with constant correlation ϱ. Then there exist ε-periodic functions $\mathscr{C}^{\varepsilon} = \mathscr{C}^{\varepsilon}(t)$, with diverging averages C_{ε}, such that a Wong–Zakai result holds of the form that $\tilde{S}^{\varepsilon} \to S$ in probability and uniformly on compacts, where*

$$\frac{\partial_t \tilde{S}^{\varepsilon}_t}{\tilde{S}^{\varepsilon}_t} = f(\widehat{W}^{\varepsilon}_t)\dot{B}^{\varepsilon}_t - \varrho\mathscr{C}^{\varepsilon}(t)f'(\widehat{W}^{\varepsilon}_t) - \tfrac{1}{2}f^2(\widehat{W}^{\varepsilon}_t)\,, \quad \tilde{S}^{\varepsilon}_0 = S_0.$$

Similar results hold for more general ("nonsimple") rough volatility models.

Remark 10.0.1. *When $H = \frac{1}{2}$, this result is an easy consequence of the well-known Itô–Stratonovich conversion formula. In the case $H < \frac{1}{2}$,* Theorem 10.1 *provides the interesting insight that genuine renormalization (in the sense of subtracting diverging quantities) is required if and only if the correlation parameter ϱ is nonzero. This is the case in equity (and many other) markets. Also note that naive approximations without renormalization (i.e., without subtracting the $\mathscr{C}^{\varepsilon}$-term) will in general diverge.*

Remark 10.0.2. *Mollification of the noise by truncation of the wavelet representation of the driving Brownian motion is natural for numerical purposes. First, it gives a simple sampling technique in terms of independent and identically distributed (i.i.d.) standard normals. Second, the construction provides a canonical hierarchy which is beneficial for QMC methods, as in the discussion in Section* 10.0.1.

In order to formulate implications for option pricing, define the Black–Scholes pricing function

$$C_{\mathrm{BS}}\left(S_0, K; \sigma^2 T\right) := \mathbb{E}\left[\left(S_0 \exp\left(\sigma\sqrt{T}Z - \frac{\sigma^2}{2}T\right) - K\right)_+\right], \tag{10.13}$$

where Z denotes a standard normal random variable. We then have the following theorem.

Theorem 10.2. *With $\mathscr{C}^\varepsilon = \mathscr{C}^\varepsilon(t)$ as in Theorem 10.1, define the renormalized integral approximation*

$$\tilde{\mathscr{I}}^\varepsilon := \tilde{\mathscr{I}}_f^\varepsilon(T) := \int_0^T f(\widehat{W}_t^\varepsilon)\mathrm{d}W_t^\varepsilon - \int_0^T \mathscr{C}^\varepsilon(t) f'(\widehat{W}_t^\varepsilon)\mathrm{d}t, \tag{10.14}$$

and also the approximate total variance

$$\mathscr{V}^\varepsilon := \mathscr{V}_f^\varepsilon(T) := \int_0^T f^2(\widehat{W}_t^\varepsilon)\mathrm{d}t.$$

Then the price of a European call option, under the pricing model (10.1), (10.3), *struck at K with time T to maturity, is given by*

$$\mathbb{E}[(S_T - K)_+] = \lim_{\varepsilon \to 0} \mathbb{E}\left[\Psi(\tilde{\mathscr{I}}^\varepsilon, \mathscr{V}^\varepsilon)\right],$$

where

$$\Psi(\mathscr{I}, \mathscr{V}) := C_{\mathrm{BS}}\left(S_0 \exp\left(\varrho \mathscr{I} - \frac{\varrho^2}{2}\mathscr{V}\right), K, \bar{\varrho}^2 \mathscr{V}\right). \tag{10.15}$$

Similar results hold for more general ("nonsimple") rough volatility models.

Let us discuss right away how to reduce the statements of Theorems 10.1 and 10.2 to the actual convergence statements that will occupy us in Section 10.2. First, note that

$$S_t = S_0 \exp\left[\int_0^t f\left(\widehat{W}_s\right)\mathrm{d}B_s - \tfrac{1}{2}\int_0^t f^2\left(\widehat{W}_s\right)\mathrm{d}s\right]. \tag{10.16}$$

The approximations W^ε, $\overline{W}^\varepsilon$, and $B^\varepsilon := \varrho W^\varepsilon + \bar{\varrho}\overline{W}^\varepsilon$ converge uniformly to the obvious limits, so that it suffices to understand the convergence of the stochastic integral. Note that $\widehat{W}$ is heavily correlated with W but independent of $\overline{W}$. The interesting part is then the convergence

$$\int_0^t f(\widehat{W}_s^\varepsilon)\mathrm{d}W_s^\varepsilon - \int_0^t \mathscr{C}^\varepsilon(s) f'(\widehat{W}_s^\varepsilon)\mathrm{d}s \to \int_0^t f(\widehat{W}_s)\mathrm{d}W_s, \tag{10.17}$$

as stated and proved in Theorem 10.22. For the other part, no correction terms arise due to independence, and it can be seen with standard methods that

$$\int_0^t f(\widehat{W}^\varepsilon)_s \mathrm{d}\overline{W}_s^\varepsilon \to \int_0^t f(\widehat{W}_s) d\overline{W}_s,$$

in the sense of convergence in probability, uniformly on compacts in t. The convergence result of Theorem 10.1 then follows readily. As for pricing, in Theorem 10.2 we consider the call payoff

$$\left(S_0 \exp\left[\int_0^T \sigma_t \mathrm{d}B_t - \frac{1}{2}\int_0^T \sigma_t^2 \mathrm{d}t\right] - K\right)_+.$$

An elementary conditioning argument with respect to W (first used by Romano and Touzi in the context of Markovian stochastic volatility models) then shows that the call price is given as expectation of

$$C_{\mathrm{BS}}\left(S_0 \exp\left(\varrho \int_0^T \sigma_t \mathrm{d}W_t - \frac{\varrho^2}{2}\int_0^T \sigma_t^2 \mathrm{d}t\right), K, \frac{\bar{\varrho}^2}{2}\int_0^T \sigma_t^2 \mathrm{d}t\right).$$

Specializing to the case $\sigma_t = f(\widehat{W}_t)$, in combination with Theorem 10.22, then yields Theorem 10.2. Note that extensions to nonsimple RV are immediate from suitable extensions of Theorem 10.22, as discussed in Section 10.4.2.

From a **mathematical perspective**, the key issue in proving the above theorems is to establish convergence of the *renormalized* approximate integrals, as $\varepsilon \to 0$,

$$\tilde{\mathscr{I}}^{\varepsilon} = \int_0^T f(\widehat{W}_t^{\varepsilon})\mathrm{d}W_t^{\varepsilon} - \int_0^T \mathscr{C}^{\varepsilon}(t) f'(\widehat{W}_t^{\varepsilon})\mathrm{d}t \to \int_0^T f(\widehat{W}_t)\mathrm{d}W_t. \tag{10.18}$$

Here we find much inspiration from singular SPDE theory, which also requires renormalized approximations for convergence to the correct Itô object. Specifically, we see that the theory of regularity structures [191], which essentially emerged from the theory of rough paths and Hairer's KPZ analysis (see [147] for a discussion and references), is a very appropriate tool here. In turn, we add an interesting new class of examples to the existing instances of regularity structures (polynomials, rough paths, many singular SPDEs, etc.). This new example avoids all considerations related to spatial structure (notably multilevel Schauder estimates [191, Chapter 5]), yet comes with the genuine need for renormalization. In fact, because we do not restrict ourselves to approximations of the white noise obtained by mollification (i.e., by convolution of $\dot{W}$ with a rescaled mollifier function, say $\delta^{\varepsilon}(x,y) = \varepsilon^{-1}\varrho(\varepsilon^{-1}(y-x)))$, our analysis naturally leads us to *renormalization functions*. In the case of mollifier approximations, which is the usual choice of Hairer and coworker [190, 191, 90]—but rules out wavelet approximations!—the renormalization function turns out to be constant (because $\dot{W}^{\varepsilon}$ is still stationary). In this case, we would obtain

$$\mathscr{C}^{\varepsilon}(t) \equiv C_{\varepsilon} = c\varepsilon^{H-\frac{1}{2}},$$

with $c = c(\varrho)$ explicitly given by integral (10.38). If, on the other hand, we consider a Haar wavelet approximation of white noise, we get[53]

$$\mathscr{C}^{\varepsilon}(t) = \frac{\sqrt{2H}}{H+\frac{1}{2}} \frac{|t - \lfloor t/\varepsilon \rfloor \varepsilon|^{H+\frac{1}{2}}}{\varepsilon} \quad \text{with mean } C_{\varepsilon} = \frac{\sqrt{2H}}{(H+\frac{1}{2})(H+\frac{3}{2})}\varepsilon^{H-\frac{1}{2}}. \tag{10.19}$$

It is natural to ask if $\mathscr{C}^{\varepsilon}(t)$ can be replaced, after all, by its mean C_{ε}. (Of course, this mean is still diverging for $\varepsilon \to 0$, as $H < \frac{1}{2}$.) For $H > 1/4$ the answer is yes, with an interesting phase transition when $H = 1/4$ (Section 10.2.2).

From a **numerical perspective**, Theorem 10.2 avoids any sampling of the independent factor $\overline{W}$. A brute-force approach then consists of simulating a scalar Brownian motion W, followed by discrete approximation of the stochastic integral $= \widehat{W}_t = \int_0^t K(t,s)\mathrm{d}W_s$. However, given the singularity of the Volterra kernel K, this is not advisable, and it is preferable to simulate the two-dimensional Gaussian process $(W_t, \widehat{W}_t : 0 \le t \le T)$, whose covariance function is readily available. A remaining problem is that the speed of convergence of

$$\sum f(\widehat{W}_s)W_{s,t} \to \int_0^T f\left(\widehat{W}_t\right)\mathrm{d}W_t\,,$$

with $[s,t]$ taken in a partition of mesh-size $\sim 1/n$, is very slow since $\widehat{W}$ has little regularity when H is small. (Gatheral and coauthors [166, 39] report $H \approx 0.05$.) Here, higher-order approximations come to our assistance, and we include quantitative estimates throughout, more precisely, *strong* rates. Such rates are essential for the design of MLMC algorithms, as was also seen in the context of general Gaussian rough differential equations [41]. (The important analysis of *weak rates* is left for future work.) Numerical aspects are further discussed in Section 10.5.

[53]Other wavelet choices are possible. In particular, in the case of fractional noise, *Alpert–Rokhlin (AR) wavelets* have been suggested for improved numerical behavior [178] where this is attributed to a series of works of A. Majda and coworkers. It would be desirable to explore this further in the context of rough volatility models.

The second set of results concerns large deviations for rough volatility models. Thanks to the contraction principle and fundamental continuity properties of Hairer's reconstruction map, the problem is reduced to understanding a large deviation principle (LDP) for a suitable enhancement of the noise. This approach requires (sufficiently) smooth coefficients, but comes with no growth restrictions, which is indeed quite suitable for financial modeling: we improve the Forde–Zhang short-time large deviations [138] for simple rough volatility models so as to include f of exponential type, a defining feature in the works of Gatheral and coauthors [166, 39]. (Such an extension is also the subject of the recent works [210, 181].)

Theorem 10.3. *Let $X_t = log(S_t/S_0)$ be the log-price under simple rough stochastic volatility model* (10.1), (10.3). *Then $(t^{H-\frac{1}{2}}X_t : t \geq 0)$ satisfies a short-time LDP with speed t^{2H} and rate function given by*

$$I(y) = \inf_{h \in L^2([0,1])} \left\{ \frac{1}{2}\|h\|_{L^2}^2 + \frac{(y - \varrho I_1(h))^2}{2I_2(h)} \right\},$$

with $I_1(h) := \int_0^1 f(\hat{h}(t))h(t)\mathrm{d}t$, $I_2(h) := \int_0^1 f(\hat{h}(t))^2\mathrm{d}t$, where $\hat{h}(t) := \int_0^t K(s,t)h(s)\mathrm{d}s$.

Theorem 10.3 will be proved as Corollary 10.25.

Remark 10.0.3. *A potential shortcoming is the nonexplicit form of the rate function. Geometric or "Hamiltonian" interpretations of the rate function, studied in a Markovian setting by many authors (e.g.,* [22, 56, 115, 116, 44]*) are then lost. A partial remedy here is to move from large deviations to (higher-order) moderate deviations. Analytic tractability is thus restored, and one still captures the main feature of the volatility smile close to the money. This method was introduced in a Markovian setting in* [145]*, and the extension to simple rough volatility models was given in* [40]*, relying on either* [138] *or the above Theorem* 10.3.

We next turn to nonsimple rough volatility models. Inspired by Rosenbaum and coworkers [126, 129, 131], we consider the stochastic Itô–Volterra equation

$$Z_t = z + \int_0^t K(s,t)(u(Z_s)\mathrm{d}W_s + v(Z_s)\mathrm{d}s),$$

with corresponding log-price process given by

$$X_t = \int_0^t f(Z_s)(\varrho \mathrm{d}W_s + \bar{\varrho}\mathrm{d}\overline{W}_s) - \frac{1}{2}\int_0^t f^2(Z_s)\mathrm{d}s.$$

(For simplicity, here we consider f, u, v to be bounded, with bounded derivatives of all orders.) For $h \in L^2([0,T])$, let z^h be the unique solution to the integral equation

$$z^h(t) = z + \int_0^t K(s,t)u(z^h(s))h(s)\mathrm{d}s,$$

and define $I_1^z(h) := \int_0^1 f(z^h(s))h(s)\mathrm{d}s$ and $I_2^z(h) := \int_0^1 f(z^h(s))^2\mathrm{d}s$. Then we have the following extension of Theorem 10.3 (and also [138, 210, 181]) to nonsimple rough volatility.

Theorem 10.4. *Let $X_t := log(S_t/S_0)$ be the log-price under nonsimple rough SV, and assume $H > 1/4$. Then $t^{H-\frac{1}{2}}X_t$ satisfies an LDP with speed t^{2H} and rate function given by*

$$I(x) = \inf_{h \in L^2([0,T])} \left\{ \frac{1}{2}\|h\|_{L^2}^2 + \frac{\left(x - \varrho I_1^z(h)\right)^2}{2I_2^z(h)} \right\}.$$

Theorem 10.4 will be proved as Corollary 10.28.

Remark 10.0.4. *We showed in* [40, *Corollary* 11]*—but see related results by Alòs, León, and Vives* [11] *and Fukasawa* [151, 153]*—that in the previously considered simple rough volatility models, now writing* $\sigma(.)$ *instead of* $f(.)$*, the implied volatility skew behaves, in the short-time limit, as* $\sim \varrho\frac{\sigma'(0)}{\sigma(0)}\langle K1,1\rangle t^{H-\frac{1}{2}}$*, where* $\langle K1,1\rangle$ *in our setting computes to* $c_H := \frac{(2H)^{\frac{1}{2}}}{(H+\frac{1}{2})(H+\frac{3}{2})}$*. (The blow-up* $t^{H-\frac{1}{2}}$ *as* $t\to 0$ *is a desired feature, in agreement with steep skews seen in the market.) To first order,* $Z_t \approx z + u(z)\int_0^t K(s,t)dW s = z + u(z)\widehat{W} =: \sigma(\widehat{W})$*, from which one obtains a skew formula in the nonsimple rough volatility case of the form*

$$\varrho u(z)\frac{f'(z)}{f(z)}c_H t^{H-\frac{1}{2}}.$$

Following the approach of [40]*, Theorem* 10.4 *allows for not only rigorous justification of these formula, but also for the computation of higher-order smile features, though this is not pursued in this chapter. In the case of classical (Markovian) stochastic volatility models,* $H = \frac{1}{2}$*, and specializing further to* $f(x) \equiv x$*, so that* Z *(resp.,* z*) models stochastic (resp., spot) volatility, this formula reduces precisely to the popular skew formula from Gatheral's book* [163, (7.5)]*, attributed therein to Medvedev and Scaillet. In the case of the rough Heston model, where* Z *models stochastic variance* (10.52)*, we have* $f(x) = \sqrt{x}$*,* $u = \eta\sqrt{x}$*, and this leads to the short-dated skew formula*

$$\frac{\varrho\eta}{2\sqrt{v_0}}c_H t^{H-\frac{1}{2}}.$$

If the above expression is multiplied with $2\sqrt{v_0}$*, we get the implied variance skew, again in agreement with Gatheral* [163, *page* 35]*. The formula may be independently verified via the characteristic function obtained in* [129]*.*

The reader may be interested in further applications of the regularity structure view on rough volatility developed in this chapter. The Stratonovich formulation opens up the possibility of constructing cubature methods (in the sense of [225, 233]*) for rough stochastic volatility methods. Indeed, our method can be seen as a level-*3 *Ninomiya–Victoir* [248] *scheme. Further, having said much about large deviations, it is not far-fetched to think of a support theorem (another classical application area of rough paths and regularity structures* [147, *Section* 9.3]*), which in turn invites one to revisit* [110] *in a setting of "rough" interest rate models. Another concrete application, the topic of the recent* [142]*, concerns precise asymptotics, allowing for considerable refinement of large deviations. (Translated into financial terms, this improvement leads to higher-order implied volatility expansions.)*

Structure of the chapter

In Section 10.1 we explain why the classical formulation of rough paths is not suitable for rough volatility models, and then go on to introduce essentials of the theory of regularity structures. We use the KPZ equation as a guiding example, which offers several similarities to rough volatility. The most basic "pricing structure" is introduced in Section 10.2. In Section 10.3 we consider a regularity structure for two-dimensional noise, which is necessary to study the asset price process in addition to the volatility process. Section 10.4 then discusses the case of nontrivial dynamics for rough volatility. Some numerical results are presented in Section 10.5, followed by several appendices with technical details. From Section 10.2 on, all our work relies on the framework of Hairer's regularity structures. There seems to be no point in repeating all the necessary definitions and terminology, which the reader can find in [190, 191, 192, 147] and a variety of survey papers on the subject. (For the reader in search of one concise reference, we recommend [147, Section 13].)

10.1 ▪ On rough paths and lessons from KPZ and singular SPDE theory

We already pointed out in Section 10.0.2 that any analysis of *correlated* ($\varrho \neq 0$) rough volatility models will involve the (Itô) integral

$$\int_0^T f(\widehat{W}_t)\mathrm{d}W_t, \tag{10.20}$$

where $(\widehat{W}_t)$ is an fBm with Hurst parameter $H < \frac{1}{2}$, itself given as an integral of a singular kernel against the Brownian motion W. Though the scalar Brownian motion W can easily be lifted to a Brownian rough path (of Itô or Stratonovich type), the integral in (10.20) *cannot* be viewed as a rough integral. Indeed, a generic integrand $f(\widehat{W})$, and even $\widehat{W}$, is not at all a rough path controlled by W (in the sense of Gubinelli). Hence, the stochastic integral cannot be defined by standard rough path theory as found in [147, Section 4].

A second attempt to view $(W, \widehat{W})$ as a two-dimensional Gaussian process—in order to use Gaussian rough path theory (e.g., [147, Section 12])—also fails. In fact, this theory would require W and $\widehat{W}$ to be independent ($\leftrightarrow$ here: fully correlated), the resulting lift to be of geometric (i.e., Stratonovich) type ($\leftrightarrow$ here: Itô integral), the Hurst parameter $H > 1/4$ ($\leftrightarrow$ here: any $H > 0$), and an enhancement of at most 3 levels (level-3 rough paths) ($\leftrightarrow$ here: $\sim 1/H$ levels).

A third attempt, in view of the nongeometric nature of Itô integration, to use *branched* rough paths is also doomed, for it requires—like classical geometric rough path theory—all iterated integrals. In our case, there is already an obstacle at level 2, before the appearance of any branching, in that the full set of second iterated integrals

$$\begin{pmatrix} \int W\mathrm{d}W, & \int W\mathrm{d}\widehat{W} \\ \int \widehat{W}\mathrm{d}W & \int \widehat{W}\mathrm{d}\widehat{W} \end{pmatrix} = \begin{pmatrix} * & ? \\ * & ? \end{pmatrix} \tag{10.21}$$

is an ill-defined object ($*$ stands for well-defined Itô integrals, ? for integrals of unclear meaning). Note that imposition of a first-order (resp., Itô) product rule would manifestly clash with Itô calculus, as $\widehat{W}$ has infinite quadratic variation when $H < \frac{1}{2}$.

On the other hand, formal expansion of (10.20) over some interval $[s, t]$ gives

$$\int_s^t f(\widehat{W}_u)\mathrm{d}W_u \approx f\left(\widehat{W}_s\right) W_{s,t} + f'\left(\widehat{W}_s\right) \int_s^t \widehat{W}_{s,u}\mathrm{d}W_u + \cdots,$$

so that the troubling terms—the "?" in (10.21)—do not appear. What is needed then, in the general case, is a higher-order "partial" branched rough path theory (for we deal with nongeometric /Itô objects), in which only partial information on the iterated integrals is stored. But even then, one faces failure of canonical (Wong–Zakai type) approximations, that is,

$$\nexists \lim_{\varepsilon \to 0} \int_0^T f(\widehat{W}_t^\varepsilon)\mathrm{d}W_t^\varepsilon . \tag{10.22}$$

Such failures are atypical for rough path theory. Having made all these observations, we see that Hairer's regularity structures (see below for more details) provide everything we desire: a tailor-made algebraic structure (which by construction only stores the required higher-order information), together with a machinery that gives continuity properties of all operations of interest *and* a consistent way to renormalize approximate stochastic integrals, such as the one appearing in (10.22).

The absence of a canonical approximation theory, as seen in (10.22), is a defining feature of the singular SPDEs recently considered by Hairer, Gubinelli, and many others. In particular, approximation of the noise (say, ε-mollification for the sake of argument) typically does *not* give rise to convergent approximations. To be specific, it is instructive to review the very example that led Hairer to regularity structures: the universal model for fluctuations of interface growth given by the Kardar–Parisi–Zhang (KPZ) equation

$$\partial_t u = \partial_x^2 u + |\partial_x u|^2 + \xi$$

with space-time white noise $\xi = \xi(x, t; \omega)$. As a matter of fact, and without going into further detail, there is a well-defined Itô solution $u = u(t, x; \omega)$ (known as the "Cole–Hopf" solution), but if one considers the equation with ε-mollified noise, then $u = u^\varepsilon$ diverges with $\varepsilon \to 0$. In this sense, there is a fundamental *lack of approximation theory*, and *no Stratonovich solution* to KPZ exists. To see the problem, take $u_0 \equiv 0$ for simplicity, and write

$$u = H \star \left(|\partial_x u|^2 + \xi\right)$$

with the space-time convolution denoted by $\star$ and the heat-kernel

$$H(t, x) = \frac{1}{\sqrt{4\pi t}} \exp\left(-\frac{x^2}{4t}\right) \mathbf{1}_{\{t>0\}}.$$

One can proceed with a Picard iteration

$$u = H \star \xi + H \star ((H' \star \xi)^2) + \cdots,$$

but there is an immediate problem with $(H' \star \xi)^2$, (naively) defined as the ε-to-zero limit of $(H' \star \xi^\varepsilon)^2$, which does not exist. However, there exists a diverging sequence of real numbers C_ε such that, in probability,

$$\exists \lim_{\varepsilon \to 0} (H' \star \xi^\varepsilon)^2 - C_\varepsilon \rightsquigarrow \left(\text{new object}\right) =: (H' \star \xi)^{\diamond 2}.$$

The idea of Hairer, following the philosophy of rough paths, was then to accept

$$H \star \xi, (H' \star \xi)^{\diamond 2} \text{ (and a few more)}$$

as enhancement of the noise ("model") upon which solution depends in pathwise robust fashion. This unlocks the seemingly fixed (and here even nonsensical) relation

$$H \star \xi \mapsto \xi \mapsto (H' \star \xi)^2.$$

Loosely speaking, one has the following theorem.

Theorem 10.5 (Hairer). *There exist diverging constants C_ε such that a Wong–Zakai[54] result holds of the form $\tilde{u}^\varepsilon \to u$, in probability and uniformly on compacts, where*

$$\partial_t \tilde{u}^\varepsilon = \partial_x^2 \tilde{u}^\varepsilon + |\partial_x \tilde{u}^\varepsilon|^2 - C_\varepsilon + \xi^\varepsilon.$$

Similar results hold for a number of other singular semilinear SPDEs.

In a sense, this can be traced back to the *Milstein-scheme* for SDEs and then *rough path theory*. Consider $dY_t = f(Y_t)dW_t$, with $Y_0 = 0$ for simplicity, and consider the second-order (Milstein) approximation

$$Y_{t_{i+1}} \approx Y_{t_i} + f\left(Y_{t_i}\right) W_{t_i, t_{i+1}} + f'\left(Y_{t_i}\right) \int_{t_i}^{t_{i+1}} W_{t_i, s} \dot{W}_s ds.$$

[54]Hairer and Pardoux [194] derive the KPZ result as special case of a Wong–Zakai result for Itô SPDEs.

One has to unlock the seemingly fixed relation

$$W \mapsto \dot{W} \mapsto \int_0^{\cdot} W_s \dot{W}_s \mathrm{d}s =: \mathbb{W},$$

for there is a choice to be made. For instance, the last term can be understood as an Itô integral $\int_0^{\cdot} W_s \mathrm{d}W_s$ or as a Stratonovich integral $\int_0^{\cdot} W_s \circ \mathrm{d}W_s$ (and in fact, there are many other choices; see, e.g., the discussion in [147]). It suffices to take this thought one step further to arrive at *rough path theory*: accept $\mathbb{W}$ as a new (analytic) object, which leads to the main (rough path) insight

$$\text{SDE theory} = \text{analysis based on } (W, \mathbb{W}).$$

In comparison,

$$\text{SPDE theory à la Hairer} \;=\; \text{analysis based on (renormalized) enhanced noise } (\xi, \dots).$$

Inside Hairer's theory[55]

As motivation, consider the Taylor expansion (at x) of a real-valued smooth function,

$$f(y) = f(x) + f'(x)(y - x) + \frac{1}{2} f''(x)(y - x)^2 + \cdots .$$

It can be written as an abstract polynomial ("jet") at x,

$$F(x) := f(x)\, \mathbf{1} + g(x) X + h(x) X^2 + \cdots ,$$

with, necessarily, $g = f'$, $h = f''/2, \dots$. If we "realize" these abstract symbols again as honest monomials, i.e., $\Pi_x : X^k \mapsto (. - x)^k$, and extend Π_x linearly, then we recover the full Taylor expansion,

$$\Pi_x[F(x)](.) = f(x) + g(x)(. - x) + \frac{1}{2} h(x)(. - x)^2 + \cdots .$$

Hairer looks for solutions of this form: at every space-time point, a jet is attached, which in the case of KPZ turns out—after solving an abstract fixed point problem—to be of the form

$$U(x, s) = u(x, s)\, \mathbf{1} + \mathsf{I} + \mathsf{Y} + v(x, s)\, X + 2 \mathsf{Y}^{\!\circ} + v(x, s) \,\mathsf{<}^{\circ} .$$

As before, every symbol is given concrete meaning by "realizing" it as an honest function (or Schwartz distribution). Naturally, one can take

$$\mathsf{I} \mapsto H \star \xi^{\epsilon} \text{(mollified noise), } \textbf{or} \quad \mathsf{I} \mapsto H \star \xi \text{(original noise)} . \tag{10.23}$$

More interestingly, Y can be mapped to any of

$$\begin{cases} H \star (H' \star \xi^{\epsilon})^2, & \text{canonically enhanced mollified noise, } \textbf{or} \\ H \star [(H' \star \xi^{\epsilon})^2 - C_{\epsilon}], & \text{renormalized} \sim, \textbf{ or} \\ H \star (H' \star \xi)^{\diamond 2}, & \text{renormalized enhanced noise.} \end{cases} \tag{10.24}$$

This realization map is called "model" and captures exactly a typical, but otherwise fixed, realization of the noise (mollified or not) together with some enhancement thereof, renormalized or

[55] In this section only, following [147], symbols are colored.

not. For instance, writing $\Pi_{x,s}$ for the realization map for renormalized enhanced noise, one has

$$\Pi_{x,s}[U(x,s)](.) = u(x,s) + H \star \xi|_{(*)} + H \star (H' \star \xi)^{\diamond 2}|_{(*)} + \cdots ,$$

where $(*)$ indicates suitable centering at (x, s). Note that U takes values in a (finite) linear space spanned by (sufficiently many) symbols,

$$U(x,s) \in \langle \dots, 1, \mathfrak{i}, \curlyvee, X, \curlyvee\!\!\!\circ, \langle, \dots \rangle =: \mathcal{T}.$$

The map $(x, s) \mapsto U(x, s)$ is an example of a **modeled distribution**; the precise definition is a mix of suitable analytic and algebraic conditions (similar to the notation of a controlled rough path).

The analysis requires keeping track of the *degree* (a.k.a. *homogeneity*) of each symbol. For instance, $|\mathfrak{i}| = \frac{1}{2} - \kappa$ (related to the Hölder regularity of the realized object one has in mind), $|X^2| = 2$, etc. All these degrees are collected in an **index set**. In order to compare jets at different points (think $(X - \delta 1)^3 = \dots$), a group of linear maps on $\mathcal{T}$ is used, called a **structure group.** Last but not least, the **reconstruction map** uniquely maps modeled distributions to functions or Schwartz distributions. (This can be seen as a generalization of the *sewing lemma*, the essence of rough integration (see, e.g., [147]), which turns a collection of sufficiently compatible local expansions into one function or Schwartz distribution.) In the KPZ context, the (Cole–Hopf or Itô) solution is then indeed obtained as a reconstruction of the abstract (modeled distribution) solution U.

10.2 ▪ The rough pricing regularity structure

In this section we develop the approximation theory for integrals of the type $\int f(\tilde{W})dW$. In the first part we present the regularity structure and the associated models we will use. In the second part we apply the reconstruction theorem from regularity structures to conclude our main result, Theorem 10.22.

10.2.1 ▪ Basic pricing setup

We are given a Hurst parameter $H \in (0, \frac{1}{2}]$, associated to an fBm (in the Riemann–Liouville sense) $\widehat{W}$, and fix an arbitrary $\kappa \in (0, H)$ and an integer

$$M \geq \max\{m \in \mathbb{N} \,|\, m \cdot (H - \kappa) - \frac{1}{2} - \kappa \leq 0\},$$

so that

$$(M + 1)(H - \kappa) - \frac{1}{2} - \kappa > 0. \tag{10.25}$$

At this stage, we can introduce the "level-$(M + 1)$" *model space*

$$\mathcal{T} = \left\langle \{\Xi, \Xi\mathcal{I}(\Xi), \dots, \Xi\mathcal{I}(\Xi)^M, \mathbf{1}, \mathcal{I}(\Xi), \dots, \mathcal{I}(\Xi)^M\} \right\rangle, \tag{10.26}$$

where $\langle \dots \rangle$ denotes the vector space generated by the (purely abstract) symbols in $\{\dots\}$. We will sometimes write

$$S = S^{(M)} := \{\Xi, \Xi\mathcal{I}(\Xi), \dots, \Xi\mathcal{I}(\Xi)^M, \mathbf{1}, \mathcal{I}(\Xi), \dots, \mathcal{I}(\Xi)^M\}$$

so that $\mathcal{T} = \mathcal{T}^{(M)} = \bigoplus_{\tau \in S} \mathbb{R}\tau$.

Remark 10.2.1. *It is useful here and in the sequel to consider as a sanity check the special case $H = \frac{1}{2}$ in which we recover the "level-2" rough path structure as introduced in* [147, *Chapter* 13]. *More specifically, if we take a Hölder exponent $\alpha := \frac{1}{2} - \kappa < \frac{1}{2}$, we may choose $M = 1$. Then condition* (10.25) *is precisely the familiar condition $\alpha > 1/3$.*

The interpretation for the symbols in S is as follows: Ξ should be understood as an abstract representation of the white noise ξ belonging to the Brownian motion W, i.e., $\xi = \dot{W}$ where the derivative is taken in the distributional sense. Note that since we set $W_x = 0$ for $x \leq 0$, we have $\dot{W}(\varphi) = 0$ for $\varphi \in C_c^\infty((-\infty, 0))$. The symbol $\mathcal{I}(\ldots)$ has the intuitive meaning of "integration against the Volterra kernel," so that $\mathcal{I}(\Xi)$ represents the integration of the white noise against the Volterra kernel, i.e.,

$$\sqrt{2H} \int_0^t |t - r|^{H - \frac{1}{2}} \mathrm{d}W_r,$$

which is nothing but the fBm $\widehat{W}_t$. Symbols like $\Xi\mathcal{I}(\Xi)^m = \Xi \cdot \mathcal{I}(\Xi) \cdot \ldots \cdot \mathcal{I}(\Xi)$ or $\mathcal{I}(\Xi)^m = \mathcal{I}(\Xi) \cdot \ldots \cdot \mathcal{I}(\Xi)$ should be read as products between the objects above. These interpretations of the symbols generating $\mathcal{T}$ will be made rigorous by the model (Π, Γ) in the next subsection. Every symbol in S is assigned a homogeneity, which we define by

$$\begin{aligned} |\Xi\mathcal{I}(\Xi)^m| &:= -\frac{1}{2} - \kappa + m(H - \kappa),\ m \geq 0, \\ |\mathcal{I}(\Xi)^m| &:= m(H - \kappa),\ m > 0, \\ |\mathbf{1}| &:= 0. \end{aligned}$$

We collect the homogeneities of elements of S in a set $A := \{|\tau| \,|\, \tau \in S\}$, whose minimum is $|\Xi| = -\frac{1}{2} - \kappa$. Note that the homogeneities are multiplicative in the sense that $|\tau\tau'| = |\tau| + |\tau'|$ for $\tau, \tau' \in S$ such that $\tau\tau' = \tau \cdot \tau' \in S$ (with the product defined in the obvious way).

At last, our regularity structure comes with a *structure group* G, an (abstract) group of linear operators on the model space $\mathcal{T}$, which should satisfy $\Gamma\tau - \tau = \bigoplus_{\tau' \in S : |\tau'| < |\tau|} \mathbb{R}\tau'$ and $\Gamma\mathbf{1} = \mathbf{1}$ for $\tau \in S$ and $\Gamma \in G$. We will choose $G = \{\Gamma_h \,|\, h \in (\mathbb{R}, +)\}$, given by

$$\Gamma_h \mathbf{1} = \mathbf{1},\ \Gamma_h \Xi = \Xi,\ \Gamma_h \mathcal{I}(\Xi) = \mathcal{I}(\Xi) + h\mathbf{1},$$

and $\Gamma_h(\tau' \cdot \tau) = \Gamma_h\tau' \cdot \Gamma_h\tau$, for $\tau', \tau \in S$, for which $\tau \cdot \tau' \in S$ is defined.

The limiting model (Π, Γ)

Let W be a Brownian motion on $\mathbb{R}_+ := [0, \infty)$, and extend it to all of $\mathbb{R}$ by requiring $W_x = 0$ for $x \leq 0$. We will frequently use the notation

$$\int_0^t f(t) \mathrm{d}W_t, \quad \int_0^t f(t) \diamond \mathrm{d}W_t \tag{10.27}$$

which denote the Itô integral and the Skorokhod integral (which boils down to an Itô integral whenever the integrand is adapted), respectively. For background on Skorokhod integration we refer the reader to [216, Section 7.3]; Nualart's ICM lecture [251] is also highly recommended. Skorokhod integrals have the distinct advantage of avoiding the need for an adapted integrand, but they coincide with Itô integrals once the integrand is adapted. For a reader unfamiliar with the (beautiful) theory of Skorokhod integration, it should be sufficient to simply think of an Itô integral with possibly nonadapted integrands for the purposes of this chapter. Whenever we make use of specific properties of Skorokhod integration we will make this explicit.

From W we now construct the fBm $\widehat{W}$ in the Riemann–Liouville sense with Hurst index $H \in (0, \frac{1}{2}]$ as

$$\widehat{W}_t := \dot{W} \star K(t) = \sqrt{2H} \int_0^t |t-r|^{H-\frac{1}{2}} \, \mathrm{d}W_r,$$

where $K(t) = \sqrt{2H}\,\mathbf{1}_{t>0}\, t^{H-\frac{1}{2}}$ denotes the Volterra kernel. We also write $K(s,t) := K(t-s)$.

To give meaning to the product terms $\Xi\mathcal{I}(\Xi)^k$ we follow the ideas from rough paths and define an "iterated integral" for $s, t \in \mathbb{R}$, $s \le t$, as

$$\mathbb{W}^m_{s,t} := \int_s^t (\widehat{W}_r - \widehat{W}_s)^m \, \mathrm{d}W_r. \tag{10.28}$$

$\mathbb{W}^m(s,t)$ satisfies the following modification of Chen's relation.

Lemma 10.6. *$\mathbb{W}^m$ as defined in* (10.28) *satisfies*

$$\mathbb{W}^m_{s,t} = \mathbb{W}^m_{s,u} + \sum_{l=0}^m \binom{m}{l} (\widehat{W}_u - \widehat{W}_s)^l \mathbb{W}^{m-l}_{u,t} \tag{10.29}$$

for $s, u, t \in \mathbb{R}$, $s \le u \le t$.

Proof. This is a direct consequence of the binomial theorem. □

We extend the domain of $\mathbb{W}^m$ to all of $\mathbb{R}^2$ by imposing Chen's relation for all $s, u, t \in \mathbb{R}$, i.e., for $t, s \in \mathbb{R}$, $t \le s$, we set

$$\mathbb{W}^m_{s,t} := -\sum_{l=0}^m \binom{m}{l} (\widehat{W}_t - \widehat{W}_s)^l \mathbb{W}^{m-l}_{t,s}. \tag{10.30}$$

We are now in position to define a model (Π, Γ) that gives rigorous meaning to the interpretation we gave above for $\Xi, \mathcal{I}(\Xi), \Xi\mathcal{I}(\Xi), \ldots$. Recall that in the theory of regularity structures a model is a collection of linear maps $\Pi_s : \mathcal{T} \to C^1_c(\mathbb{R})'$, $\Gamma_{st} \in G$ for indices $s, t \in \mathbb{R}$ that satisfy

$$\Pi_t = \Pi_s \Gamma_{st}, \tag{10.31}$$

$$|\Pi_s \tau(\varphi^\lambda_s)| \lesssim \lambda^{|\tau|}, \tag{10.32}$$

$$\Gamma_{st}\tau = \tau + \sum_{\tau' \in S : |\tau'| < |\tau|} c_{\tau'}(s,t)\tau', \text{ with } c_{\tau'}(s,t)| \lesssim |s-t|^{|\tau| - |\tau'|}, \tag{10.33}$$

where the loosely stated bounds in (10.32) and (10.33) hide a multiplicative constant, which can be chosen uniformly for $\tau \in S$, for any s, t in a compact set, and for $\varphi^\lambda_s := \lambda^{-1}\varphi(\lambda^{-1}(\cdot - s))$ with $\lambda \in (0, 1]$ and $\varphi \in C^1$ with compact support in the ball $B(0, 1)$.

We will work with the following "Itô" model (Π, Γ), which makes our interpretations of the elements of S more precise. (We will (occasionally) write $(\Pi^{\text{Itô}}, \Gamma^{\text{Itô}})$ to avoid confusion with a generic model, which we also denote by (Π, Γ).)

$$\begin{aligned}
&\Pi_s \mathbf{1} = 1, && \Gamma_{ts}\mathbf{1} = \mathbf{1}, \\
&\Pi_s \Xi = \dot{W}, && \Gamma_{ts}\Xi = \Xi, \\
&\Pi_s \mathcal{I}(\Xi)^m = \left(\widehat{W}(\cdot) - \widehat{W}(s)\right)^m, && \Gamma_{ts}\mathcal{I}(\Xi) = \mathcal{I}(\Xi) + (\widehat{W}(t) - \widehat{W}(s))\mathbf{1}, \\
&\Pi_s \Xi\mathcal{I}(\Xi)^m = \{t \mapsto \tfrac{\mathrm{d}}{\mathrm{d}t}\mathbb{W}^m(s,t)\}, && \Gamma_{ts}\tau\tau' = \Gamma_{ts}\tau \cdot \Gamma_{ts}\tau', \quad \text{for } \tau, \tau' \in S \text{ with } \tau\tau' \in S.
\end{aligned}$$

We extend both maps from S to $\mathcal{T}$ by imposing linearity.

Lemma 10.7. *The pair* (Π, Γ) *as defined above defines (a.s.) a model on* $(\mathcal{T}, A)$.

Proof. The only symbol in S for which (10.31) is not straightforward is $\Xi\mathcal{I}(\Xi)^m$, where the statement follows by Chen's relation. The bounds (10.32) and (10.33) follow for $\mathbf{1}$ trivially and for $\mathcal{I}(\Xi)^m$ by the $H-\kappa'$-Hölder regularity of $\widehat{W}$, $\kappa' \in (0, H)$. It is straightforward to check the condition (10.33) by using the rule $\Gamma_{ts}\tau\tau' = \Gamma_{ts}\tau \cdot \Gamma_{ts}\tau'$, so that we are only left with the task of bounding $\Pi_s \Xi\mathcal{I}(\Xi)^m(\varphi_s^\lambda)$. Along the lines of the proof of [147, Theorem 3.1], it follows that $|\mathbb{W}^m_{s,t}| \le C|s-t|^{mH+\frac{1}{2}-(m+1)\kappa}$ (where $C > 0$ denotes a random constant with $C \in \bigcap_{p<\infty} L^p$), so that

$$\begin{aligned}|\Pi_s \mathcal{I}(\Xi)^m \Xi(\varphi_s^\lambda)| &= \left| \int (\varphi_s^\lambda)'(t)\mathbb{W}^m(s,t)\,\mathrm{d}t \right| \\ &\le C \int \varphi'^{-1}(t-s))|s-t|^{mH+\frac{1}{2}-(m+1)\kappa} \frac{\mathrm{d}t}{\lambda^2} \\ &\le C\lambda^{mH-\frac{1}{2}-(m+1)\kappa} = C\lambda^{|\mathcal{I}(\Xi)^m\Xi|}. \qquad \square\end{aligned}$$

As we will see in Section 10.2.2 this model is the toolbox from which we can build pathwise Itô integrals of the type $\int_0^t f(r, \widehat{W}(r))\,\mathrm{d}W(r)$. For an approximation theory for such expressions we are in need of a comparable setup that describes approximations, which will be achieved by introducing a model $(\Pi^\varepsilon, \Gamma^\varepsilon)$.

The approximating model $(\Pi^\varepsilon, \Gamma^\varepsilon)$

The whole definition of the model (Π, Γ) is based on the object $\dot{W}$. It is therefore natural to build an approximating model by replacing $\dot{W}$ by some modification $\dot{W}^\varepsilon$ that converges (as a distribution) to $\dot{W}$ as $\varepsilon \downarrow 0$.

The definition of $\dot{W}^\varepsilon$ will be based on an object δ^ε which should be thought of as an approximation to the Dirac delta distribution. Our goal is to build δ^ε from wavelets, which can be as irregular as the Haar functions. We find it therefore convenient to allow δ^ε to take values in the Besov space $\mathcal{B}^\beta_{1,\infty}(\mathbb{R})$, $\beta > \frac{1}{2} + \kappa$, which includes functions like $\mathbf{1}_{[0,1]} \in \mathcal{B}^1_{1,\infty}(\mathbb{R})$.

Remark 10.2.2. *We recall the definition of the Besov space* $\mathcal{B}^\beta_{1,\infty}(\mathbb{R})$ *(see, for example, [240]), even though the definition will only be explicitly used in the proof of Lemma* 10.17 *in the appendix. Given a compactly supported wavelet basis* $\varphi_y = \varphi(\cdot - y)$, $y \in \mathbb{Z}$, $\psi_y^j = 2^{j/2}\psi(2^j(\cdot - y))$, $j \ge 0$, $y \in 2^{-j}\mathbb{Z}$, *we set*

$$\|g\|_{\mathcal{B}^\beta_{1,\infty}} := \sum_{y\in\mathbb{Z}} |(g, \varphi_y)_{L^2}| + \sup_{j\ge 0} 2^{j\beta} \sum_{y\in 2^{-j}\mathbb{Z}} 2^{-j/2}|(g, \psi_y^j)_{L^2}|.$$

Define $\mathcal{B}^\beta_{1,\infty}(\mathbb{R}) := \{g \in L^1 \,|\, \|g\|_{\mathcal{B}^\beta_{1,\infty}} < \infty\}$ *for* $\beta > 0$, *and*

$$\mathcal{B}^\beta_{1,\infty}(\mathbb{R}) := \left\{ g \in C_c^{-\lceil\beta\rceil+1}(\mathbb{R}))' \,\middle|\, \|g\|_{\mathcal{B}^\beta_{1,\infty}} < \infty \right\}$$

for $\beta \le 0$.

Definition 10.8. *Let* $\delta^\varepsilon : \mathbb{R}^2 \to \mathbb{R}$ *be a measurable, bounded function with the following properties:*

- $\delta^\varepsilon(x,y) = \delta^\varepsilon(y,x)$ *for all* $x, y \in \mathbb{R}$;
- *the map* $\mathbb{R} \ni x \mapsto \delta^\varepsilon(x, \cdot) \in \mathcal{B}^\beta_{1,\infty}(\mathbb{R})$ *is bounded and measurable for some* $\beta > -|\Xi| = \frac{1}{2} + \kappa$;

- $\int_{\mathbb{R}} \delta^{\varepsilon}(x,\cdot)\,\mathrm{d}x = 1$;
- $\sup_{x,y\in\mathbb{R}} |\delta^{\varepsilon}(x,y)| \lesssim \varepsilon^{-1}$;
- $\operatorname{supp}\delta^{\varepsilon}(x,\cdot) \subseteq B(x, c\cdot\varepsilon)$ *for any* $x \in \mathbb{R}$ *and some* $c > 0$.

Example 10.9. There are two examples which are of particular interest for our purposes:

- We say that δ^{ε} "comes from a mollifier" if there is a symmetric, compactly supported function $\varrho \in L^{\infty} \cap \mathcal{B}^{\beta}_{1,\infty}(\mathbb{R})$ integrating to 1 such that

$$\delta^{\varepsilon}(x,y) = \varepsilon^{-1}\cdot\varrho(\varepsilon^{-1}(y-x)).$$

- A further interesting example is the case where δ^{ε} "comes from a wavelet basis." Consider only $\varepsilon = 2^{-N}$, and choose compactly supported father wavelets $\varphi_{k,N} \in L^{\infty}\cap\mathcal{B}^{\beta}_{1,\infty}$, $k \in \mathbb{Z}$, (e.g., the Haar father wavelets $\varphi_{k,N} = 2^{N/2}\cdot\mathbf{1}_{[k2^{-N},(k+1)2^{-N})}$) and set

$$\delta^{\varepsilon}(x,y) := \sum_{k\in\mathbb{Z}} \varphi_{k,N}(x)\varphi_{k,N}(y).$$

  Note that we could also add some generations of mother wavelets in this choice.

Locally, $\dot{W}$ is contained in $\mathcal{B}^{|\Xi|}_{\infty,\infty}(\mathbb{R})$ (recall: $|\Xi| = -\frac{1}{2}-\kappa$), so that due to $\mathcal{B}^{|\Xi|}_{\infty,\infty}(\mathbb{R}) \subseteq (\mathcal{B}^{\beta}_{1,\infty}(\mathbb{R}))'$, we can define

$$\dot{W}^{\varepsilon}_t := \langle \dot{W}, \delta^{\varepsilon}(t,\cdot)\rangle \mathbf{1}_{\mathbb{R}_+}(t),$$

which is a Gaussian process, pathwise measurable, and locally bounded. For (maybe stochastic) integrands f we introduce the notation

$$\int_0^t f(r)\,\mathrm{d}W^{\varepsilon}_r := \int_0^t f(r)\dot{W}^{\varepsilon}_r\,\mathrm{d}r.$$

If f takes values in some (nonhomogeneous) Wiener chaos induced by $\dot{W}$, we also introduce

$$\int_0^t f(r)\diamond \mathrm{d}W^{\varepsilon}_r := \int_0^t f(r)\diamond \dot{W}^{\varepsilon}_r\,\mathrm{d}r, \tag{10.34}$$

where $\diamond$ denotes the *Wick product*. Note that these two objects do not coincide in general. A complete repetition of the definition of Wick multiplication would stray too far from the focus of this chapter. We refer the reader to [216, Section 3.1] for more details. In essence, a Wick product combines two random variables X and Y (that lie in a suitable space, namely the *Wiener chaos*) in a symmetric, bilinear manner to a new random variable $X\diamond Y$ by subtracting from the usual product $X\cdot Y$ a sum of correlation terms. If X and Y are independent, these correlations vanish, so that the Wick product just coincides with the pointwise product $X\cdot Y$. This includes the case of X being constant so that in particular $1\diamond Y = Y$. Another rather simple example arises if X, Y are both (centered) Gaussian random variables. The Wick product is then simply given by

$$X\diamond Y = X\cdot Y - \mathbb{E}[X\cdot Y].$$

In this chapter, X and Y are themselves products of Gaussian random variables. In this case, an explicit formula for $X\diamond Y = X\cdot Y - \ldots$ is given in [216, equation (3.6)].

The motivation for using the same symbol "$\diamond$" for Wick products and for Skorokhod integrals (10.27) is that Skorokhod integrals can be seen as "infinitesimal Wick products" in the sense that they are the limit of sums of Wick products (Remark 10.2.3). We sketch in Remark 10.2.3 shortly that one might want to read the Skorokhod integral as

$$\int_0^t f(r) \diamond \mathrm{d}W_r = \text{"} \int_0^t f(r) \diamond \frac{\mathrm{d}W}{\mathrm{d}r} \mathrm{d}r,\text{"} \tag{10.35}$$

where $\diamond$ should be read as Skorokhod integration on the left and "Wick multiplication" on the right-hand side (which is ill-defined, as $\frac{\mathrm{d}W}{\mathrm{d}r}$ only exists as a distribution). The ill-defined identity (10.35) can be read as a motivation for the (well-defined) definition (10.34). The close relation between Skorokhod integration and Wick multiplication plays a crucial role in the proof of Theorem 10.15 in the appendix.

Remark 10.2.3. *For the reader's convenience, we briefly comment on the close relation between the Skorokhod integral and the Wick product. Indeed, when $g = \sum X_s \mathbf{1}_{[s,t]}$, with summation over a finite partition of* $[0,T]$*, and each X_s is a (nonadapted) random variable in a finite Wiener–Itô chaos, it follows from* [216, *Theorem* 7.40] *that*

$$\int g \diamond \mathrm{d}W = \sum X_s \diamond W_{s,t},$$

where $\diamond$ denotes Skorokhod integration on the left and Wick multiplication on the right-hand side, and where $W_{s,t} = W(t) - W(s)$. Passage to L^2-limits is then standard, so that a Skorokhod integral $\int g \diamond \mathrm{d}W$ can be interpreted as the integrated Wick product "$g \diamond \frac{\mathrm{d}W}{\mathrm{d}t}$":

$$\int g \diamond \mathrm{d}W = \text{"} \int g \diamond \frac{\mathrm{d}W}{\mathrm{d}t} \mathrm{d}t,\text{"}$$

which can be seen as a motivation for our definition (10.34)*. See also* [252] *and the references therein.*

We now define an approximate fBm by setting

$$\widehat{W}_t^\varepsilon := K \star \dot{W}^\varepsilon(t) = \sqrt{2H} \int_0^t |t-r|^{H-\frac{1}{2}} \,\mathrm{d}W_r^\varepsilon,$$

which has the expected regularity as shown in the following lemma.

Lemma 10.10. *On every compact time interval* $[0,T]$ *we have the estimates*

$$|\widehat{W}_t^\varepsilon - \widehat{W}_s^\varepsilon| \lesssim C_\varepsilon |t-s|^{H-\kappa'}, \ |\widehat{W}_t^\varepsilon - \widehat{W}_s^\varepsilon - (\widehat{W}_t - \widehat{W}_s)| \lesssim C|t-s|^{H-\kappa'} \varepsilon^{\delta\kappa'}$$

uniformly in $\varepsilon \in (0,1]$ for any $\delta \in (0,1)$ and $\kappa' \in (0,H)$. Here, $C_\varepsilon, C > 0$ are random constants that are (uniformly) bounded in L^p for $p \in [1,\infty)$.

Proof. The proof is elementary but a bit bulky and therefore postponed to the appendix. □

Finally, we can give the definition of the approximating model $(\Pi^\varepsilon, \Gamma^\varepsilon)$, the "canonical" model built from the approximate (and hence regular) noise W^ε.

$$\begin{aligned}
&\Pi_s^\varepsilon \mathbf{1} = 1, && \Gamma_{st}^\varepsilon \mathbf{1} = 1,\\
&\Pi_s^\varepsilon \Xi = \dot{W}^\varepsilon, && \Gamma_{st}^\varepsilon \Xi = \Xi,\\
&\Pi_s^\varepsilon \mathcal{I}(\Xi)^m = \left(\widehat{W}_\cdot^\varepsilon - \widehat{W}_s^\varepsilon\right)^m, && \Gamma_{st}^\varepsilon \mathcal{I}(\Xi) = \mathcal{I}(\Xi) + \left(\widehat{W}_t^\varepsilon - \widehat{W}_s^\varepsilon\right)\mathbf{1},\\
&\Pi_s^\varepsilon \mathcal{I}(\Xi)^m \Xi = (\widehat{W}_\cdot^\varepsilon - \widehat{W}_s^\varepsilon)^m \,\dot{W}_\cdot^\varepsilon, && \Gamma_{st}^\varepsilon \tau\tau' = \Gamma_{st}^\varepsilon \tau \cdot \Gamma_{st}^\varepsilon \tau', \quad \text{for } \tau, \tau' \in S \text{ with } \tau\tau' \in S.
\end{aligned}$$

Lemma 10.11. *The pair* $(\Pi^\varepsilon, \Gamma^\varepsilon)$ *as defined above is a model on* $(\mathcal{T}, A)$.

Proof. The identity $\Pi_t = \Gamma_{ts}\Pi_s$ is straightforward to check. The bounds (10.32) and (10.33) on Γ_{st} and on $\Pi_s \mathcal{I}(\Xi)^m$ follow from the regularity of $\widehat{W}^\varepsilon$ as proved in Lemma 10.10. The blow-up of $\Pi_s \Xi \mathcal{I}(\Xi)^m(\varphi_s^\lambda)$, however, is even better than we need, since by the choice of δ^ε we have $|\dot{W}^\varepsilon| \leq C_\varepsilon$ on compact sets, for some random constant C_ε. □

The definition of this model is justified by the fact that application of the reconstruction operator (as in Lemma 10.20) yields integrals of the form

$$\int_0^t f(r, \widehat{W}_r^\varepsilon) \, \mathrm{d}W_r^\varepsilon. \tag{10.36}$$

As we pointed out in the introduction, there is no hope that integrals of this type will converge as $\varepsilon \downarrow 0$ if $H < \frac{1}{2}$. This can be cured by working with a renormalized model $(\hat{\Pi}^\varepsilon, \Gamma^\varepsilon)$ instead.

The renormalized model $\hat{\Pi}^\varepsilon$

From the perspective of regularity structures, the fundamental reason why integrals like (10.36) fail to converge to

$$\int_0^t f(r, \widehat{W}_r) \, \mathrm{d}W_r$$

lies in the fact that the corresponding models will not satisfy $(\Pi^\varepsilon, \Gamma^\varepsilon) \to (\Pi, \Gamma)$ in a suitable norm. To see what is going on, we will first rewrite $\Pi_s \Xi \mathcal{I}(\Xi)^k$.

Lemma 10.12. *For* $\varphi \in C_c^\infty(\mathbb{R})$, $s \in \mathbb{R}$, $m \in \{1, \ldots, M\}$ *we have*

$$\begin{aligned} \Pi_s \Xi \mathcal{I}(\Xi)^m(\varphi) = & \int_0^\infty \varphi(t) \, (\widehat{W}_t - \widehat{W}_s)^m \diamond \mathrm{d}W_t \\ & - m \int_0^\infty \varphi(t) \, K(s-t) \, (\widehat{W}_t - \widehat{W}_s)^{m-1} \, \mathrm{d}t, \end{aligned}$$

where $\diamond$ *denotes the Skorokhod integral, and* $K(t) = \sqrt{2H} \mathbf{1}_{t>0} t^{H-\frac{1}{2}}$ *denotes the Volterra kernel. Note that in the second term the domain of integration is actually* $(0, s)$.

Remark 10.2.4. *At first glance it might seem surprising (and maybe confusing) that we are in need of a Skorokhod integral in order to write the integrals in Lemma* 10.12*, especially for a reader more familiar with rough path theory. Note that the expression* $\int_0^\infty \varphi(t) \, (\widehat{W}_t - \widehat{W}_s)^m \diamond \mathrm{d}W_t$ *is* ill-defined as an Itô integral: *Since we allow* φ *to have support below* s*, the domain of integration involves* t *with* $t < s$*, in which case* $(\widehat{W}(t) - \widehat{W}(s))^m$ *is not adapted with respect to the filtration of* W *at time* t*. The concept of a Skorokhod integral is, in such cases, a natural extension of Itô's notion of integration that boils down to the classical Itô integral once the integrand is adapted. The theory of rough paths morally corresponds to the choice* $\varphi(t) = \mathbf{1}_{[s,t']}(t)$ *for* $s < t'$ *(see also* [147, *Section* 13.3.2]*), which explains why issues of this kind never arise in the rough path framework.*

There is in fact a way to write the integral $\int_0^\infty \varphi(t) \, (\widehat{W}_t - \widehat{W}_s)^m \diamond \mathrm{d}W_t$ *as an Itô integral: Expand* $(\widehat{W}_t - \widehat{W}_s)^m$ *via the binomial theorem and "pull out" all factors depending on* s *(such a point of view is essentially behind Chen's relation in* (10.30)*). However, as this seems like a rather nebulous definition, we consider it more convenient to work with a more appropriate notion of integration in this chapter.*

***Proof of Lemma* 10.12.** We prove this by re-expressing $\mathbb{W}^k_{s,t}$. For $s < t$ we already have

$$\mathbb{W}^k_{s,t} = \int_s^t \mathrm{d}W_r \diamond (\widehat{W}_r - \widehat{W}_s)^k,$$

so that it remains to see what happens for $t < s$. With relation (10.30) we then have

$$\mathbb{W}^k_{s,t} = -\sum_{l=0}^{k} \binom{k}{l} (\widehat{W}_t - \widehat{W}_s)^l \cdot \int_{\mathbb{R}} \mathrm{d}r\, \dot{W}_r \diamond (\widehat{W}_r - \widehat{W}_t)^{k-l} \mathbf{1}_{t<r<s},$$

where we use, for the sake of concision, formal notation, which is easy to translate into a rigorous formulation. Using the fact that for Gaussian random variables U_1, V, U_2 we have

$$U_1^l \cdot (V \diamond U_2^{k-l}) = V \diamond (U_1^l U_2^{k-l}) + l\mathbb{E}[VU_1]\, U_1^{l-1} U_2^{k-l} \tag{10.37}$$

(a consequence of [216, Theorems 3.15 and 7.33]), we obtain

$$\begin{aligned}\mathbb{W}^k s, t = &- \int_{\mathbb{R}} \mathrm{d}r\, \dot{W}_r \diamond (\widehat{W}_r - \widehat{W}_s)^k \mathbf{1}_{t<r<s} \\ &- \sum_{l=0}^{k} \binom{k}{l} l \cdot \int_{\mathbb{R}} \mathrm{d}r\, \mathbb{E}[\dot{W}_r \cdot (\widehat{W}_t - \widehat{W}_s)] \cdot (\widehat{W}_t - \widehat{W}_s)^{l-1} \cdot (\widehat{W}_r - \widehat{W}_t)^{k-l}.\end{aligned}$$

Using $\binom{k}{l} = k\binom{k-1}{l-1}$ and $\mathbb{E}[\dot{W}(r)\cdot(\hat{W}(t) - \widehat{W}(s))] = -K(s-r)\mathbf{1}_{r>0}$ for $t < r < s$, we can reformulate this expression and obtain

$$\mathbb{W}^k(s,t) = -\int \mathrm{d}W(r) \diamond (\widehat{W}(r) - \widehat{W}(s))^k \mathbf{1}_{t<r<s} + k \int \mathrm{d}r K(s-r)(\widehat{W}(r) - \widehat{W}(s))^{k-1} \mathbf{1}_{r>0}.$$

(An alternative derivation can be given following [253, Theorem 3.2].) Since $\Pi_s \Xi \mathcal{I}(\Xi)^m(\varphi) = \int_{\mathbb{R}} \varphi(t)\, \mathrm{d}_t \mathbb{W}^m_{s,t}$, the claim follows. □

Let us also re-express the approximating model in suitable form.

Lemma 10.13. *For $\varphi \in C_c^\infty(\mathbb{R})$, $s \in \mathbb{R}$, $m \in \{1, \dots, M\}$ we have*

$$\begin{aligned}\Pi_s^\varepsilon \Xi \mathcal{I}(\Xi)^m(\varphi) = &\int_0^\infty \varphi(t)\, (\widehat{W}_t^\varepsilon - \widehat{W}_s^\varepsilon)^m \diamond \mathrm{d}W_t^\varepsilon \\ &- m \int_0^\infty \varphi(t)\, \mathscr{K}^\varepsilon(s,t)(\widehat{W}_t^\varepsilon - \widehat{W}_s^\varepsilon)^{m-1} \mathrm{d}t \\ &+ m \int_0^\infty \varphi(t)\, \mathscr{K}^\varepsilon(t,t)(\widehat{W}_t^\varepsilon - \widehat{W}_s^\varepsilon)^{m-1}\, \mathrm{d}t,\end{aligned}$$

where $\diamond$ is defined as in (10.34)*, and where*

$$\mathscr{K}^\varepsilon(u,v) := \mathbb{E}[\widehat{W}^\varepsilon(u)\dot{W}^\varepsilon(v)] = \mathbf{1}_{u,v\geq 0} \int_0^\infty \int_0^\infty \delta^\varepsilon(v, x_1)\delta^\varepsilon(x_1, x_2) K(u - x_2)\, \mathrm{d}x_1 \mathrm{d}x_2. \tag{10.38}$$

Proof. Using that for Gaussian random variables V, U we have $VU^m = V \diamond U^m + m\mathbb{E}[VU]U^{m-1}$ (see (10.37) with $U_2 = 1$), we can rewrite as

$$\begin{aligned}\Pi_s^\varepsilon \Xi \mathcal{I}(\Xi)^m(\varphi) = &\int_0^\infty \varphi(t)\, (\widehat{W}_t^\varepsilon - \widehat{W}_s^\varepsilon)^m \diamond \mathrm{d}W_t^\varepsilon \\ &+ m \int_0^\infty \mathrm{d}t\, \varphi(t)\, \mathbb{E}[\dot{W}_t^\varepsilon\, (\widehat{W}_t^\varepsilon - \widehat{W}_s^\varepsilon)](\widehat{W}_t^\varepsilon - \widehat{W}_s^\varepsilon)^{m-1}.\end{aligned}$$

Inserting $\mathbb{E}[\dot{W}^\varepsilon(t)\, (\widehat{W}^\varepsilon(t) - \widehat{W}^\varepsilon(s))] = \mathscr{K}^\varepsilon(t,t) - \mathscr{K}^\varepsilon(s,t)$ shows the identity. □

Comparing the expressions in Lemmas 10.13 and 10.12, we see that morally we have to subtract

$$m \int \varphi(t)\, \mathscr{K}^{\varepsilon}(t,t)(\widehat{W}_t^{\varepsilon} - \widehat{W}_s^{\varepsilon})^{m-1}\, \mathrm{d}t$$

from the model, which will give us a new model $\hat{\Pi}^{\varepsilon}$. Of course we have to be careful that this step preserves "Chen's relation" $\hat{\Pi}_s^{\varepsilon}\Gamma_{st} = \hat{\Pi}_t^{\varepsilon}$; see Theorem 10.15.

If we interpret $\mathscr{K}^{\varepsilon}$ as an approximation to the Volterra kernel, we see that the expression

$$\mathscr{C}^{\varepsilon}(t) := \mathscr{K}^{\varepsilon}(t,t),\ t \geq 0,$$

will correspond to something like "$0^{H-\frac{1}{2}} = \infty$" in the limit $\varepsilon \downarrow 0$. We indeed have the following upper bound.

Lemma 10.14. *For all $s,t \in \mathbb{R}$ we have*

$$|\mathscr{K}^{\varepsilon}(s,t)| \lesssim \varepsilon^{H-\frac{1}{2}}.$$

Proof. $|\mathscr{K}^{\varepsilon}(s,t)| \lesssim \varepsilon^{-2} \int_{B(t,c\varepsilon)} \mathrm{d}x \int_{B(x,c\varepsilon)} \mathrm{d}u\, |s-u|^{H-\frac{1}{2}} \lesssim \varepsilon^{H-\frac{1}{2}}$. □

Our hope is now that the new model $\hat{\Pi}^{\varepsilon}$ converges to Π in a suitable sense. Similar to [191, equation (2.17)], we define the distance between two models (Π,Γ) and $(\tilde{\Pi},\tilde{\Gamma})$ on a compact time interval $[0,T]$ as

$$\begin{aligned} \|(\Pi,\Gamma);(\tilde{\Pi},\tilde{\Gamma})\|_T := & \sup_{\substack{\operatorname{supp}\varphi \subseteq B(0,1),\\ \lambda \in (0,1],\\ s \in [0,T], \tau \in S}} \lambda^{-|\tau|} |(\Pi_s - \tilde{\Pi}_s)\tau(\varphi_s^{\lambda})| \\ & + \sup_{\substack{t,s \in [0,T],\\ \tau \in S, A \ni \beta < |\tau|}} \frac{|\Gamma_{ts}\tau - \tilde{\Gamma}_{ts}\tau|_{\beta}}{|t-s|^{|\tau|-\beta}}, \end{aligned} \tag{10.39}$$

where $|\cdot|_{\beta}$ denotes the absolute value of the coefficient of the symbol τ' with $|\tau'| = \beta$, and where the first supremum runs over $\varphi \in C_c^1$ with $\|\varphi\|_{C^1} \leq 1$. We will also need

$$\|\Pi\|_T = \sup_{\substack{\operatorname{supp}\varphi \subseteq B(0,1),\\ \lambda \in (0,1],\\ s \in [0,T], \tau \in S}} \lambda^{-|\tau|} |\Pi_s \tau(\varphi_s^{\lambda})|.$$

We are now ready to give the fundamental result of this subsection. Recall that the (minimal) homogeneity is $|\Xi| = -\frac{1}{2} - \kappa$, which corresponds to W being Hölder with exponent $\frac{1}{2} - \kappa$.

Theorem 10.15. *For every $s \in [0,T]$, define the linear map $\hat{\Pi}_s^{\varepsilon} : \mathcal{T} \to C_c^1(\mathbb{R})'$ given by*

$$\hat{\Pi}_s^{\varepsilon}\Xi\mathcal{I}(\Xi)^m = \Pi_s^{\varepsilon}\Xi\mathcal{I}(\Xi)^m - m\mathscr{C}^{\varepsilon}(\cdot)\Pi_s^{\varepsilon}(\mathcal{I}(\Xi)^{m-1}), \quad m \in \{1,\dots,M\},$$

and $\hat{\Pi}_s^{\varepsilon} = \Pi_s^{\varepsilon}$ on all remaining symbols in S. Then

$$(\hat{\Pi}^{\varepsilon}, \hat{\Gamma}^{\varepsilon}) := (\hat{\Pi}^{\varepsilon}, \Gamma^{\varepsilon})$$

defines a ("renormalized") model on $(\mathcal{T}, A)$. On compact time intervals we have

$$\left\| \|(\hat{\Pi}^\varepsilon, \hat{\Gamma}^\varepsilon); (\Pi, \Gamma)\|_T \right\|_{L^p} \lesssim \varepsilon^{\delta\kappa}, \tag{10.40}$$

for any $\delta \in (0,1)$ and $p \in [1,\infty)$. In particular, the distance between the renormalized model and the Itô model almost decays with rate H for $M = M(\kappa, H)$ large enough.

Remark 10.2.5. *In the special case of the level-2 Brownian rough path (i.e., $H = \frac{1}{2}$, $M = 1$), the above result is in precise agreement with known results—but note that we are dealing with the simple case of scalar Brownian motion. More specifically, we don't see the usual (strong) rate "almost" $\frac{1}{2}$ but have to subtract the Hölder exponent used in the rough path/model topology (here: $\frac{1}{2} - \kappa$), which almost leads to the rate κ. Since $M = 1$ entails the condition $\frac{1}{2} - \kappa > 1/3$, we see that $\kappa < 1/6$, exactly as given, e.g., in* [147, *Exercise* 10.14]. *A better rate can be achieved by working with higher-level rough paths (here: $M > 1$), and, indeed, the special case of $H = \frac{1}{2}$ but general M can be seen as a consequence of* [148]*: at the price of working with $\sim 1/(\frac{1}{2} - \kappa)$ levels, one can choose κ arbitrarily close to $\frac{1}{2}$ and so recover the usual "almost" $\frac{1}{2}$ rate. Of course, the case $H < \frac{1}{2}$ is out of reach of rough path considerations.*

Proof. Due to Lemma 10.14, we have for fixed ε that $\sup_{t\in[0,T]} |\mathscr{C}^\varepsilon(t)| < \infty$. As $|\Pi_s \mathcal{I}(\Xi)^m| \lesssim |\cdot - s|^{mH}$, the bound (10.32) is still satisfied. The modification $\hat{\Pi}^\varepsilon_s \Xi \mathcal{I}(\Xi)^m - \Pi^\varepsilon_s \Xi \mathcal{I}(\Xi)^m$ does not lead to a violation of "Chen's relation." Indeed, using validity of (10.31) for the original model, we have

$$\begin{aligned}
\hat{\Pi}^\varepsilon_t \Gamma^\varepsilon_{ts}(\Xi \mathcal{I}(\Xi)^k) &= \hat{\Pi}^\varepsilon_t \left(\sum_{l=0}^{k} \binom{k}{l} (\widehat{W}^\varepsilon_t - \widehat{W}^\varepsilon_s)^l \Xi \mathcal{I}(\Xi)^{k-l} \right) \\
&= \Pi^\varepsilon_s(\Xi \mathcal{I}(\Xi)^k) - \sum_{l=0}^{k} \binom{k}{l} (\widehat{W}^\varepsilon_t - \widehat{W}^\varepsilon_s)^l (k-l) \mathscr{C}^\varepsilon(\cdot)(\widehat{W}^\varepsilon_\cdot - \widehat{W}^\varepsilon_t)^{k-l-1} \\
&= \Pi^\varepsilon_s(\Xi \mathcal{I}(\Xi)^k) - k\mathscr{C}^\varepsilon(\cdot) \sum_{l=0}^{k-1} \binom{k-1}{l} (\widehat{W}^\varepsilon_t - \widehat{W}^\varepsilon_s)^l (\widehat{W}^\varepsilon_\cdot - \widehat{W}^\varepsilon_t)^{k-l-1} \\
&= \Pi^\varepsilon_s(\Xi \mathcal{I}(\Xi)^k) - k\mathscr{C}^\varepsilon(\cdot)(\widehat{W}^\varepsilon_\cdot - \widehat{W}^\varepsilon_s)^k = \hat{\Pi}^\varepsilon_s(\Xi(\mathcal{I}(\Xi)^k).
\end{aligned}$$

We see that (10.31) is also satisfied after our modification, and we easily conclude that $(\hat{\Pi}^\varepsilon, \Gamma^\varepsilon)$ is still a model on $(\mathcal{T}, A)$. The bound (10.40) is a bit technical and left to Appendix 10.6.1. □

10.2.2 ▪ Approximation and renormalization theory

We now address the central question of how the integral $\int_0^t f(\widehat{W}^\varepsilon_r, r)\,\mathrm{d}W^\varepsilon_r$ has to be modified to make it convergent to $\int_0^t f(W_r, r)\mathrm{d}W_r$. The key idea is to combine the convergence result from Theorem 10.15 with Hairer's reconstruction theorem, which we state below.

We first recall the notion of a modeled distribution [191, Definition 3.1]. We say that a map $F : \mathbb{R} \to \mathcal{T}$ is in the space $\mathcal{D}^\gamma_T(\Gamma)$, $\gamma > 0$, for some time horizon $T > 0$ if

$$\|F\|_{\mathcal{D}^\gamma_T(\Gamma)} := \sup_{A\ni\beta<\gamma, s\in[0,T]} |F(s)|_\beta + \sup_{A\ni\beta<\gamma, s,t\in[0,T],\, s\neq t} \frac{|F(t) - \Gamma_{ts}F(s)|_\beta}{|t-s|^{\gamma-\beta}} < \infty, \tag{10.41}$$

where, as above, $|\cdot|_\beta$ denotes the absolute value of the coefficient of τ with $|\tau| = \beta$. Given two models (Π,Γ) and $(\overline{\Pi},\overline{\Gamma})$ and two $F,\overline{F}:\mathbb{R}\mapsto\mathcal{T}$ it is also useful to have the notion of a distance

$$\|F;\overline{F}\|_{\mathcal{D}_T^\gamma(\Gamma),\mathcal{D}_T^\gamma(\overline{\Gamma})} := \sup_{A\ni\beta<\gamma,\, t\in[0,T]} |F(t)-\overline{F}(t)|_\beta + \sup_{A\ni\beta<\gamma,\, s,t\in[0,T],\, s\neq t} \frac{|F(t)-\Gamma_{ts}F(s)-(\overline{F}(t)-\overline{\Gamma}_{ts}\overline{F}(s))|_\beta}{|t-s|^{\gamma-\beta}}.$$

The reconstruction theorem now states that for $\gamma>0$, a map $F\in\mathcal{D}_T^\gamma(\Gamma)$ can be uniquely identified with a distribution that behaves locally like $\Pi.F(\cdot)$.

Theorem 10.16 ([191, Theorem 3.10]). *Given a model* (Π,Γ), $\gamma>0$, *and* $T>0$, *there is a unique continuous operator*[56] $\mathcal{R}:\mathcal{D}_T^\gamma(\Gamma)\to C^{|\Xi|}(\mathbb{R})$ *such that for any* $s\in[0,T]$ *and* $\varphi\in C_c^1(B(0,1))$,

$$|(\mathcal{R}F-\Pi_sF(s))(\varphi_s^\lambda)|\lesssim \|\Pi\|_T\,\lambda^\gamma. \tag{10.42}$$

For two different models (Π,Γ) *and* $(\overline{\Pi},\overline{\Gamma})$ *we further have*

$$\begin{aligned}&\left|(\mathcal{R}F-\Pi_sF(s)-(\overline{\mathcal{R}}\overline{F}-\Pi_s\overline{F}(s)))(\varphi_s^\lambda)\right|\\ &\lesssim \lambda^\gamma\left(\|F\|_{\mathcal{D}_T^\gamma(\Gamma)}\,\|(\Pi,\Gamma);(\overline{\Pi},\overline{\Gamma})\|_T+\|\Pi\|_T\|F;\overline{F}\|_{\mathcal{D}_T^\gamma(\Gamma);\mathcal{D}_T^\gamma(\overline{\Gamma})}\right)\end{aligned} \tag{10.43}$$

for $F\in\mathcal{D}_T^\gamma(\Gamma)$, $\overline{F}\in\mathcal{D}_T^\gamma(\overline{\Gamma})$.

As mentioned earlier, we prefer to work with compactly supported functions $\varphi\in\mathcal{B}_{1,\infty}^\beta(\mathbb{R}^d)$, $\beta>-|\Xi|$, which include objects like the Haar wavelets. The following lemma allows us to carry over all bounds.

Lemma 10.17. *The bounds* (10.32), (10.39), (10.42), *and* (10.43) *still hold for* $\varphi\in\mathcal{B}_{1,\infty}^\beta(\mathbb{R}^d)$, $\beta>-|\Xi|$, *with compact support in* $B(0,1)$ *(after a change of constants).*

Remark 10.2.6. *This covers, in particular, functions like* $\mathbf{1}_{[0,1]}\in\mathcal{B}_{1,\infty}^1(\mathbb{R})$.

Proof. We prove this via wavelet methods in the appendix. □

In the following, we introduce $X^{(\varepsilon)}$ to denote both X and X^ε. To study objects like $\int_0^t f(\widehat{W}_r^{(\varepsilon)},r)\,\mathrm{d}W_r^{(\varepsilon)}$ with the reconstruction theorem, we first "expand" the integrand $f(\widehat{W}_r^{(\varepsilon)},r)$ in the regularity structure $\mathcal{T}$, obtaining

$$F^{(\varepsilon)}(s) := \sum_{m=0}^M \frac{1}{m!}\partial_1^m f(\widehat{W}_s^{(\varepsilon)},s)\mathcal{I}(\Xi)^m.$$

On the level of the regularity structure these objects can be multiplied with the "noise" Ξ, which gives a modeled distribution on $\mathcal{T}$. We will analyze $F^{(\varepsilon)}$ by writing it as the composition of a (random) modeled distribution with the smooth function f. To this end we need the following lemma.

Lemma 10.18. *On the regularity structure* $(\mathcal{T},A,G)$ *introduced in Section* 10.2.1, *consider a model* (Π,Γ) *which is admissible in the sense that*

$$\Pi_t\mathcal{I}(\Xi) = (K*\Pi_t\Xi)(\cdot)-(K*\Pi_t\Xi)(t)\,.$$

[56] $C^{|\Xi|}(\mathbb{R})$ denotes the space of distributions that are locally in the Besov space $\mathcal{B}_{\infty,\infty}^{|\Xi|}(\mathbb{R})$ [191, Remark 3.8].

Then

$$\mathcal{K}\Xi(t) := \mathcal{I}(\Xi) + (K * \Pi_t\Xi)(t)\mathbf{1} \tag{10.44}$$

defines a modeled distribution. More precisely, $\mathcal{K}\Xi \in \mathcal{D}_T^\infty := \bigcup_{\gamma<\infty} \mathcal{D}_T^\gamma$.

Remark 10.2.7. *Our notion of admissibility mimics* [191, *Definition* 5.9], *which, however, is not directly applicable here (due to the failure of Assumption* 5.4 *in* [191]).

Proof. By definition of the space of modeled distribution, we need to understand the action of Γ_{st} on all constituting symbols. Since $\{\mathbf{1}, \mathcal{I}(\Xi)\}$ span a sector, i.e., a space invariant by the action of the structure group, it is clear that

$$\Gamma_{st}\mathcal{I}(\Xi) = \mathcal{I}(\Xi) + (\ldots)\mathbf{1}.$$

Application of the realization map Π_s, followed by evaluation at s, immediately identifies $(\ldots)$ with

$$\Pi_t\mathcal{I}(\Xi)(s) - \Pi_s\mathcal{I}(\Xi)(s) = \Pi_t\mathcal{I}(\Xi)(s) = (K * \Pi_s\Xi)(s) - (K * \Pi_t\Xi)(t),$$

where we used admissibility and $\Pi_s\Xi = \Pi_t\Xi$ in the last step. As a consequence, $\Gamma_{st}\mathcal{K}\Xi(t) \equiv \mathcal{K}\Xi(s)$ so that trivially $\mathcal{K}\Xi \in \mathcal{D}_T^\gamma$ for any $\gamma < \infty$. □

For a given (sufficiently smooth) function f, and a generic model (Π, Γ) on our regularity structure, define

$$F^\Pi : s \mapsto \sum_{m=0}^{M} \frac{1}{m!}\partial_1^m f((\mathcal{R}\mathcal{K}\Xi(s), s)\mathcal{I}(\Xi)^m .$$

Remark that $\mathcal{K}\Xi(s)$ is function-like, i.e., takes values in the span of symbols with nonnegative degree. From [191, Proposition 3.28] we then have

$$\mathcal{R}\mathcal{K}\Xi(s) = \langle \mathcal{K}\Xi(s), \mathbf{1}\rangle = K * \Pi_s\Xi .$$

(In particular, we see that $F^{(\varepsilon)}(s)$ coincides with F^Π when Π is taken as either the approximate or the renormalized approximate model.) We can also define ΞF^Π, simply obtained by multiplying F^Π with Ξ. The properties of F^Π and ΞF^Π are summarized in the following lemma.

Lemma 10.19. *Given* $f \in C_b^{2M+3}([0,T] \times \mathbb{R})$*, there exists* $N > 0$ *such that, for all* $\gamma \in (\frac{1}{2} + \kappa, 1)$,

$$\|F^\Pi\|_{\mathcal{D}_T^\gamma(\Gamma)} \lesssim \|\Pi\|_T^N, \quad \|\Xi F^\Pi\|_{\mathcal{D}_T^{\gamma+|\Xi|}(\Gamma)} \lesssim \|\Pi\|_T^N .$$

We further have, for two given models (Π, Γ) *and* (Π', Γ'),

$$|\!|\!| F^\Pi; F^{\Pi'} |\!|\!|_{\mathcal{D}_T^\gamma(\Gamma);\mathcal{D}_T^\gamma(\Gamma')} \lesssim \left(\|\Pi\|_T^N + \|\Pi'\|_T^N\right) |\!|\!| (\Pi,\Gamma);(\Pi',\Gamma') |\!|\!|_T, \tag{10.45}$$

$$|\!|\!| \Xi F^\Pi; \Xi F^{\Pi'} |\!|\!|_{\mathcal{D}_T^{\gamma+|\Xi|}(\Gamma);\mathcal{D}_T^{\gamma+|\Xi|}(\Gamma')} \lesssim \left(\|\Pi\|_T^N + \|\Pi'\|_T^N\right) |\!|\!| (\Pi,\Gamma);(\Pi',\Gamma') |\!|\!|_T, \tag{10.46}$$

where the proportionality constants are, in particular, uniform over all f *with bounded* C^{2M+3} *norm.*

Proof. The map F^Π is simply the composition (in the sense of [191, Section 4.2]) of the function f with the modeled distributions $\mathcal{K}\Xi$ and $s \mapsto s\mathbf{1}$. The result then follows from [191, Theorem 4.16] (polynomial dependence in $\|\Pi\|_T$ is not stated there but is clear from the proof). □

Remark 10.2.8. *In the case when $f \in C^{2M+3}$, but with no global bounds, the result still holds since we only consider the values of f on the range of the continuous function $\mathcal{RK}\Xi$ (which is bounded by some $R \geq 0$). The resulting bounds then depend linearly on $\|f\|_{C^{2M+3}(B_R \times [0,T])}$.*

In the case of the Itô model (Π, Γ) and the approximating renormalized models $(\hat{\Pi}^\varepsilon, \Gamma^\varepsilon)$, we simply denote F^Π by F and F^ε, respectively. We are then allowed to apply Hairer's reconstruction theorem (see Theorem 10.16). Note that we have two reconstruction operators $\mathcal{R}$ and $\mathcal{R}^\varepsilon$, because we start with two models. $\mathcal{R}^{(\varepsilon)}\Xi F^{(\varepsilon)}$ can be written down explicitly.

Lemma 10.20. *We have (a.s.)*

$$\mathcal{R}F\Xi(\varphi) = \int_{\mathbb{R}} \varphi(t)\, f(\widehat{W}_t, t)\, \mathrm{d}W_t,$$

$$\mathcal{R}^\varepsilon F^\varepsilon \Xi(\varphi) = \int_{\mathbb{R}} \varphi(t)\, f(\widehat{W}_t^\varepsilon, t)\, \mathrm{d}W_t^\varepsilon - \int_{\mathbb{R}} \mathscr{K}^\varepsilon(t,t)\partial_1 f(\widehat{W}_t^\varepsilon, t)\varphi(t)\, \mathrm{d}t.$$

Proof. The proof is given in the appendix. □

If we take $\varphi = \mathbf{1}_{[0,T)}$, we obtain $\mathcal{R}F\Xi(\mathbf{1}_{[0,T)}) = \int_0^T f(\widehat{W}(t), t)\, \mathrm{d}W(t)$, so that it is natural to choose $\tilde{\mathscr{I}}_f^\varepsilon(T) = \mathcal{R}^\varepsilon \Xi F^\varepsilon(\mathbf{1}_{[0,T)})$ as an approximation. However, note that the key property of the reconstruction operator $\mathcal{R}^{(\varepsilon)}$ is that it is locally close to the corresponding model $\Pi^{(\varepsilon)}$, so that we, in fact, have two natural approximations.

Definition 10.21. *For F^ε, as in Lemma 10.19, and $t \geq 0$ we set*

$$\tilde{\mathscr{I}}_f^\varepsilon(t) := \mathcal{R}^\varepsilon \Xi F^\varepsilon(\mathbf{1}_{[0,t]}) = \int_0^t f(\widehat{W}_r^\varepsilon, r)\, \mathrm{d}W_r^\varepsilon - \int_0^t \mathscr{C}^\varepsilon(r)\partial_1 f(\widehat{W}_r^\varepsilon, r)\, \mathrm{d}r.$$

For a fixed partition $\{[t_l^\varepsilon, t_{l+1}^\varepsilon)\}$ of $[0,t)$ with $\left|t_{l+1}^\varepsilon - t_l^\varepsilon\right| \lesssim \varepsilon$ we further set

$$\begin{aligned}
\tilde{\mathscr{J}}_{f,M}^\varepsilon(t) &:= \sum_{[t_l^\varepsilon, t_{l+1}^\varepsilon)} \hat{\Pi}_{t_l}^\varepsilon \Xi F_{t_l}^\varepsilon(\mathbf{1}_{[t_l^\varepsilon, t_{l+1}^\varepsilon)}) \\
&= \sum_{[t_l^\varepsilon, t_{l+1}^\varepsilon)} \sum_{m=0}^{M} \frac{1}{m!} \partial_1^m f(\widehat{W}^\varepsilon(t_l^\varepsilon), t_l^\varepsilon) \int_{t_l^\varepsilon}^{t_{l+1}^\varepsilon} \left(\widehat{W}^\varepsilon(r) - \widehat{W}^\varepsilon(t_l^\varepsilon)\right)^m \mathrm{d}W^\varepsilon(r) \\
&\quad - \sum_{m=1}^{M} \frac{1}{(m-1)!} \partial_1^m f(\widehat{W}^\varepsilon(t_l^\varepsilon), t_l^\varepsilon) \int_{t_l^\varepsilon}^{t_{l+1}^\varepsilon} \mathscr{C}^\varepsilon(r) \left(\widehat{W}^\varepsilon(r) - \widehat{W}^\varepsilon(t_l^\varepsilon)\right)^{m-1} \mathrm{d}r.
\end{aligned}$$

We might drop the indices f and f, M on $\tilde{\mathscr{I}}^\varepsilon$ and $\tilde{\mathscr{J}}^\varepsilon$ if there is no risk of confusion.

The following theorem, which can be seen as the fundamental theorem of our regularity structure approach to rough pricing, shows that these approximations both do converge.

Theorem 10.22. *Fix $T > 0$. For f smooth, bounded with bounded derivatives, and $\tilde{\mathscr{I}}_f^\varepsilon$, $\tilde{\mathscr{J}}_{f,M}^\varepsilon$ as in Definition 10.21 we have that*

(i) *for any $\delta \in (0,1)$ and any $p < \infty$ there exists C such that*

$$\left\| \sup_{t \in [0,T]} \left| \tilde{\mathscr{I}}_f^\varepsilon(t) - \int_0^t f(\widehat{W}_r, r)\mathrm{d}W_r \right| \right\|_{L^p} \leq C\varepsilon^{\delta H}; \tag{10.47}$$

(ii) *for every $\delta \in (0,1)$ we can pick $M = M(\delta, H)$ large enough, such that, for any $p < \infty$, there exists C such that*

$$\left\| \sup_{t\in[0,T]} \left| \tilde{\mathscr{I}}^{\varepsilon}_{f,M}(t) - \int_0^t f(\widehat{W}_r, r)\mathrm{d}W_r \right| \right\|_{L^p} \le C\varepsilon^{\delta H}. \tag{10.48}$$

Remark 10.2.9. *With regard to* (i), *although neither $\tilde{\mathscr{I}}^{\varepsilon}_f(t)$ nor its (Itô) limit depends on the choice of M, the choice of M affects the entire regularity structure and so, implicitly also the reconstruction operator $\mathcal{R}^{\varepsilon}$ used in the definition of $\tilde{\mathscr{I}}^{\varepsilon}_f$, as well as the modeled distribution F^{ε}. The latter, in turn, requires $f \in C^M$ for the construction to make sense. If δ is chosen arbitrarily close to one, f needs to have derivatives of arbitrary order; hence we have our smoothness assumption.*

Remark 10.2.10. *By an easy localization argument one shows that for f smooth (but without any further bounds) one still has*

$$\sup_{\varepsilon\in(0,1]} \mathbb{P}\left(\sup_{t\in[0,T]} \left| \tilde{\mathscr{I}}^{\varepsilon}_f(t) - \int_0^t f(\widehat{W}_r, r)\mathrm{d}W_r \right| \le C\varepsilon^{\delta H} \right) \to 0$$

with $C \to \infty$. The original rough volatility model due to [166] *makes a point that f should be of exponential form. Now, the result with L^p-estimates still holds since we only consider the values of f on the range of the continuous function $\mathcal{R}\mathcal{K}\Xi$ (which is bounded by some $R \ge 0$). As pointed out in Remark* 10.2.8, *the bounds then depend linearly on $\|f\|_{C^{M+2}(B_R\times[0,T])}$. Since, for us, (Π,Γ) is always a Gaussian model, $\mathcal{R}\mathcal{K}\Xi$ is a Gaussian process (say $\widehat{W}$ or $\widehat{W}^{\varepsilon}$), and hence we have Gaussian concentration of Fernique-type for $\sup_{t\in[0,T]} |\mathcal{R}\mathcal{K}\Xi(t)|$. So, for instance, if f and its derivatives have exponential growth, we do have the L^p bounds of the above theorem, for all $p < \infty$. This remark justifies in particular the choice $f(x) = \exp(x)$ and $p = 2$ in the numerical discussion of Section* 10.5.

Remark 10.2.11. *In* [246] *it is shown (in a slightly different setting) that the strong rate for the standard Euler scheme (or, more precisely, left-point rule) is no better than H in general even when the fractional process is exactly simulated. In that sense, the scheme suggested in Theorem* 10.22 *is almost optimal.*

***Proof of Theorem* 10.22.** Without loss of generality $T \le 1$; otherwise, split $[0,T]$ into sufficiently many subintervals. Let us show (10.47).

$$\begin{aligned}
\tilde{\mathscr{I}}^{\varepsilon}_f(t) - \int_0^t f(\widehat{W}_r, r)\mathrm{d}W_r &= (\mathcal{R}^{\varepsilon}(F^{\varepsilon}\Xi) - \mathcal{R}(F\Xi))(\mathbf{1}_{[0,t]}) \\
&= t\big(\hat{\Pi}^{\varepsilon}_0 \Xi F^{\varepsilon}(0) - \Pi_0 \Xi F(0)\big)(t^{-1}\mathbf{1}_{[0,t]}) \\
&\quad + t\big(\mathcal{R}^{\varepsilon}\Xi F^{\varepsilon} - \hat{\Pi}^{\varepsilon}_0 \Xi F^{\varepsilon}(0) - (\mathcal{R}\Xi F - \Pi_0\Xi F(0))\big)(t^{-1}\mathbf{1}_{[0,t)}).
\end{aligned}$$

We then obtain that the error is of order $\varepsilon^{\delta\kappa}$, $\delta \in (0,1)$, using Theorem 10.15, Lemma 10.19, and (10.39) for the first term and also Theorem 10.16 for the second term. Letting $\kappa \uparrow H$ and $M \uparrow \infty$, our total rate can be chosen arbitrary close to H.

To obtain the second estimate we can bound $\tilde{\mathscr{I}}^{\varepsilon}_f(t) - \tilde{\mathscr{I}}^{\varepsilon}_{f,M}(t)$ with the first inequality in Theorem 10.16. □

10.2.3 ▪ Nonconstant vs. constant renormalization

If δ^{ε} comes from a mollifier (Example 10.9), then the renormalization $\mathscr{C}^{\varepsilon} = \mathscr{K}^{\varepsilon}(\cdot,\cdot)$ that was applied in Theorem 10.15 and thus in Definition 10.21 is a constant, which is the familiar concept one encounters in the study of singular SPDEs [190, 191, 90]. If δ^ε comes from wavelets such as the Haar basis, $\mathscr{K}^\varepsilon(\cdot,\cdot)$ is usually not constant but a periodic function with period ε. Thus we see that our analysis gives rise to a "nonconstant renormalization." It is natural to ask if one can do with constant renormalization after all. Assume that $\mathscr{C}^\varepsilon$ is periodic with mean

$$C_\varepsilon = \frac{1}{\varepsilon}\int_0^\varepsilon \mathscr{C}^\varepsilon(t)\mathrm{d}t.$$

From Lemma 10.14 it follows that $\mathscr{C}^\varepsilon$ (and its mean) are bounded by $\varepsilon^{H-\frac{1}{2}}$, uniformly in t. Putting all this together, it easily follows that $|\langle \mathscr{C}^\varepsilon - C_\varepsilon, \varphi\rangle| \lesssim \varepsilon^{\alpha+H-\frac{1}{2}}$, uniformly over all φ bounded in C^α, with convergence to zero when $\alpha > \frac{1}{2} - H$. As a consequence, taking $\varphi(t) = f(\widehat{W}^\varepsilon)$, for smooth f, we clearly can apply this estimate with any $\alpha < H$. Hence, by equating the constraints on α, we arrive at $H > 1/4$. The practical consequence regarding part (i) of Theorem 10.22 then is that we can indeed replace nonconstant renormalization by constant renormalization; however, this is at the price of restricting to $H > 1/4$ and with an according loss on the convergence rate. Interestingly, our numerical simulation suggests that no loss occurs and constant renormalization works for any $H > 0$. While we have refrained from investigating this (technical) point further,[57] we can understand the mechanism at work by looking at the following toy example: Consider the Itô integral $\int_0^1 W_t^h \mathrm{d}W_t$ where W^H is an fBm, but now with Hurst parameter $H > \frac{1}{2}$, built, say, as a Volterra process over W. Using Young integration theory, one can give a pathwise argument that shows that Riemann–Stieltjes approximations converge a.s. (with vanishing rate as $H \to \frac{1}{2}$). However, we know from stochastic theory (Itô integration) that this convergence holds in L^2 (and then in probability) for any $H > 0$. We would thus expect that constant renormalization is still valid when $H \in (0, 1/4]$, but now the difference only vanishes in a mean-square sense. This conjecture was checked numerically in Section 10.5.

10.2.4 ▪ The case of the Haar basis

The following special case of the above approximations to $\int_0^t f(\widehat{W}_r, r)\mathrm{d}W_r$ is of particular interest for our purposes. We next collect additional concrete formulas that arise in this case.

Let $\varepsilon = 2^{-N}$, $\varphi := \mathbf{1}_{[0,1)}$, and $\varphi_{l,N} = 2^{N/2}\varphi(2^N\cdot - l)$, $l \in \mathbb{Z}$. The corresponding approximation δ^ε to the Dirac delta is then

$$\delta^\varepsilon(x,y) = \sum_{l\in\mathbb{Z}} \varphi_{l,N}(x)\varphi_{l,N}(y) = 2^N \mathbf{1}_{[\lfloor x2^N\rfloor 2^{-N}, (\lfloor x2^N\rfloor+1)2^{-N})}(y), \qquad x, y \in \mathbb{R}.$$

The mollified Volterra kernel (10.38) then takes the form

$$\begin{aligned}\mathscr{K}^\varepsilon(u,v) &= \int_0^\infty\int_0^\infty \delta^\varepsilon(v,x_1)\delta^\varepsilon(x_1,x_2)K(u-x_2)\mathrm{d}x_1\mathrm{d}x_2\\ &= \sqrt{2H}\cdot 2^N \int_{[\lfloor v2^N\rfloor 2^{-N}, (\lfloor v2^N\rfloor+1)2^{-N}\wedge u)} |u-x|^{H-\frac{1}{2}}\mathbf{1}_{\lfloor v2^N\rfloor 2^{-N}\le u}\,\mathrm{d}x\\ &= \frac{\sqrt{2H}}{\frac{1}{2}+H}2^N\\ &\times\left(|u-\lfloor v2^N\rfloor 2^{-N}|^{\frac{1}{2}+H} - |u-(\lfloor v2^N\rfloor+1)2^{-N}\wedge u|^{\frac{1}{2}+H}\right)\mathbf{1}_{\lfloor v2^N\rfloor 2^{-N}\le u}.\end{aligned}$$

A special role is played by the *diagonal function* as a renormalization,

$$\mathscr{C}^\varepsilon(t) = \mathscr{K}^\varepsilon(t,t) = \frac{\sqrt{2H}\,2^N}{\frac{1}{2}+H}|t-\lfloor t2^N\rfloor 2^{-N}|^{\frac{1}{2}+H}. \tag{10.49}$$

[57] Some computations lead us to believe that this question can be settled with the aid of mixed $(1,\varrho)$-variation of the covariance function of the Volterra process [146], which we expect to hold uniformly over approximation.

We additionally have

$$\begin{aligned}\widehat{W}_t^\varepsilon &= \int_0^t K(t-r)\,\mathrm{d}W_r^\varepsilon = \sum_{l=0}^{\infty} Z_l \int_0^t K(t-r)\varphi_{k,N}(r)\mathrm{d}r \\ &= \sum_{l=0}^{\infty} 2^{-N/2}\mathscr{K}^\varepsilon(t, l2^{-N})Z_l = \sum_{l=0}^{\lfloor t2^N \rfloor} 2^{-N/2}\mathscr{K}^\varepsilon(t, l2^{-N})Z_l,\end{aligned}$$

where $Z_l = \langle \dot{W}, \varphi_{l,N}\rangle$ are i.i.d. $N(0,1)$-distributed variables. As our approximation, we can finally take $\mathscr{I}_f^\varepsilon(t)$ from Definition 10.21 with partition $\{[t_l, t_{l+1})\} = \{[l2^{-N}, (l+1)2^{-N} \wedge t)\}$, which gives us

$$\begin{aligned}\tilde{\mathscr{I}}_{f,M}^\varepsilon(t) = &\sum_{l=0}^{\lceil t2^N\rceil -1} \sum_{m=0}^{M} \frac{1}{m!}\partial_1^m f(\widehat{W}_{t_l}^\varepsilon, t_l)2^{N/2}Z_l \int_{t_l}^{t_{l+1}} \left(\widehat{W}_r^\varepsilon - \widehat{W}_{t_l}^\varepsilon\right)^m \mathrm{d}r \\ &- \sum_{m=1}^{M} \frac{1}{(m-1)!}\partial_1^m f(\widehat{W}_{t_l}^\varepsilon, t_{l+1}) \int_{t_l}^{t_{l+1}} \mathscr{C}^\varepsilon(r)\left(\widehat{W}_r^\varepsilon - \widehat{W}_{t_l}^\varepsilon\right)^{m-1} \mathrm{d}r\end{aligned}$$

and

$$\tilde{\mathscr{I}}_f^\varepsilon(t) = \sum_{l=0}^{\lceil t2^N\rceil -1} \int_{t_l}^{t_{l+1}} [2^{N/2}Z_l \cdot f(\widehat{W}_r^\varepsilon, r)\,\mathrm{d}r - \mathscr{C}^\varepsilon(r)\,\partial_1 f(\widehat{W}_r^\varepsilon, r)]\,\mathrm{d}r.$$

As explained at the end of the last section, $\mathscr{C}^\varepsilon(r)$ in these formulas could be replaced by its local mean, the constant

$$2^N \int_0^{2^{-N}} \mathscr{C}^\varepsilon(r)\,\mathrm{d}r = \frac{\sqrt{2H}}{(H+\frac{1}{2})(H+\frac{3}{2})} 2^{N(\frac{1}{2}-H)}.$$

10.3 ▪ The full rough volatility regularity structure

10.3.1 ▪ Basic setup

We want to add an independent Brownian motion so that we take an additional symbol $\bar{\Xi}$. We again fix M and define

$$\bar{S} := S \cup \{\bar{\Xi}, \bar{\Xi}\mathcal{I}(\Xi), \ldots, \bar{\Xi}\mathcal{I}(\Xi)^M\}, \quad \bar{\mathcal{T}} := \bigoplus_{\tau\in\bar{S}} \mathbb{R}\tau\,. \tag{10.50}$$

We fix $|\bar{\Xi}| = -\frac{1}{2} - \kappa$, and the homogeneity of the other symbols are defined multiplicatively as before. We also set $\widehat{W}_t = \int_0^t K(s,t)\mathrm{d}W_s$ with $K(s,t) = \sqrt{2H}|t-s|^{H-\frac{1}{2}}\mathbb{1}_{t>s}$, where W and $\overline{W}$ are independent Brownian motions.

We extend the canonical model (Π, Γ) to this regularity structure by defining

$$\Pi_s \bar{\Xi}\mathcal{I}(\Xi)^m := \left\{t \mapsto \frac{\mathrm{d}}{\mathrm{d}t}\left(\int_s^t \left(\widehat{W}(u) - \widehat{W}(s)\right)^m d\overline{W}(u)\right)\right\}$$

(the above integral being in Itô sense), and[58]

$$\Gamma_{ts}\big(\bar{\Xi}\mathcal{I}(\Xi)^m\big) = \bar{\Xi}\Gamma_{ts}(\mathcal{I}(\Xi)^m).$$

Arguments similar to the proof of Lemma 10.11 show that this indeed defines a model on $\bar{\mathcal{T}}$.

[58]Upon setting $\Gamma_{ts}(\bar{\Xi}) = \bar{\Xi}$, the given relation is precisely implied by multiplicativity of Γ.

10.3.2 ▪ Small noise model large deviation

Given $\delta > 0$ we consider the "small noise" model $(\Pi^\delta, \Gamma^\delta)$ on $\tilde{T}$ obtained by replacing $W, \overline{W}$ by $\delta W, \delta\overline{W}$, which simply means that

$$\begin{aligned}
\Pi^\delta \mathbf{1} &= 1,\\
\Pi^\delta \mathcal{I}(\Xi)^m &= \delta^m \Pi \mathcal{I}(\Xi)^m,\\
\Pi^\delta \Xi \mathcal{I}(\Xi)^m &= \delta^{m+1} \Pi \Xi \mathcal{I}(\Xi)^m,\\
\Pi^\delta \bar{\Xi} \mathcal{I}(\Xi)^m &= \delta^{m+1} \Pi \bar{\Xi} \mathcal{I}(\Xi)^m,
\end{aligned}$$

and

$$\begin{aligned}
\Gamma^\delta_{ts} \mathbf{1} = \mathbf{1}, \Gamma^\delta_{ts} \Xi &= \Xi, \Gamma^\delta_{ts} \bar{\Xi} = \bar{\Xi},\\
\Gamma^\delta_{ts} \mathcal{I}(\Xi) &= \mathcal{I}(\Xi) + \delta(\widehat{W}(t) - \widehat{W}(s))\mathbf{1},\\
\Gamma^\delta_{ts} \tau\tau' &= \Gamma^\delta_{ts}\tau \cdot \Gamma^\delta_{ts}\tau', \quad \text{for } \tau, \tau' \in \bar{S}.
\end{aligned}$$

Finally, for $h = (h_1, h_2)$ in $\mathcal{H} := L^2([0,T])^2$, we consider the deterministic model (Π^h, Γ^h) defined by

$$\begin{aligned}
\Pi^h \mathbf{1} &= 1,\\
\Pi^h_s \Xi = h_1, \ \Pi^h_s \bar{\Xi} &= h_2,\\
\Pi^h_s \mathcal{I}(\Xi)(t) &= \int_0^{t\vee s} (K(u,t) - K(u,s)) h_1(u) du,\\
\Pi^h \tau\tau' &= \Pi^h\tau \Pi^h \tau' \text{ for } \tau, \tau' \in \bar{S},
\end{aligned}$$

and

$$\begin{aligned}
\Gamma^h_{ts} \mathbf{1} = \mathbf{1}, \Gamma^h_{ts} \Xi &= \Xi, \Gamma^h_{ts} \bar{\Xi} = \bar{\Xi},\\
\Gamma^h_{ts} \mathcal{I}(\Xi) &= \mathcal{I}(\Xi) + \left(\int_0^{t\vee s} (K(u,t) - K(u,s) \right) h_1(u) du)\mathbf{1},\\
\Gamma^h_{ts} \tau\tau' &= \Gamma^h_{ts}\tau \cdot \Gamma^h_{ts}\tau', \quad \text{for } \tau, \tau' \in \bar{S}.
\end{aligned}$$

The following lemma and theorem are proved in Appendix 10.6.2.

Lemma 10.23. *For each $h \in \mathcal{H}$, Π^h does define a model. In addition, the map $h \in \mathcal{H} \mapsto \Pi^h$ is continuous.*

Theorem 10.24. *The models Π^δ satisfy a large deviation principle (LDP) in the space of models with rate δ^2 and rate function given by*

$$J(\Pi) := \begin{cases} \frac{1}{2}\|h\|^2_{\mathcal{H}} & \text{if } \Pi = \Pi^h \text{ for some } h \in \mathcal{H},\\ +\infty & \text{otherwise.} \end{cases}$$

As an immediate corollary we have the following.

Corollary 10.25. *For δ small, $\mathbb{P}(Y^\delta_1 \approx y) \approx \exp[-I(y)/\delta^2]$, in the precise sense of an LDP for*

$$Y^\delta_1 := \int_0^1 f(\delta^H \widehat{W}_s)\delta\big(\varrho dW_s + \bar{\varrho} d\overline{W}_s\big)$$

with speed δ^2*, and with rate function given by*

$$I(y) := \inf_{h_1 \in L^2([0,1])} \left\{ \frac{1}{2}\|h_1\|_{L^2}^2 + \frac{(y - I_1(h_1))^2}{2I_2(h_1)} \right\}, \tag{10.51}$$

where

$$I_1(h_1) = \varrho \int_0^1 f\left(\int_0^s K(u,s)h_1(u)du\right)h_1(s)\mathrm{d}s,$$

$$I_2(h_1) = \int_0^1 f\left(\int_0^s K(u,s)h_1(u)du\right)^2 \mathrm{d}s.$$

Remark 10.3.1. *This improves a similar result obtained in* [138]. *In fact, we now cover functions* f *of exponential form, as required in rough volatility modeling* [166, 39, 40].

Proof. Note that

$$Y_1^\delta = \left\langle \mathcal{R}^\delta F^\delta \cdot (\varrho\Xi + \bar{\varrho}\bar{\Xi}), 1_{[0,1]} \right\rangle,$$

where $F^\delta \equiv F^{\Pi^\delta}$ as defined in Lemma 10.19. By the contraction principle and the continuity estimate from Theorem 10.16, it holds that Y_1^δ satisfies an LDP, with rate function given by

$$I(y) = \inf\left\{ \frac{1}{2}\left(\|h_1\|_{L^2}^2 + \|h_2\|_{L^2}^2\right), \quad y = \left\langle \mathcal{R}^h F^h \cdot (\varrho\Xi + \bar{\varrho}\bar{\Xi}), 1_{[0,1]} \right\rangle \right\},$$

where we used $F^h \equiv F^{\Pi^h}$. It then suffices to note that

$$\left\langle \mathcal{R}^h\left(F^h \cdot (\varrho\Xi + \bar{\varrho}\bar{\Xi})\right), 1_{[0,1]} \right\rangle = \int_0^1 f\left(\int_0^s K(u,s)h_1(u)\mathrm{d}u\right)(\varrho h_1(s)\mathrm{d}s + \bar{\varrho}h_2(s)\mathrm{d}s),$$

and optimizing over h_2 for fixed h_1, we obtain (10.51). □

We note that thanks to Brownian (resp., fractional Brownian) scaling, small noise large deviations translate immediately into short-time large deviations [138].

Although the rate function here is not given in a very useful form, it is possible to expand it in small y and thus compute (explicitly in terms of the model parameters) higher-order moderate deviations. In [40] this was related to implied volatility skew expansions.

10.4 ▪ Rough Volterra dynamics for volatility

10.4.1 ▪ Motivation from market microstructure

Rosenbaum and coworkers, [126, 129, 131] show that stylized facts of modern market microstructure naturally give rise to fractional dynamics and leverage effects. Specifically, they construct a sequence of Hawkes processes, suitably rescaled in time and space, that converges in law to a rough volatility model of rough Heston form

$$\mathrm{d}S_t/S_t = \sqrt{V_t}\mathrm{d}B_t \equiv \sqrt{v}\left(\varrho\mathrm{d}W_t + \bar{\varrho}\mathrm{d}\overline{W}_t\right), \tag{10.52}$$

$$V_t = V_0 + \int_0^t \frac{a - bV_s}{(t-s)^{\frac{1}{2}-H}}\mathrm{d}s + \int_0^t \frac{c\sqrt{V_s}}{(t-s)^{\frac{1}{2}-H}}\mathrm{d}W_s.$$

(As earlier, $W, \overline{W}$ are independent Brownian motions.) Similar to the case of the classical Heston model, the square-root provides both pain (with regard to any methods that rely on sufficient smooth coefficients) and comfort (an *affine structure*, here infinite-dimensional, which allows for closed-form computations of moment generating functions). Arguably, there is no real financial reason for the square-root dynamics,[59] and ongoing work attempts to modify the above square-root dynamics, so as to obtain (something close to) *lognormal volatility*. We note that lognormal volatility was a key feature of the rough volatility model discussed in Gatheral et al. [166]. This motivates the study of more general dynamic rough volatility models of the form

$$dS_t/S_t = f(Z_t)dB_t \equiv f(Z_t)\big(\varrho dW_t + \bar{\varrho}d\overline{W}_t\big), \tag{10.53}$$

$$Z_t = z + \int_0^t K(s,t)v(Z_s)ds + \int_0^t K(s,t)u(Z_s)dW_s, \tag{10.54}$$

with sufficiently nice functions f, u, v. (While $f(x) = \sqrt{x}$ is a possible choice in what follows, we assume $u, v \in C^3$ for a local solution theory and then in fact impose $u, v \in C^3_b$ for global existence. One clearly expects nonexplosion under, e.g., linear growth, but in order not to stray too far from our main line of investigation we refrain from a discussion.) Note that $f(z)$ plays the role of spot volatility. Further note that the choice $z = 0$, $v \equiv 0$, $u \equiv 1$ brings us back to the "simple" case with (rough stochastic) volatility $f(Z_t) = f(\widehat{W}_t)$ considered in earlier sections.

Equation (10.53) fits, with some good will, into the existing theory of stochastic Volterra equations with singular kernels (e.g., [255] or [106]).[60]

10.4.2 ▪ Regularity structure approach

We insist that (10.53) is not a classical Itô SDE (solutions will not be semimartingales) nor a rough differential equation (in the sense of rough paths, driven by a Gaussian rough path as in [147, Chapter 10]). If rough paths have established themselves as a powerful tool for analyzing classical Itô SDE, we here make the point that Hairer's theory is an equally powerful tool for analyzing stochastic Volterra (resp., mixed Itô–Volterra) equations in the singular regime of interest.

As a preliminary step, we have to find the correct model space, spanned by symbols which arise by formal Picard iteration. To this end, rewrite (10.53) as an equation for modeled distributions,

$$\mathcal{Z} = \mathcal{I}(U(\mathcal{Z}) \cdot \Xi) + (\ldots)\mathbf{1}\,, \tag{10.55}$$

from which one can guess (or rigorously derive along [191, Section 8.1]) the need for the symbols

$$\mathbf{1}, \mathcal{I}(\Xi), \mathcal{I}(\Xi)^2, \mathcal{I}(\Xi\mathcal{I}(\Xi)), \ldots .$$

We have degrees $|\mathbf{1}| = 0$, $|\mathcal{I}(\Xi)| = H - \kappa$. For subsequent symbols, the degree is computed as

$$\left(\frac{1}{2} + H\right) \times \{\text{number of } \mathcal{I}\} + \left(-\frac{1}{2} + \kappa\right) \times \{\text{number of } \Xi\}.$$

For a modeled distribution, $\mathcal{Z}(t)$ takes values in the linear span of sufficiently many symbols, the (minimal) number of which is dictated by the Hurst parameter H. Loosely speaking, $\mathcal{Z} \in \mathcal{D}^\gamma$ indicates an expansions with γ-error estimate, which in practice is easy to see from the degree of the lowest degree symbols that do not figure in the expansion. For example, in the case of a "level-2 expansion" we can expect

$$\mathcal{Z}(t) = (\ldots)\mathbf{1} + (\ldots)\mathcal{I}(\Xi) \in \mathcal{D}_0^{2(H-\kappa)}$$

[59] This is also a frequent remark for the classical Heston model.

[60] We are not aware of any literature on mixed Itô–Volterra systems (although expect no difficulties). Here, of course, it suffices to first solves for Z and then construct S as a stochastic exponential.

since $|\mathcal{I}(\Xi)^2| = |\mathcal{I}(\Xi\mathcal{I}(\Xi))| = 2H - 2\kappa$. It follows from general theory [191, Theorem 4.16] that if $\mathcal{Z} \in \mathcal{D}_0^\gamma$, then so is $U(\mathcal{Z})$, the composition with a smooth function, and by [191, Theorem 4.7] the product with $\Xi \in \mathcal{D}^\infty_{-\frac{1}{2}-\kappa}$ is a modeled distribution in $\mathcal{D}^{\gamma-\frac{1}{2}-\kappa}$. For both reconstruction and convolution with singular kernels, one needs modeled distributions with positive degree $\gamma - \frac{1}{2} - \kappa > 0$. Given $H \in (0, \frac{1}{2}]$ we can then determine which symbols (up to which degree) are required in the expansion. As earlier, fix an integer

$$M \geq \max\left\{m \in \mathbb{N} | m \cdot (H-\kappa) - \frac{1}{2} - \kappa \leq 0\right\}$$

(so that $(M+1).(H-\kappa) - \frac{1}{2} - \kappa > 0$) and see that $\mathcal{Z} \in \mathcal{D}_0^{(M+1).(H-\kappa)}$ will do. When $H > 1/4$, and by choosing $\kappa > 0$ small enough, we see that $M = 1$ will do. That is, the symbols required to describe $\mathcal{Z}$ are $\{\mathbf{1}, \mathcal{I}(\Xi)\}$, and if one adds the symbols required to describe the right-hand side, one ends up with the level-2 model space spanned by

$$\{\Xi, \Xi\mathcal{I}(\Xi), \mathbf{1}, \mathcal{I}(\Xi)\}$$

which is exactly the model space for the "simple" rough pricing regularity structure, (10.26) in case $M = 1$. When $H \leq 1/4$ this precise correspondence is no longer true. To wit, in case $H \in (1/3, 1/4]$, taking $M = 2$ accordingly, solving (10.54) on the level of modeled distributions will require a ("level-3") model space given by

$$\langle \Xi, \Xi\mathcal{I}(\Xi), \Xi\mathcal{I}(\Xi)^2, \Xi\mathcal{I}(\Xi\mathcal{I}(\Xi)), \mathbf{1}, \mathcal{I}(\Xi), \mathcal{I}(\Xi)^2, \mathcal{I}(\Xi\mathcal{I}(\Xi))\rangle$$

which is strictly larger than the corresponding level-3 simple model space given in (10.26). In general, one needs to consider an extended model space $\hat{\mathcal{T}} = \langle \hat{S} \rangle$, so as to have

$$\tau \in \hat{S} \Rightarrow \Xi\mathcal{I}(\tau)^m, \mathcal{I}(\tau)^m \in \hat{S}, m \geq 0,$$

(with the understanding that only finitely many such symbols are needed, depending on H as explained above). As a result, symbols such as

$$\Xi\mathcal{I}(\Xi(\mathcal{I}(\Xi))^m),\ m \geq 0,\ \ \mathcal{I}(\Xi(\mathcal{I}(\Xi(\mathcal{I}(\Xi))^m)^{m'})), m, m' \geq 0,\ \ \ldots$$

will appear. At this stage a tree notation (omnipresent in the works of Hairer) would come in handy, and we refer the reader to [71] (and the references therein) for a recent attempt to reconcile the tree formalism of branched rough path [179, 193] and for the most recent algebraic formalism of regularity structures. (In a nutshell, the simple case (10.26) corresponds to trees where one node has m branches; in the present nonsimple case, symbol branching can happen everywhere.) Carrying out the following construction in the general case of fixed $H > 0$ is certainly possible;[61] however, the algebraic complexity is essentially the one from branched rough paths, and hence the general case requires a Hopf algebraic (Connes–Kreimer, Grossman–Larson, etc.) construction of the structure group (a.k.a. positive renormalization). Although this, and negative renormalization, is well understood [191, 72] (see also [71] for a rough path perspective), all complete exposition would lead us far astray from the main topic of this chapter. Hence, for simplicity only, we shall restrict our attention from here on to the level-2 case $H > 1/4$ (with $M = 1$ accordingly) but will mention general results whenever useful.

[61] We note that as $H \downarrow 0$, the number of symbols tends to infinity. In comparison, among all recently studied singular SPDEs, only the sine-Gordon equation [195] exhibits the similar feature of requiring arbitrarily many symbols.

10.4.3 ▪ Solving for rough volatility

We rewrite (10.54) as an equation for modeled distributions in $\mathcal{D}^\gamma$,

$$\mathcal{Z} = z\mathbf{1} + \mathcal{K}(U(\mathcal{Z}) \cdot \Xi + V(\mathcal{Z})). \tag{10.56}$$

(Here U, V are the operators associated to composition with $u, v \in C^{M+2}$, respectively.) We also impose

$$\gamma \in \left(\frac{1}{2} + \kappa, 1\right),$$

which is clearly necessary so as to have the product $U(\mathcal{Z}) \cdot \Xi$ in a modeled distribution space of positive parameter, so that reconstruction, convolution, etc. make sense. Let $H > 1/4, M = 1$ and pick $\kappa \in (0, \frac{4H-1}{6})$ so that $(M + 1).(H - \kappa) - \frac{1}{2} - \kappa > 0$. As explained in the previous section, this exactly allows us to work in the familiar structure of Section 10.2.1. That is, with $M = 1$,

$$\mathcal{T} = \langle \Xi, \Xi\mathcal{I}(\Xi), \mathbf{1}, \mathcal{I}(\Xi) \rangle,$$

with index set and structure group as given in that section. This structure is equipped with the Itô model and its (renormalization) approximations. Equation (10.56) critically involves the convolution operator $\mathcal{K}$ acting on $\mathcal{D}^\gamma$. The general construction [191, Section 5] is among the most technical in Hairer's work, and in fact is not directly applicable (our kernel K, although β-regularizing with $\beta = \frac{1}{2} + H$) fails Assumption 5.4 in [191]) so we shall be rather explicit.

Lemma 10.26. *On the regularity structure $(\mathcal{T}, A, G)$ of Section* 10.2.1 *with $M = 1$, consider a model (Π, Γ) which is admissible in the sense that*

$$\Pi_t \mathcal{I}(\Xi) = (K * \Pi_t \Xi)(\cdot) - (K * \Pi_t \Xi)(t) .$$

Let $\gamma > 0$, $F \in \mathcal{D}^\gamma$ and set[62]

$$\mathcal{K}F : s \in [0, T] \mapsto \mathcal{I}(F(s)) + (K * \mathcal{R}F)(s)\mathbf{1}.$$

Then (i) *$\mathcal{K}$ maps $\mathcal{D}^\gamma \to \mathcal{D}^{\min\{\gamma+\beta,1\}}$, and* (ii) *$\mathcal{R}(\mathcal{K}F) = K * \mathcal{R}F$, i.e., convolution commutes with reconstruction.*

Remark 10.4.1. [191, *Theorem* 5.2] *suggests that the estimate $\mathcal{K}$ maps $\mathcal{D}^\gamma \to \mathcal{D}^{\gamma+\beta}$. The difference to our baby Schauder estimate stems from the fact that, unlike Assumption* 5.3 *in* [191, *page* 64], *we do not assume that our regularity structure contains the polynomial structure.*

Proof. (Sketch) The special case $F \equiv \Xi \in \mathcal{D}^\infty$ was already treated in Lemma 10.18. We only show that, in the general case, $\mathcal{K}$ necessarily has the stated form but will not check the properties. It is enough to consider F with values in $\langle \Xi, \Xi\mathcal{I}\Xi \rangle$ and make the ansatz

$$(\mathcal{K}F)(s) := \mathcal{I}F(s) + (\ldots)\mathbf{1}.$$

Applying reconstruction, together with [191, Proposition 3.28] we see that $\mathcal{R}(\mathcal{K}F) \equiv (\ldots)$, which in turn must equal $K * \mathcal{R}F$, provided we postulate validity of (ii). This is the given definition of $\mathcal{K}F$. □

We return to our goal of solving

$$\mathcal{Z} = z\mathbf{1} + \mathcal{K}(U(\mathcal{Z}) \cdot \Xi + V(\mathcal{Z})), \tag{10.57}$$

[62] $\mathcal{I}$ is extended linearly to all of $\mathcal{T}$ by taking $\mathcal{I}\tau = 0$ for symbols $\tau \neq \Xi$).

noting perhaps that $U(\mathcal{Z})$ makes sense for every function-like modeled distribution, say $F(t) = F_0(t)\mathbf{1} + \sum_{k=1}^{M} F_k(t)(\mathcal{I}\Xi)^k \in \mathcal{T}_+ := \langle \mathbf{1}, \mathcal{I}(\Xi), \ldots, (\mathcal{I}\Xi)^M \rangle$, in which case

$$U(F)(t) = u(F_0(t))\mathbf{1} + u'(F_0(t)) \sum_{k=1}^{M} F_k(t)\mathcal{I}(\Xi)^k. \tag{10.58}$$

(Similar remarks apply to V, the composition operator associated to $v \in C^{M+2}$). Recall $M = 1$.

Theorem 10.27. *For any admissible model* (Π, Γ) *and* $u, v \in C_b^{M+2}(\mathbb{R})$, *for any* $T > 0$, *the equation* (10.57) *has a unique solution in* $\mathcal{D}^\gamma(\mathcal{T}_+)$, *and the map* $(u, v, \Pi) \mapsto \mathcal{Z}$ *is locally Lipschitz in the sense that if* $\mathcal{Z}$ *and* $\tilde{\mathcal{Z}}$ *are the solutions corresponding, respectively, to* (u, v, Π) *and* $(\tilde{u}, \tilde{v}, \tilde{\Pi})$, *then*

$$\|\mathcal{Z}; \tilde{\mathcal{Z}}\|_{\mathcal{D}_T^\gamma} \lesssim \|u - \tilde{u}\|_{C_b^{M+2}} + \|v - \tilde{v}\|_{C_b^{M+2}} + \|(\Pi, \Gamma); (\tilde{\Pi}, \tilde{\Gamma})\|_T,$$

with the proportionality constant being bounded when the (resp., C_b^{M+2} *and model) norms of the arguments stay bounded.*

In addition, if (Π, Γ) *is the canonical Itô model (associated to Brownian, resp., fractional Brownian motion,* $H > 1/4$*), then* $Z = \mathcal{R}\mathcal{Z}$ *is solves* (10.53) *in the Itô sense.*

Remark 10.4.2. $Z = \mathcal{R}\mathcal{Z}$ *is clearly the (unique) reconstruction of the (unique) solution to the abstract problem. We also checked that* Z *is indeed a solution for the Itô–Volterra equation. However, if one desires to know that* Z *is the unique strong solution to the stochastic Itô–Volterra equation, it is clear that one has to resort to uniqueness results of the stochastic theory* [106].

Proof. The well-posedness and continuous dependence on the parameters essentially follow from results of [191]; details are spelled out in Appendix 10.6.3.

The fact that the reconstruction of the solution solves the Itô equation can be obtained by considering approximations, as is done in [194, Theorem 6.2] or [147, Chapter 5]. □

Using the large deviation results obtained in the previous subsection, we can directly obtain an LDP for the log-price

$$X_t = \int_0^t f(Z_s)(\varrho \mathrm{d}W_s + \bar{\varrho}\mathrm{d}\overline{W}_s) - \frac{1}{2}\int_0^t f^2(Z_s)\mathrm{d}s.$$

For square integrable h, let z^h be the unique solution to the integral equation

$$z^h(t) = z + \int_0^t K(s,t)u(z^h(s))h(s)\mathrm{d}s.$$

Corollary 10.28. *Let* $H \in (1/4, \frac{1}{2}]$ *and* f *smooth (without boundedness assumption). Then* $t^{H-\frac{1}{2}}X_t$ *satisfies an LDP with speed* t^{2H} *and rate function given by*

$$I(x) = \inf_{h \in L^2([0,1])} \left\{ \frac{1}{2}\|h\|_{L^2}^2 + \frac{\left(x - I_1^z(h)\right)^2}{2I_2^z(h)} \right\}$$

where

$$I_1^z(h) = \varrho \int_0^1 f(z^h(s))h(s)\mathrm{d}s, \qquad I_2^z(h) = \int_0^1 f(z^h(s))^2\mathrm{d}s.$$

Remark 10.4.3. *Concerning the case $H \leq 1/4$, the following proof extends to any $H > 0$ provided one builds the correct regularity structure as discussed at the end of Section* 10.4.2. *(In particular, the proof of Theorem* 10.24 *for obtaining Schilder-type large deviations for the appropriate Itô model extends immediately.)*

Proof. We ignore the second part $\int_0^t(\dots)\mathrm{d}s$ in X_t, which is $O(t) = o(t^{\frac{1}{2}-H})$ since f is bounded. Let $\hat{X}_t = \int_0^t f(Z_s)(\varrho \mathrm{d}W_s + \bar{\varrho}\mathrm{d}\overline{W}_s)$. By scaling, we see that $t^{H-\frac{1}{2}}\hat{X}_t$ is equal in law to $\hat{X}_1^\delta$, where $\delta = t^H$ and X^δ, Z^δ are defined in the same way as X, Z with $W, \overline{W}$ replaced by $\delta W, \delta\overline{W}$ and v replaced by $v^\delta = \delta^{1+\frac{1}{2H}}h$. We then note that

$$X_1^\delta = \left\langle \mathcal{R}^\delta F(\mathcal{Z}^\delta)(\varrho\Xi + \bar{\varrho}\bar{\Xi}), 1_{[0,1]}\right\rangle =: \Psi(\Pi^\delta, v^\delta)$$

where Ψ is locally Lipschitz by Theorem 10.27. We can then directly combine the fact that Π^δ satisfies an LDP (Theorem 10.24) with a contraction principle such as Lemma 3.3 in [196], to obtain that X_1^δ satisfies an LDP with rate function

$$I(x) = \inf\left\{\frac{1}{2}(\|h\|^2_{L^2} + \|\bar{h}\|^2_{L^2}, \quad x = \Psi(\Pi^{(h,\bar{h})}, 0)\right\}.$$

It then suffices to note that z^h is exactly $\mathcal{R}\mathcal{Z}$ for $\mathcal{Z}$, the solution to (10.57) corresponding to a model $\Pi^{(h,\bar{h})}$ and with $h \equiv 0$, and to optimize separately over $\bar{h}$ as in the proof of Corollary 10.25. □

We also have an approximation result. (Assume u, v to be smooth with three bounded derivatives.)

Corollary 10.29. *Let $H > 1/4$ (but see remark). Then $Z = \lim Z^\varepsilon$, uniformly on compacts and in probability, where*

$$Z_t^\varepsilon = z + \int_0^t K(s,t)\big(u(Z_s^\varepsilon)dW_s^\varepsilon + (v(Z_s^\varepsilon) - \mathscr{C}^\varepsilon(s)uu'(Z_s^\varepsilon))\mathrm{d}s\big)\,. \tag{10.59}$$

Remark 10.4.4. *Replacing the renormalization function $\mathscr{C}^\varepsilon$ by its mean is possible, provided $H >$* 1/4. *However, unlike the discussion at the end of Section* 10.2.2, *this is no longer a consequence of quantifying the distributional convergence. In the present context, this is achieved by directly checking mode-convergence, which, fortunately, is not much harder. We leave the details to the interested reader.*

Remark 10.4.5. *In contrast to the previous statement, the above result is more involved for $H \in (0, 1/4]$, and additional renormalization terms appear, the general description of which would benefit from pre-Lie products, as recently introduced* [71].

Proof. Thanks to Theorems 10.15 and 10.27, it follows from continuity of reconstruction that

$$Z = \mathcal{R}\mathcal{Z} = \lim_{\varepsilon\to 0}\mathcal{R}^\varepsilon\mathcal{Z}^\varepsilon,$$

so that the only thing to do is check that Z^ε solves (10.59). Note that (10.57) implies that one has (omitting upper ε's at all normal and calligraphic Z ...)

$$\mathcal{Z}(t) = Z_t\mathbf{1} + u(Z_t)\mathcal{I}(\Xi),$$

and, with (10.58),

$$U(\mathcal{Z}(t))\Xi = u(Z_t)\Xi + u'u(Z_t)\mathcal{I}(\Xi)\Xi.$$

But then since $\hat{\Pi}^\varepsilon$ is a "smooth" model, in the sense of Remark 3.15 in [191], one has

$$\begin{aligned}\mathcal{R}^\varepsilon(U(\mathcal{Z}^\varepsilon)\Xi)(t) &= \hat{\Pi}^\varepsilon_t(U(\mathcal{Z}^\varepsilon(t))\Xi)(t)\\ &= u(Z^\varepsilon_t)(\hat{\Pi}^\varepsilon_t\Xi)(t) + u'u(Z^\varepsilon_t)(\hat{\Pi}^\varepsilon_t\Xi\mathcal{I}(\Xi))(t)\\ &= u(Z^\varepsilon_t)\dot{W}^\varepsilon(t) - u'u(Z^\varepsilon_t)\mathcal{K}^\varepsilon(t,t)\,.\end{aligned}$$

Since convolution commutes with reconstruction (Lemma 10.26), it follows that Z^ε is indeed a solution to (10.59). □

10.5 ▪ Numerical results

We will revisit the case of European option pricing under rough volatility. Building on the theoretical underpinnings of Section 10.2, we present a concise, but self-contained, description of the central algorithm of this paper—for simplicity, restricted to the unit time interval—and complement the theoretical convergence rates obtained in previous sections with numerical counterparts. The code used to run the simulations is available at https://www.github.com/RoughStochVol.

10.5.1 ▪ Implementation

Without loss of generality, set time to maturity $T = 1$. We are interested in pricing a European call option with spot S_0 and strike K under rough volatility—the risk-free interest rate is assumed to satisfy $r = 0$. From Theorem 10.2, we have

$$C(S_0, K, 1) = \mathbb{E}\left[C_{\mathrm{BS}}\left(S_0 \exp\left(\varrho\mathcal{I} - \frac{\varrho^2}{2}\mathcal{V}\right), K, \frac{\bar{\varrho}^2}{2}\mathcal{V}\right)\right], \tag{10.60}$$

where the computational challenge obviously lies in the efficient simulation of

$$(\mathcal{I},\mathcal{V}) = \left(\int_0^1 f(\widehat{W}_t,t)\mathrm{d}W_t, \int_0^1 f^2(\widehat{W}_t,t)\mathrm{d}t\right).$$

As explored in Subsection 10.2.4, we take a Wong–Zakai-style approach to simulating $\mathcal{I}$, that is, we approximate the white noise process $\dot{W}$ on the grid associated with the Haar basis as follows. Let $\{Z_i\}_{i=1,\dots,2^N-1} \sim$ i.i.d. $\mathcal{N}(0,1)$ and choose a Haar grid level $N \in \mathbb{N}$ such that the step size of the grid satisfies $\varepsilon = 2^{-N}$. Then, for all $t \in [0,1]$, we set

$$\dot{W}^\varepsilon(t) := \sum_{i=0}^{2^N-1} Z_i e^\varepsilon_i(t), \quad \text{where} \quad e^\varepsilon_i(t) := 2^{N/2}\mathbf{1}_{[i2^{-N},(i+1)2^{-N})}(t), \tag{10.61}$$

which induces an approximation of the fBm

$$\widehat{W}^\varepsilon(t) = \sum_{i=0}^{2^N-1} Z_i \hat{e}^\varepsilon_i(t) \quad \text{with} \tag{10.62}$$

$$\hat{e}^\varepsilon_i(t) := \mathbf{1}_{t>i2^{-N}} \frac{\sqrt{2H}2^{N/2}}{H+\frac{1}{2}}\left(|t-i2^{-N}|^{H+\frac{1}{2}} - |t-\min((i+1)2^{-N},t)|^{H+\frac{1}{2}}\right). \tag{10.63}$$

As outlined above, the central issue is that the object $\int_0^1 f(\widehat{W}^\varepsilon(t),t)\dot{W}^\varepsilon(t)\mathrm{d}t$ does *not* converge in an appropriate sense to the object of interest $\mathcal{I}$ as $\varepsilon \to 0$. This is overcome by *renormalizing* the object, two possible approaches of which are explored in Subsection 10.2.4. For the remainder, we will consider the "simpler" renormalized object given by

$$\tilde{\mathscr{I}}^{\varepsilon} = \int_0^1 f(\widehat{W}_t^{\varepsilon}, t)\dot{W}_t^{\varepsilon}\mathrm{d}t - \int_0^1 \mathscr{C}^{\varepsilon}(t)\partial_1 f(\widehat{W}_t^{\varepsilon}, t)\mathrm{d}t, \tag{10.64}$$

where the renormalization object $\mathscr{C}^{\varepsilon}(t)$ can be one of

$$\mathscr{C}^{\varepsilon}(t) = \begin{cases} \frac{2^N\sqrt{2H}}{H+\frac{1}{2}}|t - \lfloor t2^N \rfloor 2^{-N}|^{H+\frac{1}{2}}, \\ \frac{\sqrt{2H}}{(H+\frac{1}{2})(H+\frac{3}{2})}2^{N(\frac{1}{2}-H)}. \end{cases} \tag{10.65}$$

Inserting the nonconstant version of (10.65) into (10.64), we obtain

$$\tilde{\mathscr{I}}^{\varepsilon} = \sum_{i=0}^{2^N-1} \int_{i2^{-N}}^{(i+1)2^{-N}} \left[Z_i 2^{N/2} f(\widehat{W}^{\varepsilon}(t), t) - \frac{\sqrt{2H}2^N}{H+\frac{1}{2}}|t - i2^{-N}|^{H+\frac{1}{2}}\partial_1 f(\widehat{W}^{\varepsilon}(t), t)\right]\mathrm{d}t, \tag{10.66}$$

$$\mathscr{V}^{\varepsilon} := \sum_{i=0}^{2^N-1} \int_{i2^{-N}}^{(i+1)2^{-N}} f^2(\widehat{W}^{\varepsilon}(t), t)\mathrm{d}t. \tag{10.67}$$

Algorithm 1: Simulation

Parameters: $M \in \mathbb{N}$: # Monte Carlo simulations
$N \in \mathbb{N}$: Haar grid "level" such that $\varepsilon = 2^{-N}$
$d \in \mathbb{N}$: # discretization points of trapezoidal rule in each Haar subinterval

Output: M samples of bivariate object $(\tilde{\mathscr{I}}^{\varepsilon}, \mathscr{V}^{\varepsilon})$

initialize $\tilde{\mathscr{I}}^{\varepsilon} = \mathscr{V}^{\varepsilon} = \mathbf{0} \in \mathbb{R}^M$;
simulate array $\mathbf{Z} \in \mathbb{R}^{M\times 2^N}$ of i.i.d. standard normals;
for each Haar subinterval $[i2^{-N}, (i+1)2^{-N})$ where $i \in \{0, \dots, 2^N - 1\}$ **do**
  choose discretization grid $\mathcal{D}^i$ with d points on the Haar subinterval;
  evaluate functions $\hat{e}_k^{\epsilon}, k = 0, \dots, i$, from (10.63) on $\mathcal{D}^i$ to obtain $\hat{\mathbf{e}}^{\epsilon} \in \mathbb{R}^{(i+1)\times d}$;
  compute $\hat{\mathbf{W}}^{\epsilon} = \mathbf{Z}^* \times \hat{\mathbf{e}}^{\epsilon} \in \mathbb{R}^{M\times d}$ where $\mathbf{Z}^* \in \mathbb{R}^{M\times(i+1)}$ is the truncation of $\mathbf{Z}$ to its first $i+1$ columns such that $\hat{\mathbf{W}}^{\epsilon}$ is an approximation of the fBm on $\mathcal{D}^i$;
  evaluate integrands from equations (10.66), (10.67) on $\mathcal{D}^i$ using $\hat{\mathbf{W}}^{\epsilon}$ and the last column of $\mathbf{Z}^*$;
  approximate respective integrals on subinterval by trapezoidal rule;
  add obtained estimates to running sums $\tilde{\mathscr{I}}^{\varepsilon}$ and $\mathscr{V}^{\varepsilon}$;
end
return $\tilde{\mathscr{I}}^{\varepsilon}, \mathscr{V}^{\varepsilon}$

10.5.2 • Observed convergence rates

In this subsection, we will discuss strong convergence of the approximating object $\tilde{\mathscr{I}}^{\varepsilon}$ to the actual object of interest $\mathscr{I}$ as well as weak convergence of the option price itself as the Haar grid interval size $\varepsilon \to 0$. Specifically, we will be looking at Monte Carlo estimates of our errors; that is, in order to approximate some quantity $\mathbb{E}[X]$ for some random variable X, we will instead be looking at $\frac{1}{M}\sum_{i=1}^{M} X_i$ where the X_i are M i.i.d. samples drawn from the same distribution as X.

In other words, we need to generate M realizations of the bivariate stochastic object $\left(\tilde{\mathscr{I}}^{\varepsilon}, \mathscr{V}^{\varepsilon}\right)$, a task that can be vectorized as described below, thus avoiding expensive looping through realizations.

Strong convergence

We verify Theorem 10.22(i) numerically, albeit in the $L^2(\Omega)$ sense and—for simplicity—with $f(x,t) := \exp(x)$, i.e., with no explicit time dependence. That is, we are concerned with Monte Carlo approximations of

$$\left\| \tilde{\mathscr{I}}^{\varepsilon} - \int_0^1 \exp(\widehat{W}_t)\mathrm{d}W_t \right\|_{L^2(\Omega)},$$

and we expect an error "almost" of order H.

Remark 10.5.1. *We choose $f(x,t) = \exp(x)$ because this closely resembles the* rough Bergomi *model (see* [39] *and below). Also, for the simplest nontrivial choice $f(x,t) = x$, the discretization error is overshadowed by the Monte Carlo error, even for very coarse grids.*

Since $(W, \widehat{W})$ is a two-dimensional Gaussian process with known covariance structure, it is possible to use the Cholesky algorithm [39, 40] to simulate the joint paths on some grid and then use standard Riemann sums to approximate the integral. The value obtained in this way could serve as a reference value for our scheme. However, for strong convergence, we need both objects to be based on the same stochastic sample. For this reason, we find it easier to construct a reference value by the wavelet-based scheme itself, i.e., we simply pick some $\varepsilon' \ll \varepsilon$ and consider

$$\|\tilde{\mathscr{I}}^{\varepsilon} - \tilde{\mathscr{I}}^{\varepsilon'}\|_{L^2(\Omega)} \tag{10.68}$$

as $\varepsilon \to \varepsilon'$. As can be seen in Figures 10.1 and 10.2, both renormalization approaches stated in (10.65) are consistent with a theoretical strong rate of almost H even for $H < 1/4$ (see discussion at the end of Section 10.2.2).

Remark 10.5.2 (weak convergence). *As the model does not have the Markov property, a proper weak convergence analysis proves to be subtle. Indeed, the rate of convergence of*

$$\left| \mathbb{E}\left[\varphi(\tilde{\mathscr{I}}^{\varepsilon})\right] - \mathbb{E}\left[\varphi\left(\int_0^1 \exp(\widehat{W}_t)\mathrm{d}W_t\right)\right]\right|$$

as $\epsilon \to 0$ (for suitable test functions φ) remains an open problem. However, picking $\varphi(x) = x^2$, Itô's isometry yields

$$\mathbb{E}\left[\left(\int_0^1 \exp(\widehat{W}_t)\mathrm{d}W_t\right)^2\right] = \int_0^1 \mathbb{E}\left[\exp\left(2\widehat{W}_t\right)\right]\mathrm{d}t = \int_0^1 \exp\left(2t^{2H}\right)\mathrm{d}t \tag{10.69}$$

which can be approximated numerically. So, we can consider

$$\left|\mathbb{E}\left[\left(\tilde{\mathscr{I}}^{\varepsilon}\right)^2\right] - \int_0^1 \exp\left(2t^{2H}\right)\mathrm{d}t\right| \tag{10.70}$$

as $\varepsilon \to 0$. Our preliminary results indicate that for both renormalization approaches, the weak rate seems to be around the strong rate H.

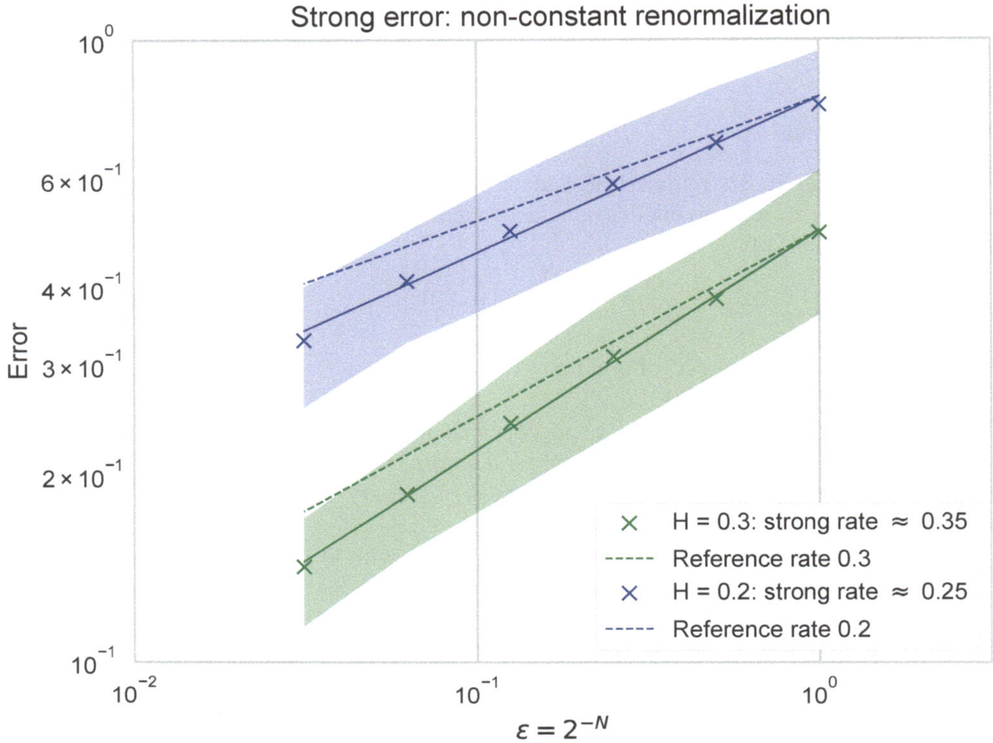

Figure 10.1: Empirical strong errors in the sense of (10.68) on a log-log-scale under a *nonconstant* renormalization, obtained by $M = 10^5$ Monte Carlo samples with a trapezoidal rule delta of $\Delta = 2^{-17}$ and fineness of the reference Haar grid $\varepsilon' = 2^{-8}$. Solid lines represent the empirical rates of convergence obtained by least squares regression, and dashed lines provide visual reference rates. Shaded color bands show interpolated 95% confidence regions based on normality of Monte Carlo estimator.

Option pricing

We pick a simplified version of the *rough Bergomi* model [39], for which the instantaneous variance is given by

$$f^2(x) = \sigma_0^2 \exp(\eta x)$$

with σ_0 and η denoting spot volatility and volatility of volatility, respectively. Let C^ε denote the approximation of the call price (10.60) based on $\left(\tilde{\mathscr{I}}^\varepsilon, \mathscr{V}^\varepsilon\right)$, fix some $\varepsilon' \ll \varepsilon$, and consider

$$\left|C^\varepsilon(S_0, K, T = 1) - C^{\varepsilon'}(S_0, K, T = 1)\right| \tag{10.71}$$

as $\varepsilon \to \varepsilon'$. Empirical results displayed in Figure 10.3 indicate a weak rate of $2H$ across the full range of $0 < H < \frac{1}{2}$.

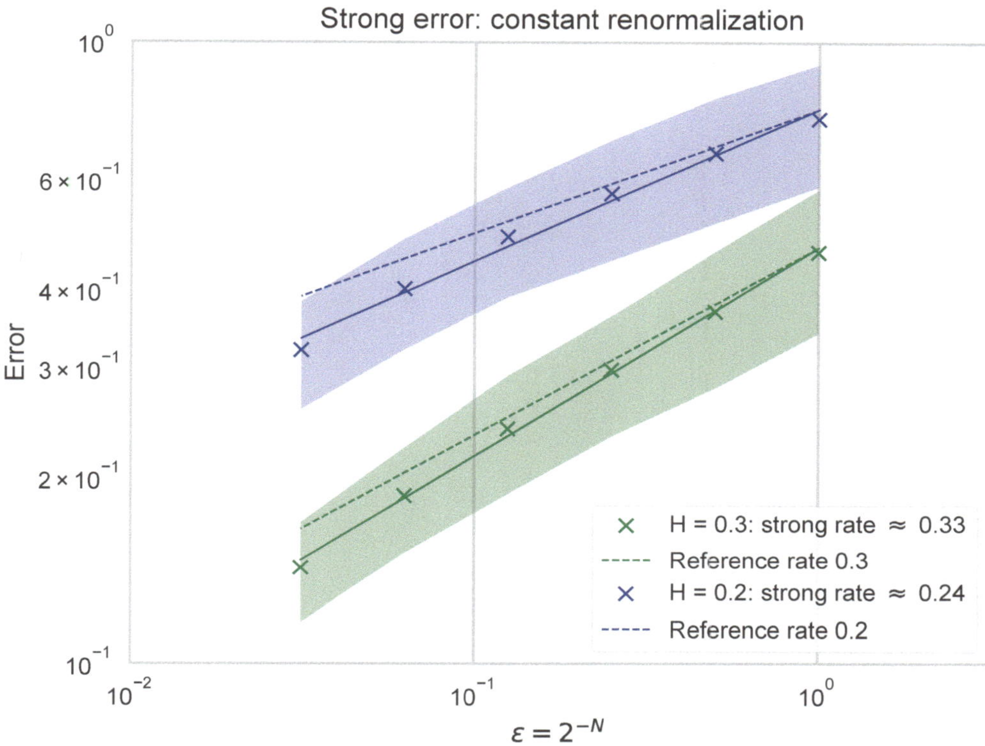

Figure 10.2: Empirical strong errors in the sense of (10.68) on a log-log-scale under a *constant* renormalization, obtained by $M = 10^5$ Monte Carlo samples with a trapezoidal rule delta of $\Delta = 2^{-17}$ and fineness of the reference Haar grid $\varepsilon' = 2^{-8}$. Solid lines represent the empirical rates of convergence obtained by least squares regression, and dashed lines provide visual reference rates. Shaded color bands show interpolated 95% confidence regions based on normality of Monte Carlo estimator.

10.6 ▪ Appendix

10.6.1 ▪ Approximation and renormalization (proofs)

Lemma 10.30. *Given $a, b > 0$ and $\delta \in [0,1]$, for $x \notin [0,1)$ we have*

$$|a^x - b^x| \le 2^{1-\delta}|x|^\delta (a^{x-\delta} \vee b^{x-\delta}) \cdot |a-b|^\delta ,$$

and for $x \in (0,1)$,

$$|a^x - b^x| \le 2^{1-\delta}|x|^\delta (a^{(x-1)\delta} b^{x(1-\delta)} \vee b^{(x-1)\delta} a^{x(1-\delta)}) \cdot |a-b|^\delta .$$

Proof. This follows from interpolation between $|a^x - b^x| \le |x| \sup_{z \in [a,b]} z^{x-1}|a-b| \le |x| a^{x-1} \vee b^{x-1}|a-b|$ and $|a^x - b^x| \le a^x + b^x \le 2a^x \vee b^x$. □

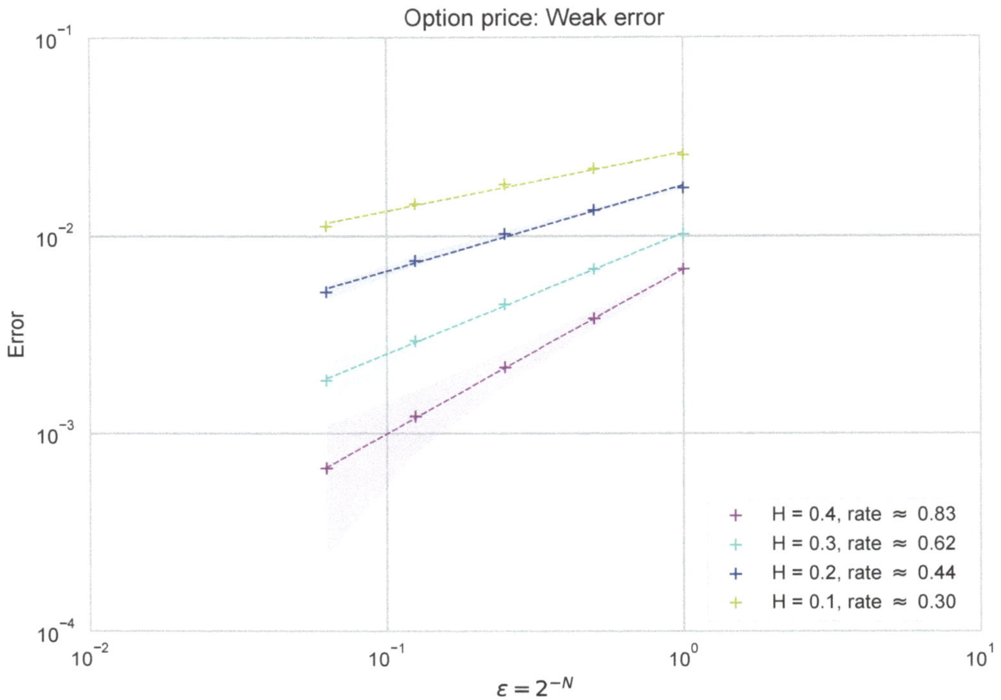

Figure 10.3: Empirical weak errors in the sense of (10.71) on a log-log-scale as $\varepsilon \to \varepsilon' = 2^{-8}$, obtained with $M = 10^5$ Monte Carlo samples with spot $S_0 = 1$, strike $K = 1$, correlation $\varrho = -0.8$, spot vol $\sigma_0 = 0.2$, vvol $\eta = 2$, and trapezoidal rule delta $\Delta = 2^{-17}$. Dashed lines represent least squares estimates for rate estimation, and shaded color bands show confidence regions based on normality of Monte Carlo estimator.

***Proof of Lemma* 10.10.** Rewriting $\widehat{W}^\varepsilon(t) = \sqrt{2H}\int_0^\infty \mathrm{d}W(u) \int_0^\infty \mathrm{d}r\, \delta^\varepsilon(r,u)\, |t-r|^{H-\frac{1}{2}} \mathbf{1}_{r<t}$, we have

$$\begin{aligned}
\mathbb{E}\left|\widehat{W}^\varepsilon(t) - \widehat{W}^\varepsilon(s)\right|^2 &= 2H \int_0^\infty \mathrm{d}u \left(\int_0^\infty \mathrm{d}r \delta^\varepsilon(r,u)(\mathbf{1}_{r<t}|t-r|^{H-\frac{1}{2}} - \mathbf{1}_{r<s}|s-r|^{H-\frac{1}{2}}) \right)^2 \\
&\lesssim \int_0^\infty \mathrm{d}u \int_0^\infty \mathrm{d}r |\delta^\varepsilon(r,u)| \left(\mathbf{1}_{r<t}|t-r|^{H-\frac{1}{2}} - \mathbf{1}_{r<s}|s-r|^{H-\frac{1}{2}} \right)^2 \\
&\lesssim \int_0^{s\vee t} \mathrm{d}r \left(\mathbf{1}_{r<t}|t-r|^{H-\frac{1}{2}} - \mathbf{1}_{r<s}|s-r|^{H-\frac{1}{2}} \right)^2,
\end{aligned}$$

where we used the Itô isometry in the first step and Jensen's inequality in the second. Assuming $s < t$, we can split the integral in domains $[0,s]$ and $[s,t]$ which yields the bound $|t-s|^{2H} \int_0^s |s-r|^{4H-1} + |t-s|^{2H} \lesssim |t-s|^{2H}$. Application of equivalence of moments for Gaussian random variables and Kolmogorov's criterion then shows the first inequality.

The second estimate follows by interpolation (and once more Kolmogorov), if we can prove that

$$\mathbb{E}|\widehat{W}^\varepsilon(t) - \widehat{W}(t)|^2 \lesssim \varepsilon^{2H-\kappa'}. \tag{10.72}$$

We have, by Itô isometry,

$$\mathbb{E}\left[\left|\widehat{W}^{\varepsilon}(t)-\widehat{W}(t)\right|^2\right]=2H\int_0^{\infty}\mathrm{d}u\left(\int_0^{\infty}\mathrm{d}r\,\delta^{\varepsilon}(r,u)\,|t-r|^{H-\frac{1}{2}}\mathbf{1}_{r<t}-|t-u|^{H-\frac{1}{2}}\mathbf{1}_{u<t}\right)^2.$$

We can enlarge the inner integral such that $\int\delta^{\varepsilon}(r,u)dr=1$ by neglecting an error term which can be estimated by $\int_{B(0,c\varepsilon)}\mathrm{d}u(\int_{B(0,c\varepsilon)}\mathrm{d}r\,\varepsilon^{-1}|t-r|^{H-\frac{1}{2}})^2\lesssim\varepsilon^{2H}$. Application of Jensen's inequality then yields

$$\int_0^{\infty}\mathrm{d}u\int_{-\infty}^{\infty}\mathrm{d}r\,|\delta^{\varepsilon}(r,u)|\left(|t-r|^{H-\frac{1}{2}}\mathbf{1}_{r<t}-|t-u|^{H-\frac{1}{2}}\mathbf{1}_{u<t}\right)^2.$$

The cases where either $r>u$ or $u>t$ yield an error term of order ε^{2H}. Hence, an application of Lemma 10.30,

$$\left||t-r|^{H-\frac{1}{2}}-|t-u|^{H-\frac{1}{2}}\right|\lesssim(|t-r|^{-\frac{1}{2}+\kappa}+|t-u|^{-\frac{1}{2}+\kappa})\cdot|u-r|^{H-\kappa},$$

finishes the proof of (10.72). □

***Proof of* (10.40).** We only consider the symbols $\Xi\mathcal{I}^m(\Xi)$; the symbols $\mathcal{I}(\Xi)^m$ can be handled with Lemma 10.10. In view of Lemmas 10.12 and 10.13, we have to control (for $m\geq 0$ in the first equation and $m>0$ in the second equation)

$$\mathbb{E}\left|\int_0^{\infty}\mathrm{d}W^{\varepsilon}(t)\diamond\varphi_s^{\lambda}(t)(\widehat{W}^{\varepsilon}_{st})^m-\int_0^{\infty}\mathrm{d}W(t)\diamond\varphi_s^{\lambda}(t)(\widehat{W}_{st})^m\right|^2\lesssim\varepsilon^{2\delta\kappa'}\lambda^{2mH-1-2\kappa'},\tag{10.73}$$

$$\mathbb{E}\left|\int_0^{\infty}\mathrm{d}t\,\varphi_s^{\lambda}(t)\left(\mathscr{K}^{\varepsilon}(s,t)(\widehat{W}^{\varepsilon}_{st})^{m-1}-K(s-t)(\widehat{W}_{st})^{m-1}\right)\right|^2\lesssim\varepsilon^{2\delta\kappa'}\lambda^{2mH-1-2\kappa'},\tag{10.74}$$

where $\widehat{W}^{(\varepsilon)}_{st}=\widehat{W}^{(\varepsilon)}(t)-\widehat{W}^{(\varepsilon)}(s)$, and where $\delta\in(0,1)$, $\kappa'\in(0,H)$ is arbitrary. Equivalence of norms in the Wiener chaos and a version of Kolmogorov's criterion for models ([191, Proposition 3.32]) then gives (10.40) (note that this gives a better homogeneity than we actually need since we only subtract $2\kappa'$ and not $2m\kappa'$ in the exponent of $\lambda\in(0,1]$). We can rewrite the random variable in the expectation of (10.73) using [216, Theorem 7.40] as

$$\int_0^{T+1}\mathrm{d}W(t)\diamond\int\mathrm{d}u\,\delta^{\varepsilon}(t,u)\left(\mathbf{1}_{u\geq 0}\varphi_s^{\lambda}(u)(\hat{W}^{\varepsilon}_{su})^m-\varphi_s^{\lambda}(t)(\widehat{W}_{st})^m\right).$$

Using [216, Theorem 7.39] and Jensen's inequality, we can estimate the second moment of this Skorokhod integral by

$$\mathbb{E}|(10.73)|^2\lesssim\int_0^{T+1}\mathrm{d}t\int\mathrm{d}u\,|\delta^{\varepsilon}(t,u)|\,\mathbb{E}\left(\mathbf{1}_{u\geq 0}\varphi_s^{\lambda}(u)(\widehat{W}^{\varepsilon}_{su})^m-\varphi_s^{\lambda}(t)(\widehat{W}_{st})^m\right)^2.$$

In the regime $\lambda\leq\varepsilon$, every term in the parentheses can simply be bounded (using Lemma 10.10) by $\lambda^{2H-1}\lesssim\lambda^{2H-1-2\kappa'}\varepsilon^{\kappa'}$. If, on the other hand, $\varepsilon<\lambda$, we can split off a term of order $\int_{B(0,c\varepsilon)}\mathrm{d}t\int_{B(0,c\varepsilon)}\frac{\mathrm{d}u}{\varepsilon}\lesssim\lambda^{2mH-1-2\kappa'}\varepsilon^{2\kappa'}$, drop the indicator $\mathbf{1}_{u\geq 0}$, and bound on the support of $\delta^{\varepsilon}(t,u)$,

$$\begin{aligned}|\varphi_s^{\lambda}(u)(\widehat{W}^{\varepsilon}_{su})^m-\varphi_s^{\lambda}(t)(\widehat{W}_{ts})^m|&\leq|(\varphi_s^{\lambda}(u)-\varphi_s^{\lambda}(t))\cdot|\widehat{W}^{\varepsilon}_{su}|^m+|\varphi_s^{\lambda}(t)|\cdot\left|(\widehat{W}^{\varepsilon}_{su})^m-(\widehat{W}_{st})^m\right|\\&\lesssim C_{\varepsilon}\mathbf{1}_{B(s,(1+2c)\lambda)}(t)\,\lambda^{-1-\kappa'}\varepsilon^{\kappa'}\lambda^{mH}+C_{\varepsilon}\mathbf{1}_{B(s,\lambda)}(t)\lambda^{-1}\lambda^{mH-\kappa'}\varepsilon^{\kappa'},\end{aligned}$$

where $C_\varepsilon > 0$ denote random constants that are uniformly bounded in L^p for $p \in [1,\infty)$. This shows (10.73). To estimate (10.74) we first note that due to $\mathbb{E}|(\widehat{W})^{m-1}_{st} - (\widehat{W}^\varepsilon_{st})^{m-1}|^2 \lesssim |t-s|^{2(m-1)H-2\kappa'}\varepsilon^{\delta 2\kappa'}$, we are only left with

$$\mathbb{E}\left|\int_0^\infty \mathrm{d}t\, \varphi_s^\lambda(t)(\mathscr{K}^\varepsilon(s,t) - K(s-t))(\widehat{W}^\varepsilon_{st})^{m-1}\right|^2 \lesssim \int_0^\infty \mathrm{d}t\, \varphi_s^\lambda(t)|\mathscr{K}^\varepsilon(s,t) - K(s-t)|^2\,|s-t|^{2(m-1)H},$$

which is straightforward to bound with Lemma 10.14 if $\lambda \le \varepsilon$. For $\lambda < \varepsilon$ and $t > 2c\varepsilon$ with $c > 0$ as in Definition 10.8, the desired bound follows from Lemma 10.31. The remaining case, however, contributes with

$$\begin{aligned}&\int_{B(0,2c\varepsilon)} \mathrm{d}t\, \varphi_s^\lambda(t)|t-s|^{2(m-1)H}(\varepsilon^{2H-1} + |t-s|^{2H-1})\\ &\lesssim \int_{B(s,\lambda^{-1}2c\varepsilon)} \mathrm{d}t\,(\lambda^{2(m-1)H}\varepsilon^{2H-1} + \lambda^{2mH-1}|t|^{2mH-1})\\ &\lesssim \lambda^{2(m-1)H-1}\varepsilon^{2H} + \lambda^{2mH-1}(\lambda^{-1}\varepsilon)^{2mH} \le \lambda^{2kH-\kappa'}\varepsilon^{\kappa'},\end{aligned}$$

which completes the proof. □

Lemma 10.31. *For c as in Definition 10.8 and $t > 2c\varepsilon$ and $s \in \mathbb{R}$, for $\kappa' \in (0,H)$ we have*

$$|K(s-t) - \mathscr{K}^\varepsilon(s,t)| \lesssim |s-t|^{H-\frac{1}{2}-\kappa'}\varepsilon^{\kappa'}.$$

Proof. If $2c\varepsilon \ge |s-t|/2$, the bound easily follows from Lemma 10.14. If $2c\varepsilon \ge |s-t|/2$, we can reshape

$$|K(s-t) - \mathscr{K}^\varepsilon(s,t)| = \left|\int_{-\infty}^\infty \mathrm{d}u\, \delta^{2,\varepsilon}(t,u)(\mathbb{1}_{t<s}|s-t|^{H-\frac{1}{2}} - \mathbb{1}_{s<u}|s-u|^{H-\frac{1}{2}})\right|,$$

where $\delta^{2,\varepsilon}(t,\cdot) := \int_{-\infty}^\infty \mathrm{d}x_1 \int_{-\infty}^\infty \mathrm{d}x_2 \delta^\varepsilon(t,x_1)\delta^\varepsilon(x_1,\cdot)$ satisfies the properties in Definition 10.8 with support in $B(t,2c\varepsilon)$. Note that for $2c\varepsilon \ge |s-t|/2$ either both indicator functions or none vanish so that we only have to consider $t < s$ where we obtain, with Lemma 10.30 up to a constant, $\int_{-\infty}^\infty |\delta^{2,\varepsilon}(t,u)||t-s|^{H-\frac{1}{2}-\kappa'}\varepsilon^{\kappa'} \lesssim |t-s|^{H-\frac{1}{2}-\kappa'}\varepsilon^{\kappa'}$. □

***Proof of Lemma* 10.17.** We restrict ourselves to the proof of (10.42); the other three inequalities follow by similar arguments. We fix a wavelet basis $\varphi_y = \varphi(\cdot - y)$, $y \in \mathbb{Z}$, $\psi_y^j = 2^{j/2}\psi(2^j(\cdot - y))$, $j \ge 0$, $y \in 2^{-j}\mathbb{Z}$, and use in the following the notation $\varphi_y^j = 2^{j/2}\varphi(2^j(\cdot - y))$, $j \ge 0$, $y \in 2^{-j}\mathbb{Z}$. Within this basis we can express the $\mathscr{B}^\beta_{1,\infty}$ regularity of φ by

$$\sum_{y\in\mathbb{Z}} |(\varphi,\varphi_y)_{L^2}| + \sup_{j\ge 0} 2^{j\beta} \sum_{y\in 2^{-j}\mathbb{Z}} 2^{-dj/2}|(\varphi,\psi_y^j)_{L^2}| \lesssim \|\varphi\|_{\mathscr{B}^\beta_{1,\infty}}.$$

Without loss of generality, we can assume that $\lambda = 2^{-j_0}$ is dyadic, so that by scaling,

$$\sum_{y\in 2^{-j_0}\mathbb{Z}} |(\varphi_s^\lambda,\varphi_y^{j_0})_{L^2}| + \sup_{j\ge j_0} 2^{(j-j_0)\beta} \sum_{y\in 2^{-j}\mathbb{Z}} 2^{-(j-j_0)d/2}|(\varphi_s^\lambda,\psi_y^j)_{L^2}| \lesssim 2^{j_0 d/2}\|\varphi\|_{\mathscr{B}^\beta_{1,\infty}}. \qquad (10.75)$$

We can now rewrite

$$
\begin{aligned}
&(\mathcal{R}F - \Pi_s F_s)(\varphi_s^\lambda) \\
&= \sum_{y\in 2^{-j_0}\mathbb{Z}} (\mathcal{R}F - \Pi_s F_s)(\varphi_y^{j_0}) \cdot (\varphi_y^{j_0}, \varphi_s^\lambda)_{L^2} + \sum_{j\geq j_0} \sum_{y\in 2^{-j}\mathbb{Z}} (\mathcal{R}F - \Pi_s F_s)(\psi_y^j) \cdot (\psi_y^j, \varphi_s^\lambda)_{L^2} \\
&= \sum_{y\in 2^{-j_0}\mathbb{Z}} (\mathcal{R}F - \Pi_y F_y)(\varphi_y^{j_0}) (\varphi_y^{j_0}, \varphi_s^\lambda)_{L^2} + \sum_{y\in 2^{-j_0}\mathbb{Z}} \Pi_y(F_y - \Gamma_{ys}F_s)(\varphi_y^{j_0}) (\varphi_y^{j_0}, \varphi_s^\lambda) \qquad (10.76) \\
&+ \sum_{\substack{j\geq j_0,\\ y\in 2^{-j}\mathbb{Z}}} (\mathcal{R}F - \Pi_y F_y)(\psi_y^j) (\psi_y^j, \varphi_s^\lambda)_{L^2} + \sum_{\substack{j\geq j_0,\\ y\in 2^{-j}\mathbb{Z}}} \Pi_y(F_y - \Gamma_{ys}F_s)(\psi_y^j) (\psi_y^j, \varphi_s^\lambda)_{L^2}\,. \qquad (10.77)
\end{aligned}
$$

Only finitely many terms in (10.76) contribute, which all can be bounded a constant times $2^{-j_0\gamma} = \lambda^\gamma$. Moreover,

$$
\begin{aligned}
(10.77) &\lesssim \sum_{j\geq j_0} 2^{-j\gamma} + \sum_{j\geq j_0} \sum_{A\ni\alpha<\gamma} 2^{-j\alpha} 2^{-(\gamma-\alpha)j_0} \sum_{y\in 2^{-j}\mathbb{Z}} 2^{jd/2} |(\varphi_s^\lambda, \psi_y^j)_{L^2}| \\
&\lesssim \sum_{j\geq j_0} 2^{-j\gamma} + 2^{-\gamma j_0} \sum_{A\ni\alpha<\gamma} \sum_{j\geq j_0} 2^{-(j-j_0)\alpha} 2^{-(j-j_0)\beta} \lesssim 2^{-j_0\gamma} = \lambda^\gamma\,,
\end{aligned}
$$

where we used $\beta + \alpha > 0$, $\alpha \in A$, in the last line. □

***Proof of Lemma* 10.20.** It easy to check, using Taylor's formula, that for scaled Haar wavelets φ_s^λ and $\gamma \in (0, (M+1)H)$,

$$
\mathbb{E}\left[\left|\int \varphi_s^\lambda(t)\, f(\widehat{W}(t), t)\mathrm{d}W(t) - \Pi_s F\Xi(s)(\varphi_s^\lambda)\right|^2\right]^{\frac{1}{2}} \lesssim \lambda^{(\gamma-\frac{1}{2}-\kappa)}\,, \qquad (10.78)
$$

uniformly for s in compact sets. The same argument as in the proof of Lemma 10.17 then implies that (10.78) actually holds for compactly supported smooth function φ (or even compactly supported functions in $\mathcal{B}^\beta_{1,\infty}(\mathbb{R}^d)$). Proceeding now as in [191] we choose test functions $\eta, \psi \in C_c^\infty$ with η even and $\operatorname{supp}\eta \subseteq \mathrm{B}(0,1)$, $\int \eta(t)\,\mathrm{d}t = 1$. We then obtain, for $\psi^\delta(s) = \langle \psi, \eta_s^\delta \rangle$,

$$
\begin{aligned}
&\mathbb{E}\left[|\mathcal{R}F\Xi(\psi^\delta) - \int \psi^\delta(t)\, f(\widehat{W}(t), t))\mathrm{d}W(t)|^2\right]^{\frac{1}{2}} \\
&\quad = \mathbb{E}\left[\left|\int \mathrm{d}x\, \psi(x) \left(\mathcal{R}F\Xi(\eta_x^\delta) - \int \eta_x^\delta(t)\, f(\widehat{W}(t), t)\mathrm{d}W(t)\right)\right|^2\right]^{\frac{1}{2}} \\
&\quad \lesssim \int \mathrm{d}x\, \psi^2(x)\, \delta^{\gamma-\frac{1}{2}-\kappa} \overset{\delta\to 0}{\to} 0,
\end{aligned}
$$

where we included a term $\Pi_x \Xi F(x)$ in the second step. It remains to note that

$$
\int \psi^\delta(t)\, f(\widehat{W}(t), t))\mathrm{d}W(t) \overset{\delta\to 0}{\to} \int \psi(t)\, f(\widehat{W}(t), t)\mathrm{d}W(t)
$$

in $L^2(\mathbb{P})$ and further that $\mathcal{R}F\Xi(\psi^\delta) \to \mathcal{R}F\Xi(\psi)$ a.s. and thus in $L^2(\mathbb{P})$. Putting everything together we obtain

$$\mathbb{E}\left[|\mathcal{R}F\Xi(\psi) - \int \psi(t)\, f(\widehat{W}(t), t)\mathrm{d}W(t)|^2\right] = 0,$$

which implies the first statement. For the second identity, we proceed in the same way but now use Lemma 10.32. □

Lemma 10.32. *For $F \in L^2(\mathbb{P} \times \mathrm{Leb})$ we have*

$$\mathbb{E}\left[\left|\int F(t)\mathrm{d}W^{\varepsilon}(t)\right|^2\right] \lesssim \int \mathbb{E}\left[|F(t)|^2\right]\mathrm{d}t\,.$$

Proof. As a consequence of Definition 10.8 we see that $\int |\delta^{\varepsilon}(x,y)\mathrm{d}x|$ is bounded uniformly in ε and y. We can thus normalize $|\delta^{\varepsilon}(\cdot, r)|$ to a probability density and apply Itô's isometry and Jensen's inequality to obtain

$$\int F(t)\mathrm{d}W^{\varepsilon}(t) = \int_0^{\infty}\int_0^{\infty} \delta^{\varepsilon}(t,r)\, F(t)\mathrm{d}t\, \mathrm{d}W(r). \qquad \square$$

10.6.2 ▪ Large deviations proofs

***Proof of Lemma* 10.23.** The fact that Π^h satisfies the algebraic constraints is obvious, so we focus on the analytic ones. The Sobolev embedding $L^2 \subset C^{-\frac{1}{2}}$ yields that $\Pi\Xi$, $\Pi\bar{\Xi}$ satisfy the right bounds. Noting that (by, e.g., [267, section 3.1]) $\|K * h\|_{C^H} \le C\|h\|_{C^{-\frac{1}{2}}}$ gives the bound for $\Pi\mathcal{I}(\Xi)^m$. Finally, we note that by using the Cauchy–Schwarz inequality,

$$\begin{aligned}
\left|\left\langle \Pi_t \Xi \mathcal{I}(\Xi)^m, \varphi_x^{\lambda}\right\rangle\right| &= \left|\int h_1(s)(K * h_1(s) - K * h_1(t))^m \varphi_x^{\lambda}(s)ds\right| \\
&\le \left(\sup_{|s-t|\le\lambda} |K * h_1(s) - K * h_1(t)|\right)^m \|h_1\|_{L^2}\| \, \|\varphi_x^{\lambda}\|_{L^2} \\
&\lesssim \lambda^{mH-\frac{1}{2}}.
\end{aligned}$$

The inequality for $\Pi\bar{\Xi}\mathcal{I}(\Xi)^m$ follows in the same way, and the bounds for Γ also follow. Continuity in h is proved by similar arguments which we leave to the reader. □

***Proof of Theorem* 10.24.** The theorem is a special case of results in Hairer and Weber [196] for large deviations of Banach-valued Gaussian polynomials. Let us recall the setting.

Let $(B, \mathcal{H}, \mu)$ be an abstract Wiener space, and let us call ξ the associated B-valued Gaussian random variable and call (e_i) an orthonormal basis of $\mathcal{H}$ with $e_i \in B^*$. For a multi-index $\alpha \in \mathbb{N}^{\mathbb{N}}$ with only finitely many nonzero entries, define $H_{\alpha}(\xi) = \Pi_{i\ge 0} H_{\alpha_i}(\langle \xi, e_i\rangle)$, where the H_n, $n \ge 0$, are the usual Hermite polynomials. For a given Banach space E, the homogeneous Wiener chaos $\mathcal{H}^{(k)}(E)$ is defined as the closure in $L^2(E, \mu)$ of the linear space generated by elements of the form

$$H_{\alpha}(\xi)y, \quad |\alpha| = k,\ y \in E.$$

Also define the inhomogeneous Wiener chaos $\mathcal{H}^k(E) = \oplus_{i=0}^k \mathcal{H}^{(i)}(E)$. Finally, for $\Psi \in \mathcal{H}^{(k)}(E)$ and $h \in \mathcal{H}$ we define $\Psi^{hom}(h) = \int \Psi(\xi + h)\mu(d\xi)$, and for $\Psi = \sum_{i\le k} \Psi_i \in \mathcal{H}^k(E)$, we let $\Psi^{hom} = (\Psi_k)^{hom}$.

Now let $E = \oplus_{\tau\in\mathcal{W}} E_\tau$, where $\mathcal{W}$ is a finite set and each E_τ is a separable Banach space. Let $\Psi = \oplus_{\tau\in\mathcal{W}}\Psi_\tau$ be a random variable such that each Ψ_τ is in $\mathcal{H}^{K_\tau}(E_\tau)$. Letting $\Psi^\delta = \oplus_\tau \delta^{K_\tau}\Psi_\tau$, Theorem 3.5 in [196] states that Ψ^δ satisfies an LDP with rate function given by

$$I(\Psi) = \inf\left\{\frac{1}{2}\|h\|_{\mathcal{H}}^2,\quad \Psi = \oplus_{\tau\in\mathcal{W}}\Psi_\tau^{hom}(h)\right\}.$$

In our case, we apply this result with $\mathcal{W} = \left\{\Xi\mathcal{I}(\Xi)^m, \bar{\Xi}\mathcal{I}(\Xi)^m, 0 \le m \le M\right\}$, and each E_τ is the closure of smooth functions $(t,s) \mapsto \Pi_t\tau(s)$ under the norms

$$\|\Pi\tau\| = \sup_{\lambda,t,\varphi} \lambda^{-|\tau|}\left|\left\langle \Pi_t\tau, \varphi_t^\lambda\right\rangle\right|.$$

In order to obtain Theorem 10.24, it suffices then to identify $(\Pi\tau)^{hom}(h)$, which is done in the following lemma. □

Lemma 10.33. *For each $\tau \in \mathcal{W}$ and $h \in \mathcal{H}$, $(\Pi\tau)^{hom}(h) = \Pi^h\tau$.*

Proof. We prove it for $\tau = \Xi\mathcal{I}(\Xi)^m$; the other cases are similar. Note that $\Psi \mapsto \Psi^{hom}(h)$ is continuous from $\mathcal{H}^k$ to $\mathbb{R}$ for fixed h (by an application of the Cameron–Martin formula), and so it is enough to prove that

$$\lim_{\varepsilon\to 0}\left(\hat{\Pi}^\varepsilon\tau\right)^{hom}(h) = \Pi^h\tau, \tag{10.79}$$

where $\hat{\Pi}^\varepsilon$ corresponds to the (renormalized model) with piecewise-linear approximation of ξ. For any test function φ, by definition one has

$$\left\langle \Pi_t^\varepsilon\tau, \varphi\right\rangle = -\left\langle I^\varepsilon, \varphi'\right\rangle,$$

where

$$I^\varepsilon(s) = \int_t^s ((K*\xi^\varepsilon)(u) - (K*\xi^\varepsilon)(t))^m \xi^\varepsilon(u)\mathrm{d}u - C_\varepsilon R_m^\varepsilon,$$

where R_m^ε is a renormalization term which is valued in the lower-order chaos $\mathcal{H}^m$, so that by definition it does not play a role in the value of $(\Pi\tau)^{hom}$. Now note that if Φ is a Wiener polynomial whose leading order term is given by $\Pi_{i=1}^k\langle\xi, g_i\rangle$ (where the g_i are in $\mathcal{H}$), then $\Phi^{hom}(h) = \Pi_{i=1}^k\langle h, g_i\rangle$. In our case this means that

$$(I^\varepsilon)^{hom}(s) = \int_t^s \left((K*h_1^\varepsilon)(u) - (K*h_1^\varepsilon)(t)\right)^m h_1^\varepsilon(u)\mathrm{d}u$$

where $h_1^\varepsilon = \varrho^\varepsilon * h_1$. In other words, we have $(\hat{\Pi}^\varepsilon\tau)^{hom} = \Pi_\tau^{h^\varepsilon}$, and by continuity of $h \mapsto \Pi^h$ we obtain (10.79). □

10.6.3 ▪ Proofs of Section 10.4

The proof of Theorem 10.27 follows from the estimates in the following lemmas, using the standard procedure of taking a time horizon T small enough to obtain a contraction and then iterating the procedure. Note that due to the global boundedness of u and v, the estimates are uniform in the starting point z, so that one obtains global existence (unlike the typical situation in SPDEs where the theory only gives local in time existence).

By translating u and v, we can assume without loss of generality that the initial condition is $z = 0$. Then the solution will be an element of $\mathcal{D}_{0,T}^\gamma(\Gamma) := \left\{F \in \mathcal{D}_T^\gamma(\Gamma),\ \ F(0) = 0\right\}$.

Lemma 10.34. *Let F and $\tilde{F}$ in $\mathcal{D}^{\gamma}_{0,T}(\mathcal{T})$ for the respective models (Π,Γ) and $(\tilde{\Pi},\tilde{\Gamma})$. For each $\gamma < 1$ and $T \in (0,1]$ one has*

$$\|\mathcal{K}F;\mathcal{K}\tilde{F}\|_{\mathcal{D}^{\gamma}_{T}(\Gamma),\mathcal{D}^{\gamma}_{T}(\tilde{\Gamma})} \lesssim T^{\eta}\|F;\tilde{F}\|_{\mathcal{D}^{\gamma+|\Xi|}_{T}(\Gamma),\mathcal{D}^{\gamma+|\Xi|}_{T}(\tilde{\Gamma})}$$

for some $\eta > 0$. The constant of proportionality only depends on γ and the norms of (Π,Γ) and $(\tilde{\Pi},\tilde{\Gamma})$.

Proof. The choice $\gamma < 1$ avoids the appearance of polynomial terms [191, Section 5]. Note that if F belongs to $\mathcal{D}^{\gamma}_{0,T}$, so does $\mathcal{K}F$. Since K is a regularizing kernel of order $\beta := \frac{1}{2} + H$ in the sense of [191], it follows along the lines of [191, Section 5] that

$$\|\mathcal{K}F;\mathcal{K}\tilde{F}\|_{\mathcal{D}^{\bar{\gamma}}_{T}(\Gamma),\mathcal{D}^{\bar{\gamma}}_{T}(\tilde{\Gamma})} \lesssim \|F;\tilde{F}\|_{\mathcal{D}^{\gamma+|\Xi|}_{T}(\Gamma),\mathcal{D}^{\gamma+|\Xi|}_{T}(\tilde{\Gamma})},$$

where we pick $\bar{\gamma} \in (\gamma, 1)$ such that $\bar{\gamma} \le \gamma + |\Xi| + \beta = \gamma + H - \kappa$. On the other hand, it is clear from the definition of $\|\cdot;\cdot\|$ that since $\mathcal{K}F$ and $\mathcal{K}\tilde{F}$ vanish at $t = 0$, it holds that

$$\|\mathcal{K}F;\mathcal{K}\tilde{F}\|_{\mathcal{D}^{\gamma}_{T}(\Gamma),\mathcal{D}^{\gamma}_{T}(\tilde{\Gamma})} \lesssim T^{\eta}\|\mathcal{K}F;\mathcal{K}\tilde{F}\|_{\mathcal{D}^{\bar{\gamma}}_{T}(\Gamma),\mathcal{D}^{\bar{\gamma}}_{T}(\tilde{\Gamma})}$$

for $\eta = \bar{\gamma} - \gamma$. □

Lemma 10.35. *Let G (resp., $\tilde{G}$) be the composition operator corresponding to g (resp., $\tilde{g}$) $\in C^{M+2}_b$. Then one has*

$$\|G(F);\tilde{G}(\tilde{F})\|_{\mathcal{D}^{\gamma}_{T}(\Gamma),\mathcal{D}^{\gamma}_{T}(\tilde{\Gamma})} \lesssim \|G - \tilde{G}\|_{C^{M+2}} + \|F;\tilde{F}\|_{\mathcal{D}^{\gamma}_{T}(\Gamma),\mathcal{D}^{\gamma}_{T}(\tilde{\Gamma})}$$

with a proportionality constant depending only on γ and the norms of (Π,Γ), $(\tilde{\Pi},\tilde{\Gamma})$, F, $\tilde{F}$, g, $\tilde{g}$.

Proof. This follows from the estimate in [191, Theorem 4.16]. The joint continuity is not stated there but is clear from the triangle inequality. □

Acknowledgments

The authors acknowledge financial support from DFG research grants BA5484/1 (CB, BS) and FR2943/2 (PKF, BS), the ERC via grant CoG-683164 (PKF), the ANR via grant ANR-16-CE40-0020-01 (PG), and the DFG Research Training Group RTG 1845 (JM).

Participants of Global Derivatives 2017 (Barcelona) and Gatheral 60th birthday conference (CIMS, NYU) are thanked for their valuable feedback. The authors are also very thankful to the anonymous referees for their very constructive feedback.

Bibliography

[1] Eduardo Abi Jaber, *Lifting the Heston model*, Quantitative Finance, 19 (2019), pp. 1995–2013. (Cited on p. 89)

[2] Eduardo Abi Jaber and Omar El Euch, *Markovian structure of the Volterra Heston model*, Statistics & Probability Letters, 149 (2019), pp. 63–72. (Cited on p. 109)

[3] ———, *Multifactor approximation of rough volatility models*, SIAM Journal on Financial Mathematics, 10 (2019), pp. 309–349. (Cited on pp. 89, 90, 112)

[4] Eduardo Abi Jaber, Martin Larsson, and Sergio Pulido, *Affine volterra processes*, The Annals of Applied Probability, 29 (2019), pp. 3155–3200. (Cited on pp. 68, 103, 104, 106, 107, 191)

[5] Peter A. Acworth, Mark Broadie, and Paul Glasserman, *A comparison of some Monte Carlo and quasi-Monte Carlo techniques for option pricing*, in Monte Carlo and Quasi-Monte Carlo Methods 1996 (Salzburg), Harald Niederreiter, Peter Hellekalek, Gerhard Larcher, and Peter Zinterhof, eds., Springer-Verlag, New York, 1998, pp. 1–18. (Cited on p. 201)

[6] Robert Almgren and Neil Chriss, *Optimal execution of portfolio transactions*, Journal of Risk, 3 (2001), pp. 5–40. (Cited on p. 63)

[7] Elisa Alòs, *A generalization of the Hull and White formula with applications to option pricing approximation*, Finance and Stochastics, 10 (2006), pp. 353–365. (Cited on p. 160)

[8] ———, *A decomposition formula for option prices in the Heston model and applications to option pricing approximation*, Finance and Stochastics, 16 (2012), pp. 403–422. (Cited on p. 93)

[9] Elisa Alòs, Jim Gatheral, and Radoš Radoičić, *Exponentiation of conditional expectations under stochastic volatility*, Quantitative Finance, 20 (2020), pp. 13–27. (Cited on pp. 88, 93, 183, 185, 186, 189)

[10] Elisa Alòs and Jorge A. León, *An intuitive introduction to fractional and rough volatilities*, Mathematics, 9 (2021), article 994. (Cited on p. 159)

[11] Elisa Alòs, Jorge A León, and Josep Vives, *On the short-time behavior of the implied volatility for jump-diffusion models with stochastic volatility*, Finance and Stochastics, 11 (2007), pp. 571–589. (Cited on pp. 35, 128, 159, 160, 162, 199, 206)

[12] Elisa Alòs and David Garcia Lorite, *Malliavin Calculus in Finance: Theory and Practice*, CRC Press, Boca Raton, 2021. (Cited on p. 160)

[13] Elisa Alòs and Yan Yang, *A fractional Heston model with* $H > 1/2$, Stochastics, 89 (2017), pp. 384–399. (Cited on p. 200)

[14] Leif B.G. Andersen and Vladimir V. Piterbarg, *Moment explosions in stochastic volatility models*, Finance and Stochastics, 11 (2007), pp. 29–50. (Cited on p. 96)

[15] Torben G. Andersen and Tim Bollerslev, *Heterogeneous information arrivals and return volatility dynamics: Uncovering the long-run in high frequency returns*, The Journal of Finance, 52 (1997), pp. 975–1005. (Cited on p. 18)

[16] ———, *Intraday periodicity and volatility persistence in financial markets*, Journal of Empirical Finance, 4 (1997), pp. 115–158. (Cited on p. 2)

[17] Torben G. Andersen, Tim Bollerslev, Francis X. Diebold, and Heiko Ebens, *The distribution of realized stock return volatility*, Journal of Financial Economics, 61 (2001), pp. 43–76. (Cited on p. 9)

[18] Torben G. Andersen, Tim Bollerslev, Francis X. Diebold, and Paul Labys, *The distribution of realized exchange rate volatility*, Journal of the American Statistical Association, 96 (2001), pp. 42–55. (Cited on pp. 2, 9, 17, 18)

[19] ———, *Modeling and forecasting realized volatility*, Econometrica, 71 (2003), pp. 579–625. (Cited on pp. 17, 18, 20)

[20] Jean-Pascal Ansel and Christophe Stricker, *Couverture des actifs contingents et prix maximum*, Annales de l'Institut Henri Poincaré Probabilités et Statistiques, 30(2) (1994), pp. 303–315. (Cited on p. 115)

[21] Søren Asmussen and Peter W. Glynn, *Stochastic simulation: Algorithms and Analysis*, Stoch. Model. Appl. Probab. 57, Springer, New York, 2007. (Cited on pp. 128, 140)

[22] Marco Avellaneda, Dash Boyer-Olson, Jérôme Busca, and Peter K. Friz, *Application of large deviation methods to the pricing of index options in finance*, Comptes Rendus Mathematique, 336 (2003), pp. 263–266. (Cited on p. 205)

[23] Emmanuel Bacry, Sylvain Delattre, Marc Hoffmann, and Jean-Francois Muzy, *Some limit theorems for Hawkes processes and application to financial statistics*, Stochastic Processes and Their Applications, 123 (2013), pp. 2475–2499. (Cited on p. 63)

[24] Emmanuel Bacry, Adrian Iuga, Matthieu Lasnier, and Charles-Albert Lehalle, *Market impacts and the life cycle of investors orders*, Market Microstructure and Liquidity, 1 (2015), 1550009. (Cited on pp. 60, 61, 64, 69)

[25] Emmanuel Bacry, Thibault Jaisson, and Jean-François Muzy, *Estimation of slowly decreasing Hawkes kernels: Application to high-frequency order book dynamics*, Quantitative Finance, 16 (2016), pp. 1179–1201. (Cited on p. 61)

[26] Emmanuel Bacry, Iacopo Mastromatteo, and Jean-François Muzy, *Hawkes processes in finance*, Market Microstructure and Liquidity, 1 (2015), article 1550005. (Cited on p. 61)

[27] Emmanuel Bacry and Jean-François Muzy, *Log-infinitely divisible multifractal processes*, Communications in Mathematical Physics, 236 (2003), pp. 449–475. (Cited on pp. 17, 20)

[28] Emmanuel Bacry and Jean-François Muzy, *First-and second-order statistics characterization of Hawkes processes and non-parametric estimation*, IEEE Transactions on Information Theory, 62 (2016), pp. 2184–2202. (Cited on p. 65)

[29] Ole Barndorff-Nielsen, *Notes on the Gamma Kernel*, Thiele Research Report 3, TN Thiele Centre, University of Aarhus, 2012. (Cited on pp. 130, 131)

[30] Ole E. Barndorff-Nielsen and Andreas Basse-O'Connor, *Quasi Ornstein-Uhlenbeck processes*, Bernoulli, 17 (2011), pp. 916–941. (Cited on p. 118)

[31] Ole E. Barndorff-Nielsen, Fred Espen Benth, and Almut E.D. Veraart, *Modelling energy spot prices by volatility modulated Lévy-driven Volterra processes*, Bernoulli, 19 (2013), pp. 803–845. (Cited on p. 127)

[32] Ole E. Barndorff-Nielsen, José Manuel Corcuera, and Mark Podolskij, *Multipower variation for Brownian semistationary processes*, Bernoulli, 17 (2011), pp. 1159–1194. (Cited on p. 149)

[33] ———, *Limit theorems for functionals of higher order differences of Brownian semi-stationary processes*, in Prokhorov and Contemporary Probability Theory, A.N. Shiryaev, S.R.S. Varadhan, and E.L. Presman, eds., Springer, Heidelberg, 2013, pp. 69–96. (Cited on pp. 128, 142)

[34] Ole E. Barndorff-Nielsen, Mikko S. Pakkanen, and Jürgen Schmiegel, *Assessing relative volatility/intermittency/energy dissipation*, Electronic Journal of Statistics, 8 (2014), pp. 1996–2021. (Cited on p. 127)

[35] Ole E. Barndorff-Nielsen and Jürgen Schmiegel, *Ambit processes; with applications to turbulence and tumour growth*, in Stochastic Analysis and Applications, F.E. Benth, G. Di Nunno, T. Lindstrøm, B. Øksendal, and T. Zhang, eds., Springer, Berlin, 2007, pp. 93–124. (Cited on p. 127)

[36] ———, *Brownian Semistationary Processes and Volatility/Intermittency*, Walter de Gruyter GmbH & Co. KG, Berlin, 2009. (Cited on p. 127)

[37] Harry Bateman, *Higher Transcendental Functions, Vol.* 1, McGraw-Hill, New York, 1953. (Cited on pp. 139, 154)

[38] Christian Bayer, Peter K. Friz, Paul Gassiat, Jorg Martin, and Benjamin Stemper, *A regularity structure for rough volatility*, Mathematical Finance, 30 (2020), pp. 782–832. (Cited on pp. xxvii, 173)

[39] Christian Bayer, Peter K. Friz, and Jim Gatheral, *Pricing under rough volatility*, Quantitative Finance, 16 (2016), pp. 887–904. (Cited on pp. xxvi, 4, 23, 105, 116, 118, 128, 129, 144, 169, 194, 199, 200, 204, 205, 227, 235, 236)

[40] Christian Bayer, Peter K. Friz, Archil Gulisashvili, Blanka Horvath, and Benjamin Stemper, *Short-time near-the-money skew in rough fractional volatility models*, Quantitative Finance, 19 (2019), pp. 779–798. (Cited on pp. 170, 172, 205, 206, 227, 235)

[41] Christian Bayer, Peter K. Friz, Sebastian Riedel, and John Schoenmakers, *From rough path estimates to multilevel Monte Carlo*, SIAM Journal on Numerical Analysis, 54 (2016), pp. 1449–1483. (Cited on p. 204)

[42] Christian Bayer, Fabian A. Harang, and Paolo Pigato, *Log-modulated rough stochastic volatility models*, SIAM Journal on Financial Mathematics, 12 (2021), pp. 1257–1284. (Cited on pp. 166, 179, 180, 181, 182)

[43] Christian Bayer, Blanka Horvath, Aitor Muguruza, Benjamin Stemper, and Mehdi Tomas, *On Deep Calibration of (Rough) Stochastic Volatility Models*, arXiv:1908.08806, 2019. (Cited on p. 116)

[44] Christian Bayer and Peter Laurence, *Asymptotics beats Monte Carlo: The case of correlated local vol baskets*, Communications on Pure and Applied Mathematics, 67 (2014), pp. 1618–1657. (Cited on p. 205)

[45] Shalom Benaim and Peter K. Friz, *Smile asymptotics* II: *Models with known moment generating functions*, Journal of Applied Probability, 45 (2008), pp. 16–32. (Cited on p. 96)

[46] Shalom Benaim and Peter K. Friz, *Regular variation and smile asymptotics*, Mathematical Finance, 19 (2009), pp. 1–12. (Cited on p. 171)

[47] Shalom Benaim, Peter K. Friz, and Roger Lee, *On Black-Scholes implied volatility at extreme strikes*, Frontiers in Quantitative Finance: Volatility and Credit Risk Modeling, (2008), pp. 19–45. (Cited on p. 96)

[48] Mikkel Bennedsen, *A rough multi-factor model of electricity spot prices*, Energy Economics, 63 (2017), pp. 301–313. (Cited on p. 127)

[49] Mikkel Bennedsen, Asger Lunde, and Mikko S. Pakkanen, *Discretization of Lévy Semistationary Processes with Application to Estimation.* arXiv:1407.2754, 2014. (Cited on p. 142)

[50] ———, *Hybrid scheme for Brownian semistationary processes*, Finance and Stochastics, 21 (2017), pp. 931–965. (Cited on pp. xxvii, 116, 164, 177)

[51] ———, *Decoupling the short-and long-term behavior of stochastic volatility*, Journal of Financial Econometrics 20, (2022) pp. 961–1006. (Cited on pp. 22, 59, 200)

[52] Sónia R. Bentes and M. Mendes da Cruz, *Is stock market volatility persistent? A fractionally integrated approach.* Unpublished, 2011. (Cited on p. 2)

[53] Fred Espen Benth, Heidar Eyjolfsson, and Almut E.D. Veraart, *Approximating Lévy semistationary processes via Fourier methods in the context of power markets*, SIAM Journal on Financial Mathematics, 5 (2014), pp. 71–98. (Cited on pp. 128, 136)

[54] Jan Beran, *Statistics for Long-Memory Processes*, Monogr. Statist. Appl. Probab. 61, Chapman and Hall, New York, 1994. (Cited on p. 2)

[55] Henri Berestycki, Jérôme Busca, and Igor Florent, *Computing the implied volatility in stochastic volatility models*, Communications on Pure and Applied Mathematics, 57 (2004), pp. 1–22. (Cited on p. 171)

[56] Henri Berestycki, Jérôme Busca, and Igor Florent, *Computing the implied volatility in stochastic volatility models*, Communications on Pure and Applied Mathematics, 57 (2004), pp. 1352–1373. (Cited on p. 205)

[57] Lorenzo Bergomi, *Smile dynamics* II, Risk, October (2005), pp. 67–73. (Cited on pp. 33, 39, 116)

[58] Lorenzo Bergomi and Julien Guyon, *Stochastic volatility's orderly smiles*, Risk, 25 (2012), p. 60. (Cited on pp. 33, 48, 105, 162, 188)

[59] Alexandros Beskos and Gareth O. Roberts, *Exact simulation of diffusions*, The Annals of Applied Probability, 15 (2005), pp. 2422–2444. (Cited on p. 139)

[60] Rabi N. Bhattacharya and R. Ranga Rao, *Normal Approximation and Asymptotic Expansions*, Wiley, 1979. (Cited on p. 167)

[61] Rabi N. Bhattacharya and R. Ranga Rao, *Normal Approximation and Asymptotic Expansions*, Classics Appl. Math. 64, SIAM, 2010. (Cited on p. 167)

[62] Bo Martin Bibby, Ib Michael Skovgaard, and Michael Sorensen, *Diffusion-type models with given marginal distribution and autocorrelation function*, Bernoulli, 11 (2005), pp. 191–220. (Cited on p. 23)

[63] Nicholas H. Bingham, Charles M. Goldie, and Jozef L. Teugels, *Regular Variation*, Encyclopedia Math. Appl. 27, Cambridge University Press, Cambridge, 1989. (Cited on pp. 81, 127, 129, 130, 145)

[64] Pierre Blanc, Jonathan Donier, and Jean-Philippe Bouchaud, *Quadratic Hawkes processes for financial prices*, Quantitative Finance, 17 (2017), pp. 171–188. (Cited on p. 99)

[65] Jean-Philippe Bouchaud, *Price impact*, in Encyclopedia of Quantitative Finance, R. Cont, ed., John Wiley & Sons, Ltd, New York, 2010. (Cited on pp. 61, 64)

[66] Jean-Philippe Bouchaud and Marc Potters, *More statistical properties of order books and price impact*, Physica A: Statistical Mechanics and Its Applications, 324 (2003), pp. 133–140. (Cited on p. 64)

[67] ———, *Theory of Financial Risk and Derivative Pricing: From Statistical Physics to Risk Management*, Cambridge University Press, 2003. (Cited on p. 17)

[68] Michelle Boué and Paul Dupuis, *A variational representation for certain functionals of Brownian motion*, The Annals of Probability, 26 (1998), pp. 1641–1659. (Cited on pp. 173, 174)

[69] Florian Bourgey, Stefano De Marco, Peter K. Friz, and Paolo Pigato, *Local volatility under rough volatility*, Mathematical Finance, 33 (2023), pp. 1119–1145. (Cited on pp. 177, 178)

[70] Alexandre Brouste and Stefano M Iacus, *Parameter estimation for the discretely observed fractional Ornstein–Uhlenbeck process and the Yuima R package*, Computational Statistics, 28 (2013), pp. 1529–1547. (Cited on p. 11)

[71] Yvain Bruned, Ilya Chevyrev, Peter K. Friz, and Rosa Preiss, *A rough path perspective on renormalization*, Journal of Functional Analysis, 277 (2019), 108283. (Cited on pp. 229, 232)

[72] Yvain Bruned, Martin Hairer, and Lorenzo Zambotti, *Algebraic renormalisation of regularity structures*, Inventiones Mathematicae, 215 (2019), pp. 1039–1156. (Cited on p. 229)

[73] Gerard Brunick and Steven Shreve, *Mimicking an Itô process by a solution of a stochastic differential equation*, The Annals of Applied Probability, 23 (2013), pp. 1584–1628. (Cited on p. 177)

[74] Amarjit Budhiraja, Jiang Chen, and Paul Dupuis, *Large deviations for stochastic partial differential equations driven by a Poisson random measure*, Stochastic Processes and Their Applications, 123 (2013), pp. 523–560. (Cited on p. 174)

[75] Amarjit Budhiraja and Paul Dupuis, *A variational representation for positive functionals of infinite-dimensional Brownian motion*, Probability and Mathematical Statistics, 20 (2000), pp. 39–61. (Cited on p. 173)

[76] ———, *Analysis and Approximation of Rare Events: Representation and Weak Convergence Methods*, Springer, New York, 2019. (Cited on p. 174)

[77] Amarjit Budhiraja, Paul Dupuis, and Arnab Ganguly, *Large deviations for small noise diffusions in a fast Markovian environment*, Electronic Journal of Probability, 23 (2018), pp. 1–33. (Cited on p. 174)

[78] Amarjit Budhiraja, Paul Dupuis, and Vasileios Maroulas, *Large deviations for infinite dimensional stochastic dynamical systems*, The Annals of Probability, 36 (2008), pp. 1390–1420. (Cited on p. 174)

[79] ———, *Variational representations for continuous time processes*, in Annales de l'IHP Probabilités et Statistiques, 47(3), 2011, pp. 725–747. (Cited on p. 174)

[80] Hans Buehler, Lukas Gonon, Josef Teichmann, and Ben Wood, *Deep hedging*, Quantitative Finance, 19 (2019), pp. 1271–1291. (Cited on p. 116)

[81] Hans Bühler, *Volatility Markets – Consistent Modeling, Hedging and Practical Implementation*, PhD thesis, TU Berlin, 2006. (Cited on p. 105)

[82] Giorgia Callegaro, Martino Grasselli, and Gilles Pagès, *Fast hybrid schemes for fractional Riccati equations (rough is not so tough)*, Mathematics of Operations Research, 46 (2021), pp. 221–255. (Cited on pp. 92, 193)

[83] Francesco Caravenna and Jacopo Corbetta, *General smile asymptotics with bounded maturity*, SIAM Journal on Financial Mathematics, 7 (2016), pp. 720–759. (Cited on p. 171)

[84] P. Carmona, L. Coutin, and G. Montseny, *Approximation of some Gaussian processes*, Statistical Inference for Stochastic Processes, 3 (2000), pp. 161–171. (Cited on pp. 90, 112)

[85] Peter Carr and Dilip Madan, *Option valuation using the fast Fourier transform*, Journal of Computational Finance, 2 (1999), pp. 61–73. (Cited on pp. 84, 86, 189)

[86] ——, *Towards a theory of volatility trading*, in Option Pricing, Interest Rates and Risk Management, E. Jouini, J. Cvitanić, and M. Musiela, eds., Handbooks in Mathematical Finance, Cambridge University Press, Cambridge, 2001, pp. 458–476. (Cited on p. 88)

[87] Peter Carr and Liuren Wu, *What type of process underlies options? A simple robust test*, Journal of Finance, 58 (2003), pp. 2581–2610. (Cited on pp. 4, 42)

[88] Miriana Cellupica and Barbara Pacchiarotti, *Pathwise asymptotics for Volterra type stochastic volatility models*, Journal of Theoretical Probability, 34 (2020), pp. 682–727. (Cited on p. 173)

[89] Narasinga R Chaganty, *Large deviations for joint distributions and statistical applications*, Sankhyā: The Indian Journal of Statistics, Series A, 59 (1997), pp. 147–166. (Cited on p. 173)

[90] Ajay Chandra and Martin Hairer, *An Analytic BPHZ Theorem for Regularity Structures*. arXiv:1612.08138, 2016. (Cited on pp. 204, 224)

[91] Zhiyao Chen, Robert T. Daigler, and Ali M. Parhizgari, *Persistence of volatility in futures markets*, Journal of Futures Markets, 26 (2006), pp. 571–594. (Cited on p. 2)

[92] Patrick Cheridito, Hideyuki Kawaguchi, and Makoto Maejima, *Fractional Ornstein-Uhlenbeck processes*, Electronic Journal of Probability, 8 (2003), p. 14. (Cited on pp. 2, 11, 24, 31)

[93] Alberto Chiarini and Markus Fischer, *On large deviations for small noise Itô processes*, Advances in Applied Probability, 46 (2014), pp. 1126–1147. (Cited on p. 174)

[94] Alexandra Chronopoulou, *Parameter estimation and calibration for long-memory stochastic volatility models*, in Handbook of Modeling High-Frequency Data in Finance, Frederi G. Viens, Maria C. Mariani, and Ionut Florescu, eds., John Wiley & Sons, New York, 2011, pp. 219–231. (Cited on p. 2)

[95] Alexandra Chronopoulou and Frederi G. Viens, *Estimation and pricing under long-memory stochastic volatility*, Annals of Finance, 8 (2012), pp. 379–403. (Cited on p. 14)

[96] Jean-François Coeurjolly, *Estimating the parameters of a fractional Brownian motion by discrete variations of its sample paths*, Statistical Inference for Stochastic Processes, 4 (2001), pp. 199–227. (Cited on p. 11)

[97] Fabienne Comte, Laure Coutin, and Eric Renault, *Affine fractional stochastic volatility models*, Annals of Finance, 8 (2012), pp. 337–378. (Cited on pp. 2, 12, 36, 85, 103, 200)

[98] Fabienne Comte and Eric Renault, *Long memory continuous time models*, Journal of Econometrics, 73 (1996), pp. 101–149. (Cited on pp. 40, 85)

[99] Fabienne Comte and Eric Renault, *Long memory in continuous-time stochastic volatility models*, Mathematical Finance, 8 (1998), pp. 291–323. (Cited on pp. 2, 4, 11, 23, 85, 200)

[100] Rama Cont, *Volatility clustering in financial markets: Empirical facts and agent-based models*, in Long Memory in Economics, Gilles Teyssière and Alan P. Kirman, eds., Springer, Berlin, 2007, pp. 289–309. (Cited on p. 2)

[101] Rama Cont and Adrien de Larrard, *Price dynamics in a Markovian limit order market*, SIAM Journal on Financial Mathematics, 4 (2013), pp. 1–25. (Cited on p. 61)

[102] Rama Cont, Sasha Stoikov, and Rishi Talreja, *A stochastic model for order book dynamics*, Operations Research, 58 (2010), pp. 549–563. (Cited on p. 61)

[103] José Manuel Corcuera, Emil Hedevang, Mikko S. Pakkanen, and Mark Podolskij, *Asymptotic theory for Brownian semi-stationary processes with application to turbulence*, Stochastic Processes and Their Applications, 123 (2013), pp. 2552–2574. (Cited on pp. 36, 127, 128, 131, 142)

[104] Fulvio Corsi, *A simple approximate long-memory model of realized volatility*, Journal of Financial Econometrics, 7 (2009), pp. 174–196. (Cited on p. 20)

[105] Fulvio Corsi, Nicola Fusari, and Davide La Vecchia, *Realizing smiles: Options pricing with realized volatility*, Journal of Financial Economics, 107 (2013), pp. 284–304. (Cited on p. 52)

[106] Laure Coutin and Laurent Decreusefond, *Stochastic Volterra equations with singular kernels*, in Stochastic Analysis and Mathematical Physics, A.B. Cruzeiro and J.-C. Zambrini, eds., Birkhäuser Boston, Inc., Boston, MA, 2001, pp. 39–50. (Cited on pp. 228, 231)

[107] Christa Cuchiero and Josef Teichmann, *Generalized Feller processes and Markovian lifts of stochastic Volterra processes: The affine case*, Journal of Evolution Equations, 20 (2020), pp. 1301–1348. (Cited on pp. 111, 112)

[108] Giuseppe Da Prato and Jerzy Zabczyk, *Stochastic Equations in Infinite Dimensions*, Cambridge University Press, Cambridge, 2014. (Cited on p. 109)

[109] Aditi Dandapani, Paul Jusselin, and Mathieu Rosenbaum, *From quadratic Hawkes processes to super-Heston rough volatility models with Zumbach effect*, Quantitative Finance, 21 (2021), pp. 1235–1247. (Cited on p. 99)

[110] Mark H.A. Davis and Vicente Mataix-Pastor, *Negative Libor rates in the swap market model*, Finance and Stochastics, 11 (2007), pp. 181–193. (Cited on pp. 201, 202, 206)

[111] Khalil Dayri and Mathieu Rosenbaum, *Large tick assets: Implicit spread and optimal tick size*, Microstructure and Liquidity, 1 (2015), 155003. (Cited on p. 6)

[112] Fredd Y. Delbaen and Walter Schachermayer, *The no-arbitrage property under a change of numéraire*, Stochastics and Stochastic Reports, 53 (1995), pp. 213–226. (Cited on p. 115)

[113] A. Dembo and O. Zeitouni, *Large Deviations Techniques and Applications*, Stoch. Model. Appl. Probab. 38, Springer-Verlag, Berlin, 2010. (Cited on pp. 169, 173)

[114] Kresimir Demeterfi, Emanuel Derman, Michael Kamal, and Joseph Zou, *More than You Ever Wanted to Know About Volatility Swaps*, Quantitative Strategies Research Notes, Goldman Sachs, 1999. (Cited on p. 88)

[115] Jean-Dominique Deuschel, Peter K. Friz, Antoine Jacquier, and Sean Violante, *Marginal density expansions for diffusions and stochastic volatility* I: *Theoretical foundations*, Communications on Pure and Applied Mathematics, 67 (2014), pp. 40–82. (Cited on pp. 95, 205)

[116] Jean-Dominique Deuschel, Peter K. Friz, Antoine Jacquier, and Sean Violante, *Marginal density expansions for diffusions and stochastic volatility* II: *Applications*, Communications on Pure and Applied Mathematics, 67 (2014), pp. 321–350. (Cited on p. 205)

[117] Francis X. Diebold and Atsushi Inoue, *Long memory and regime switching*, Journal of Econometrics, 105 (2001), pp. 131–159. (Cited on p. 18)

[118] Kai Diethelm, Neville J. Ford, and Alan D. Freed, *Detailed error analysis for a fractional Adams method*, Numerical Algorithms, 36 (2004), pp. 31–52. (Cited on p. 91)

[119] Zhuanxin Ding, Clive W.J. Granger, and Robert F. Engle, *A long memory property of stock market returns and a new model*, Journal of Empirical Finance, 1 (1993), pp. 83–106. (Cited on p. 2)

[120] Jonathan Donier and Julius Bonart, *A million metaorder analysis of market impact on the bitcoin*, Market Microstructure and Liquidity, 1 (2015), 1550008. (Cited on p. 60)

[121] Jean Duchon, Raoul Robert, and Vincent Vargas, *Forecasting volatility with the multifractal random walk model*, Mathematical Finance, 22 (2012), pp. 83–108. (Cited on p. 20)

[122] Darrell Duffie, D. Filipović, and Walter Schachermayer, *Affine processes and applications in finance*, The Annals of Applied Probability, 13 (2003), pp. 984–1053. (Cited on pp. 87, 103)

[123] Bruno Dupire, *Model art*, Risk, (1993). (Cited on p. 88)

[124] ———, *Pricing with a smile*, Risk Magazine, 7 (1994), pp. 18–20. (Cited on pp. 1, 176)

[125] Paul Dupuis and Konstantinos Spiliopoulos, *Large deviations for multiscale diffusion via weak convergence methods*, Stochastic Processes and Their Applications, 122 (2012), pp. 1947–1987. (Cited on p. 174)

[126] Omar El Euch, Masaaki Fukasawa, and Mathieu Rosenbaum, *The microstructural foundations of leverage effect and rough volatility*, Finance and Stochastics, 22 (2018), pp. 241–280. (Cited on pp. 24, 60, 61, 200, 205, 227)

[127] Omar El Euch, Jim Gatheral, Radoš Radoičić, and Mathieu Rosenbaum, *The Zumbach effect under rough Heston*, Quantitative Finance, 20 (2020), pp. 235–241. (Cited on p. 98)

[128] Omar El Euch, Jim Gatheral, and Mathieu Rosenbaum, *Roughening Heston*, Risk, (2019), pp. 84–89. (Cited on p. 68)

[129] Omar El Euch and Mathieu Rosenbaum, *The characteristic function of rough Heston models*, Mathematical Finance, 29 (2019), pp. 3–38. (Cited on pp. 24, 59, 60, 68, 71, 72, 76, 77, 78, 84, 86, 92, 103, 104, 108, 190, 200, 205, 206, 227)

[130] Omar El Euch, Masaaki Fukasawa, Jim Gatheral, and Mathieu Rosenbaum, *Short-term at-the-money asymptotics under stochastic volatility models*, SIAM Journal on Financial Mathematics, 10 (2019), pp. 491–511. (Cited on pp. 167, 168, 176)

[131] Omar El Euch and Mathieu Rosenbaum, *Perfect hedging in rough Heston models*, The Annals of Applied Probability, 28 (2018), pp. 3813–3856. (Cited on pp. 24, 60, 72, 84, 86, 87, 99, 116, 200, 205, 227)

[132] Vladimir Filimonov and Didier Sornette, *Apparent criticality and calibration issues in the Hawkes self-excited point process model: Application to high-frequency financial data*, Quantitative Finance, 15 (2015), pp. 1293–1314. (Cited on pp. 62, 65)

[133] Martin Forde, Masaaki Fukasawa, Stefan Gerhold, and Benjamin Smith, *The Riemann-Liouville field and its GMC as $H \to 0$, and skew flattening for the rough Bergomi model*, Statistics and Probability Letters, 181 (2022) article 109265. (Cited on pp. 179, 180)

[134] Martin Forde, Stefan Gerhold, and Benjamin Smith, *Small-time, large-time, and asymptotics for the rough Heston model*, Mathematical Finance, 31 (2021), pp. 203–241. (Cited on pp. 97, 98, 172)

[135] Martin Forde and Antoine Jacquier, *Small-time asymptotics for implied volatility under the Heston model*, International Journal of Theoretical and Applied Finance, 12 (2009), pp. 861–876. (Cited on p. 97)

[136] Martin Forde, Antoine Jacquier, and Roger Lee, *The small-time smile and term structure of implied volatility under the Heston model*, SIAM Journal on Financial Mathematics, 3 (2012), pp. 690–708. (Cited on p. 176)

[137] Martin Forde, Benjamin Smith, and Lauri Viitasaari, *Rough volatility and CGMY jumps with a finite history and the rough Heston model—small-time asymptotics in the regime*, Quantitative Finance, 21 (2021), pp. 541–563. (Cited on pp. 98, 168)

[138] Martin Forde and Hongzhong Zhang, *Asymptotics for rough stochastic volatility models*, SIAM Journal on Financial Mathematics, 8 (2017), pp. 114–145. (Cited on pp. 170, 171, 172, 173, 174, 176, 199, 201, 205, 227)

[139] Jean-Pierre Fouque, George Papanicolaou, and K. Ronnie Sircar, *Derivatives in Financial Markets with Stochastic Volatility*, Cambridge University Press, Cambridge, 2000. (Cited on p. xxvi)

[140] Jean-Pierre Fouque, George Papanicolaou, Ronnie Sircar, and Knut Sølna, *Multiscale Stochastic Volatility for Equity, Interest Rate, and Credit Derivatives*, Cambridge University Press, Cambridge, 2011. (Cited on p. 92)

[141] Eric Fournié, Jean-Michel Lasry, Jérôme Lebuchoux, and Pierre-Louis Lions, *Applications of Malliavin calculus to Monte-Carlo methods in Finance*, Finance and Stochastics, 5 (2001), pp. 201–236. (Cited on p. 177)

[142] Peter K. Friz, Paul Gassiat, and Paolo Pigato, *Precise asymptotics: Robust stochastic volatility models*, The Annals of Applied Probability, 31 (2021), pp. 896–940. (Cited on pp. 170, 172, 175, 206)

[143] ———, *Short-dated smile under rough volatility: Asymptotics and numerics*, Quantitative Finance, 22 (2022), pp. 463–480. (Cited on pp. 172, 175, 176, 177)

[144] Peter K. Friz, Jim Gatheral, and Radoš Radoičić, *Forests, cumulants, martingales*, The Annals of Probability, 50 (2022), pp. 1418–1445. (Cited on pp. 183, 184, 185, 186)

[145] Peter K. Friz, Stefan Gerhold, and Arpad Pinter, *Option pricing in the moderate deviations regime*, Mathematical Finance, 28 (2018), pp. 962–988. (Cited on pp. 172, 205)

[146] Peter K. Friz, Benjamin Gess, Archil Gulisashvili, and Sebastian Riedel, *The Jain–Monrad criterion for rough paths and applications to random Fourier series and non-Markovian Hörmander theory*, The Annals of Probability, 44 (2016), pp. 684–738. (Cited on p. 224)

[147] Peter K. Friz and Martin Hairer, *A Course on Rough Paths*, Springer, Cham, 2020. (Cited on pp. 202, 204, 206, 207, 209, 210, 211, 213, 216, 219, 228, 231)

[148] Peter K. Friz and Sebastian Riedel, *Convergence rates for the full Brownian rough paths with applications to limit theorems for stochastic flows*, Bulletin des Sciences Mathématiques, 135 (2011), pp. 613–628. (Cited on p. 219)

[149] Peter K. Friz and Thomas Wagenhofer, *Reconstructing volatility: Pricing of index options under rough volatility*, Mathematical Finance, 33 (2023), pp. 19–40. (Cited on p. 170)

[150] Masaaki Fukasawa, *Asymptotic analysis for stochastic volatility: Edgeworth expansion*, Electronic Journal of Probability, 16 (2011), pp. 764–791. (Cited on p. 167)

[151] Masaaki Fukasawa, *Asymptotic analysis for stochastic volatility: Martingale expansion*, Finance and Stochastics, 15 (2011), pp. 635–654. (Cited on pp. 4, 23, 35, 163, 199, 206)

[152] ———, *Volatility derivatives and model-free implied leverage*, International Journal of Theoretical and Applied Finance, 17 (2014), 1450002. (Cited on p. 187)

[153] ———, *Short-time at-the-money skew and rough fractional volatility*, Quantitative Finance, 17 (2017), pp. 189–198. (Cited on pp. 103, 128, 199, 206)

[154] ———, *Volatility has to be rough*, Quantitative Finance, 21 (2021), pp. 1–8. (Cited on pp. 164, 166)

[155] Masaaki Fukasawa and Asuto Hirano, *Refinement by reducing and reusing random numbers of the hybrid scheme for Brownian semistationary processes*, Quantitative Finance, 21 (2021), pp. 1127–1146. (Cited on p. 116)

[156] Arnab Ganguly, *Large deviation principle for stochastic integrals and stochastic differential equations driven by infinite-dimensional semimartingales*, Stochastic Processes and Their Applications, 128 (2018), pp. 2179–2227. (Cited on p. 173)

[157] Kun Gao and Roger Lee, *Asymptotics of implied volatility to arbitrary order*, Finance and Stochastics, 18 (2014), pp. 349–392. (Cited on p. 176)

[158] Jorge Garcia, *A large deviation principle for stochastic integrals*, Journal of Theoretical Probability, 21 (2008), pp. 476–501. (Cited on p. 173)

[159] Josselin Garnier and Knut Sølna, *Correction to Black–Scholes formula due to fractional stochastic volatility*, SIAM Journal on Financial Mathematics, 8 (2017), pp. 560–588. (Cited on p. 163)

[160] ———, *Option pricing under fast-varying and rough stochastic volatility*, Annals of Finance, 14 (2018), pp. 489–516. (Cited on p. 163)

[161] ———, *Option pricing under fast-varying long-memory stochastic volatility*, Mathematical Finance, 29 (2019), pp. 39–83. (Cited on p. 163)

[162] Paul Gassiat, *On the martingale property in the rough Bergomi model*, Electronic Communications in Probability, 24 (2019), pp. 1–9. (Cited on pp. 116, 158, 171)

[163] Jim Gatheral, *The Volatility Surface: A Practitioner's Guide*, John Wiley & Sons, New York, 2006. (Cited on pp. xxvi, 1, 3, 41, 45, 51, 56, 84, 88, 94, 144, 178, 189, 194, 206)

[164] ———, *No-dynamic-arbitrage and market impact*, Quantitative Finance, 10 (2010), pp. 749–759. (Cited on p. 60)

[165] Jim Gatheral and Antoine Jacquier, *Arbitrage-free SVI volatility surfaces*, Quantitative Finance, 14 (2014), pp. 59–71. (Cited on pp. 3, 32, 41, 51)

[166] Jim Gatheral, Thibault Jaisson, and Mathieu Rosenbaum, *Volatility is rough*, Quantitative Finance, 18 (2018), pp. 933–949. (Cited on pp. xxvi, 31, 35, 51, 55, 59, 69, 103, 124, 128, 199, 204, 205, 223, 227, 228)

[167] Jim Gatheral, Paul Jusselin, and Mathieu Rosenbaum, *The quadratic rough Heston model and the joint S&P 500/VIX smile calibration problem*, Risk.net, May (2020). (Cited on p. 100)

[168] Jim Gatheral and Martin Keller-Ressel, *Affine forward variance models*, Finance and Stochastics, 23 (2019), pp. 501–533. (Cited on pp. 103, 108, 191)

[169] Jim Gatheral and Roel C.A. Oomen, *Zero-intelligence realized variance estimation*, Finance and Stochastics, 14 (2010), pp. 249–283. (Cited on pp. 5, 55)

[170] Jim Gatheral and Radoš Radoičić, *Rational approximation of the rough Heston solution*, International Journal of Theoretical and Applied Finance, 22 (2019), 1950010. (Cited on pp. 93, 193)

[171] Stefan Gerhold, Christoph Gerstenecker, and Arpad Pinter, *Moment explosions in the rough Heston model*, Decisions in Economics and Finance, 42 (2019), pp. 575–608. (Cited on pp. 87, 92, 95)

[172] Stefan Gerhold, Antoine Jacquier, Mikko S. Pakkanen, Henry Stone, and Thomas Wagenhofer, *Pathwise large deviations for the rough Bergomi model: Corrigendum*, Journal of Applied Probability, 58 (2021), pp. 849–850. (Cited on p. 173)

[173] JOHN GEWEKE AND SUSAN PORTER-HUDAK, *The estimation and application of long memory time series models*, Journal of Time Series Analysis, 4 (1983), pp. 221–238. (Cited on p. 18)

[174] TILMANN GNEITING AND MARTIN SCHLATHER, *Stochastic models that separate fractal dimension and the Hurst effect*, SIAM Review, 46 (2004), pp. 269–282. (Cited on p. 128)

[175] CARLA GOMES AND HENRI WAELBROECK, *Is market impact a measure of the information value of trades? Market response to liquidity vs. informed metaorders*, Quantitative Finance, 15 (2015), pp. 773–793. (Cited on p. 64)

[176] CLIVE W.J. GRANGER AND NAMWON HYUNG, *Occasional structural breaks and long memory with an application to the S&P* 500 *absolute stock returns*, Journal of Empirical Finance, 11 (2004), pp. 399–421. (Cited on p. 18)

[177] CLIVE W.J. GRANGER AND TIMO TERÄSVIRTA, *A simple nonlinear time series model with misleading linear properties*, Economics Letters, 62 (1999), pp. 161–165. (Cited on p. 18)

[178] DENIS S. GREBENKOV, DMITRY BELYAEV, AND PETER W. JONES, *A multiscale guide to Brownian motion*, Journal of Physics A: Mathematical and Theoretical, 49 (2016), 043001. (Cited on p. 204)

[179] MASSIMILIANO GUBINELLI, *Ramification of rough paths*, Journal of Differential Equations, 248 (2010), pp. 693–721. (Cited on p. 229)

[180] HAMZA GUENNOUN, ANTOINE JACQUIER, PATRICK ROOME, AND FANGWEI SHI, *Asymptotic behavior of the fractional Heston model*, SIAM Journal on Financial Mathematics, 9 (2018), pp. 1017–1045. (Cited on pp. 85, 103)

[181] ARCHIL GULISASHVILI, *Large deviation principle for Volterra type fractional stochastic volatility models*, SIAM Journal on Financial Mathematics, 9 (2018), pp. 1102–1136. (Cited on pp. 173, 205)

[182] ARCHIL GULISASHVILI, *Gaussian stochastic volatility models: Scaling regimes, large deviations, and moment explosions*, Stochastic Processes and Their Applications, 130 (2020), pp. 3648–3686. (Cited on pp. 158, 172)

[183] ARCHIL GULISASHVILI, *Time-inhomogeneous Gaussian stochastic volatility models: Large deviations and super roughness*, Stochastic Processes and Their Applications, 139 (2021), pp. 37–79. (Cited on p. 181)

[184] ARCHIL GULISASHVILI, FREDERI VIENS, AND XIN ZHANG, *Extreme-strike asymptotics for general Gaussian stochastic volatility models*, Annals of Finance, 15 (2019), pp. 59–101. (Cited on p. 173)

[185] ———, *Small-time asymptotics for Gaussian self-similar stochastic volatility models*, Applied Mathematics & Optimization, 82 (2020), pp. 183–223. (Cited on p. 173)

[186] JULIEN GUYON, *The joint S&P* 500*/VIX smile calibration puzzle solved*, Risk, April (2020). (Cited on p. 100)

[187] ISTVÁN GYÖNGY, *Mimicking the one-dimensional marginal distributions of processes having an Itô differential*, Probability Theory and Related Fields, 71 (1986), pp. 501–516. (Cited on p. 177)

[188] PATRICK S. HAGAN, DEEP KUMAR, ANDREW S. LESNIEWSKI, AND DIANA E. WOODWARD, *Managing smile risk*, Wilmott Magazine, (2002), pp. 84–108. (Cited on p. 1)

[189] PAUL HAGER AND EYAL NEUMAN, *The multiplicative chaos of* $H = 0$ *fractional Brownian fields*, The Annals of Applied Probability, 32 (2022), pp. 2139–2179. (Cited on p. 180)

[190] MARTIN HAIRER, *Solving the KPZ equation*, Annals of Mathematics (2), 178 (2013), pp. 559–664. (Cited on pp. 202, 204, 206, 224)

[191] Martin Hairer, *A theory of regularity structures*, Inventiones Mathematicae, 198 (2014), pp. 269–504. (Cited on pp. 202, 204, 206, 218, 219, 220, 221, 224, 228, 229, 230, 231, 233, 239, 241, 244)

[192] ———, *Introduction to regularity structures*, Brazilian Journal of Probability and Statistics, 29 (2015), pp. 175–210. (Cited on p. 206)

[193] Martin Hairer and David Kelly, *Geometric versus non-geometric rough paths*, Annales de l'Institut Henri Poincaré, Probabilités et Statistiques, 51 (2015), pp. 207–251. (Cited on p. 229)

[194] Martin Hairer and Étienne Pardoux, *A Wong-Zakai theorem for stochastic PDEs*, J. Math. Soc. Japan, 67 (2015), pp. 1551–1604. (Cited on pp. 202, 208, 231)

[195] Martin Hairer and Hao Shen, *The dynamical Sine-Gordon model*, Communications in Mathematical Physics, 341 (2016), pp. 933–989. (Cited on p. 229)

[196] Martin Hairer and Hendrik Weber, *Large deviations for white-noise driven, nonlinear stochastic PDEs in two and three dimensions*, Annales de la Faculté des Sciences de Toulouse, Mathématiques, Série 6, 24 (2015), pp. 55–92. (Cited on pp. 232, 242, 243)

[197] Stephen J. Hardiman, Nicolas Bercot, and Jean-Philippe Bouchaud, *Critical reflexivity in financial markets: A Hawkes process analysis*, The European Physical Journal B, 86 (2013), article 442. (Cited on pp. 62, 69)

[198] Philipp Harms and David Stefanovits, *Affine representations of fractional processes with applications in mathematical finance*, Stochastic Processes and Their Applications, 129 (2019), pp. 1185–1228. (Cited on p. 112)

[199] Steven L Heston, *A closed-form solution for options with stochastic volatility with applications to bond and currency options*, Review of Financial Studies, 6 (1993), pp. 327–343. (Cited on pp. 1, 83, 84, 103)

[200] Blanka Horvath, Antoine Jacquier, and Chloé Lacombe, *Asymptotic behaviour of randomised fractional volatility models*, Journal of Applied Probability, 56 (2019), pp. 496–523. (Cited on pp. 95, 173)

[201] Blanka Horvath, Antoine Jacquier, and Peter Tankov, *Volatility options in rough volatility models*, SIAM Journal on Financial Mathematics, 11 (2020), pp. 437–469. (Cited on p. 117)

[202] Blanka Horvath, Aitor Muguruza, and Mehdi Tomas, *Deep learning volatility: A deep neural network perspective on pricing and calibration in (rough) volatility models*, Quantitative Finance, 21 (2021), pp. 11–27. (Cited on p. 116)

[203] Blanka Horvath, Josef Teichmann, and Žan Žurič, *Deep hedging under rough volatility*, Risk, 9 (2021), p. 138. (Cited on p. 116)

[204] Weibing Huang, Charles-Albert Lehalle, and Mathieu Rosenbaum, *Simulating and analyzing order book data: The queue-reactive model*, Journal of the American Statistical Association, 110 (2015), pp. 107–122. (Cited on p. 63)

[205] Gur Huberman and Werner Stanzl, *Price manipulation and quasi-arbitrage*, Econometrica, 72 (2004), pp. 1247–1275. (Cited on p. 60)

[206] John Hull and Alan White, *One-factor interest-rate models and the valuation of interest-rate derivative securities*, Journal of Financial and Quantitative Analysis, 28 (1993), pp. 235–254. (Cited on p. 1)

[207] Jacques Istas and Gabriel Lang, *Quadratic variations and estimation of the local Hölder index of a Gaussian process*, Annales de l'Institut Henri Poincaré, Probabilités et Statistiques, 33 (1997), pp. 407–436. (Cited on p. 11)

[208] Saul D. Jacka, *A martingale representation result and an application to incomplete financial markets*, Mathematical Finance, 2 (1992), pp. 239–250. (Cited on p. 115)

[209] Jean Jacod and Albert Shiryaev, *Limit Theorems for Stochastic Processes*, Grundlehren Math. Wiss. 288, Springer-Verlag, Berlin, 2013. (Cited on pp. 163, 179)

[210] Antoine Jacquier, Mikko S. Pakkanen, and Henry Stone, *Pathwise large deviations for the rough Bergomi model*, Journal of Applied Probability, 55 (2018), pp. 1078–1092. (Cited on pp. 172, 173, 205)

[211] Antoine Jacquier and Alexandre Pannier, *Large and moderate deviations for stochastic Volterra systems*, Stochastic Processes and Their Applications, 149 (2022), pp. 142–187. (Cited on pp. 172, 173, 174)

[212] Antoine Jacquier and Konstantinos Spiliopoulos, *Pathwise moderate deviations for option pricing*, Mathematical Finance, 30 (2020), pp. 426–463. (Cited on p. 174)

[213] Thibault Jaisson, *Market impact as anticipation of the order flow imbalance*, Quantitative Finance, 15 (2015), pp. 1123–1135. (Cited on pp. 60, 61, 62)

[214] Thibault Jaisson and Mathieu Rosenbaum, *Limit theorems for nearly unstable Hawkes processes*, The Annals of Applied Probability, 25 (2015), pp. 600–631. (Cited on pp. 62, 65, 77)

[215] ———, *Rough fractional diffusions as scaling limits of nearly unstable heavy tailed Hawkes processes*, The Annals of Applied Probability, 26 (2016), pp. 2860–2882. (Cited on pp. 31, 61, 62, 65, 66, 67, 71, 74, 79)

[216] Svante Janson, *Gaussian Hilbert Spaces*, Cambridge University Press, Cambridge, 1997. (Cited on pp. 211, 214, 215, 217, 239)

[217] Paul Jusselin and Mathieu Rosenbaum, *No-arbitrage implies power-law market impact and rough volatility*, Mathematical Finance, 30 (2020), pp. 1309–1336. (Cited on pp. xxvi, 178)

[218] Olav Kallenberg, *Foundations of Modern Probability*, Probab. Appl. (N.Y.) 2, Springer Verlag, New York, 1997. (Cited on p. 149)

[219] Jan Kallsen, *A didactic note on affine stochastic volatility models*, in From Stochastic Calculus to Mathematical Finance, Springer, 2006, pp. 343–368. (Cited on p. 103)

[220] Ioannis Karatzas and Steven E. Shreve, *Brownian Motion and Stochastic Calculus*, Grad. Texts in Math. 113, Springer-Verlag, New York, 2012. (Cited on p. 83)

[221] Samuel Karlin and Howard E. Taylor, *A Second Course in Stochastic Processes*, Academic Press, Inc. [Harcourt Brace Jovanovich, Publishers], New York-London, 1981. (Cited on p. 83)

[222] Martin Keller-Ressel, *Moment explosions and long-term behavior of affine stochastic volatility models*, Mathematical Finance, 21 (2011), pp. 73–98. (Cited on p. 87)

[223] Martin Keller-Ressel and Assad Majid, *A comparison principle between rough and non-rough Heston models—with applications to the volatility surface*, Quantitative Finance, 20 (2020), pp. 919–933. (Cited on pp. 87, 95, 96)

[224] Anatoliy Aleksandrovich Kilbas, Hari M. Srivastava, and Juan J. Trujillo, *Theory and Applications of Fractional Differential Equations*, North-Holland Math. Stud. 204, Elsevier Science B.V., Amsterdam, 2006. (Cited on p. 97)

[225] Shigeo Kusuoka, *Approximation of expectation of diffusion processes and mathematical finance*, in Proceedings of Final Taniguchi Symposium, Nara 1998, Adv. Stud. Pure Math. 31, Mathematical Society of Japan, 2001, pp. 147–165. (Cited on pp. 201, 206)

[226] Hubert Lacoin, Rémi Rhodes, and Vincent Vargas, *A probabilistic approach of ultraviolet renormalization in the boundary Sine-Gordon model*, Probability Theory and Related Fields, 185 (2023), pp. 1–40. (Cited on pp. 185, 186)

[227] Roger Lee, *Gamma swap*, in Encyclopedia of Quantitative Finance, John Wiley & Sons, Ltd., New York, 2010. (Cited on p. 187)

[228] Roger W. Lee, *The moment formula for implied volatility at extreme strikes*, Mathematical Finance, 14 (2004), pp. 469–480. (Cited on p. 95)

[229] Alan L. Lewis, *Option Valuation under Stochastic Volatility* II, Finance Press, 2016. (Cited on pp. 86, 189)

[230] Fabrizio Lillo, J. Doyne Farmer, and Rosario N. Mantegna, *Master curve for price-impact function*, Nature, 421 (2003), pp. 129–130. (Cited on pp. 60, 61, 64)

[231] Pierre-Louis Lions and Marek Musiela, *Correlations and bounds for stochastic volatility models*, Annales de l'Institut Henri Poincaré C, Analyse Non Linéaire, 24 (2007), pp. 1–16. (Cited on p. 158)

[232] Giulia Livieri, Saad Mouti, Andrea Pallavicini, and Mathieu Rosenbaum, *Rough volatility: Evidence from option prices*, IISE Transactions, 50 (2018), pp. 767–776. (Cited on pp. 59, 124)

[233] Terry Lyons and Nicolas Victoir, *Cubature on Wiener space*, Proceedings of the Royal Society of London A: Mathematical, Physical and Engineering Sciences, 460 (2004), pp. 169–198. (Cited on pp. 201, 206)

[234] Stephane Mallat, *A Wavelet Tour of Signal Processing*, Elsevier/Academic Press, Amsterdam, 2009. (Cited on p. 140)

[235] Benoit B. Mandelbrot, Adlai J. Fisher, and Laurent E. Calvet, *A Multifractal Model of Asset Returns*. Unpublished, 1997. (Cited on p. 180)

[236] Benoit B. Mandelbrot and John W. Van Ness, *Fractional Brownian motions, fractional noises and applications*, SIAM Review, 10 (1968), pp. 422–437. (Cited on pp. 2, 23)

[237] Rosario N Mantegna and H. Eugene Stanley, *Introduction to Econophysics: Correlations and Complexity in Finance*, Cambridge University Press, Cambridge, 2000. (Cited on p. 17)

[238] Michael McAleer and Marcelo C Medeiros, *Realized volatility: A review*, Econometric Reviews, 27 (2008), pp. 10–45. (Cited on p. 55)

[239] Ryan McCrickerd and Mikko S. Pakkanen, *Turbocharging Monte Carlo pricing for the rough Bergomi model*, Quantitative Finance, 18 (2018), pp. 1877–1886. (Cited on p. 116)

[240] Yves Meyer, *Wavelets and Operators*, Cambridge Stud. Adv. Math. 37, Cambridge University Press, Cambridge, 1992. (Cited on p. 213)

[241] Aleksandar Mijatović and Peter Tankov, *A new look at short-term implied volatility in asset price models with jumps*, Mathematical Finance, 26 (2016), pp. 149–183. (Cited on p. 199)

[242] Thomas Mikosch and Catalin Starica, *Is it really long memory we see in financial returns*, in Extremes and Integrated Risk Management, Paul Embrechts, ed., Risk Books, London, 2000, pp. 149–168. (Cited on p. 2)

[243] Yuliya S. Mishura, *Stochastic Calculus for Fractional Brownian Motion and Related Processes*, Lecture Notes in Math. 1929, Springer-Verlag, Berlin, 2008. (Cited on p. 131)

[244] Marek Musiela and Marek Rutkowski, *Martingale Methods in Financial Modelling*, Stoch. Model. Appl. Probab. 36, Springer-Verlag, Berlin, 2005. (Cited on pp. xxvi, 1)

[245] Anthony Neuberger, *Volatility Trading*, Institute of Finance and Accounting, London Business School, 1990. (Cited on p. 88)

[246] Andreas Neuenkirch and Taras Shalaiko, *The Order Barrier for Strong Approximation of Rough Volatility Models*. arXiv:1606.03854, 2016. (Cited on p. 223)

[247] Eyal Neuman and Mathieu Rosenbaum, *Fractional Brownian motion with zero Hurst parameter: A rough volatility viewpoint*, Electronic Communications in Probability, 23 (2018), pp. 1–12. (Cited on p. 180)

[248] Syoiti Ninomiya and Nicolas Victoir, *Weak approximation of stochastic differential equations and application to derivative pricing*, Applied Mathematical Finance, 15 (2008), pp. 107–121. (Cited on pp. 201, 206)

[249] Diaa Noureldin, Neil Shephard, and Kevin Sheppard, *Multivariate high-frequency-based volatility (heavy) models*, Journal of Applied Econometrics, 27 (2012), pp. 907–933. (Cited on p. 52)

[250] Alexander Novikov and Esko Valkeila, *On some maximal inequalities for fractional Brownian motions*, Statistics & Probability Letters, 44 (1999), pp. 47–54. (Cited on p. 24)

[251] David Nualart, *Fractional Brownian motion: Stochastic calculus and applications*, in International Congress of Mathematicians, vol. 3, Eur. Math. Soc., Zürich, 2006, pp. 1541–1562. (Cited on p. 211)

[252] ———, *The Malliavin Calculus and Related Topics*, Springer-Verlag, Berlin, 2013. (Cited on p. 215)

[253] David Nualart and Étienne Pardoux, *Stochastic calculus with anticipating integrands*, Probability Theory and Related Fields, 78 (1988), pp. 535–581. (Cited on pp. 160, 217)

[254] Carl J. Nuzman and Vincent H. Poor, *Linear estimation of self-similar processes via Lamperti's transformation*, Journal of Applied Probability, 37 (2000), pp. 429–452. (Cited on pp. 19, 21, 50)

[255] Etienne Pardoux and Philip Protter, *Stochastic Volterra equations with anticipating coefficients*, The Annals of Probability, 18 (1990), pp. 1635–1655. (Cited on p. 228)

[256] Jan Pedersen and Orimar Sauri, *On Lévy semistationary processes with a Gamma kernel*, in XI Symposium on Probability and Stochastic Processes, R.H. Mena, J.C. Pardo, V. Rivero, and G.U. Bravo, eds., Birkhäuser/Springer, Cham, 2015, pp. 217–239. (Cited on p. 131)

[257] Mathias Pohl, Alexander Ristig, Walter Schachermayer, and Ludovic Tangpi, *The amazing power of dimensional analysis: Quantifying market impact*, Market Microstructure and Liquidity, 3 (2017), 1850004. (Cited on p. 61)

[258] Andrei D. Polyanin and Valentin F. Zaitsev, *Handbook of Exact Solutions for Ordinary Differential Equations*, 2nd Edition, Chapman and Hall/CRC, Boca Raton, 2003. (Cited on p. 87)

[259] Rémi Rhodes and Vincent Vargas, *Gaussian multiplicative chaos and applications: A review*, Probability Surveys, 11 (2014), pp. 315–392. (Cited on p. 179)

[260] Christian Y. Robert and Mathieu Rosenbaum, *A new approach for the dynamics of ultra-high-frequency data: The model with uncertainty zones*, Journal of Financial Econometrics, 9 (2011), pp. 344–366. (Cited on pp. 5, 6, 15)

[261] Christian Y. Robert and Mathieu Rosenbaum, *Volatility and covariation estimation when microstructure noise and trading times are endogenous*, Mathematical Finance, 22 (2012), pp. 133–164. (Cited on pp. 5, 6)

[262] Mathieu Rosenbaum, *Estimation of the volatility persistence in a discretely observed diffusion model*, Stochastic Processes and Their Applications, 118 (2008), pp. 1434–1462. (Cited on p. 2)

[263] Mathieu Rosenbaum, *First order p-variations and Besov spaces*, Statistics & Probability Letters, 79 (2009), pp. 55–62. (Cited on p. 5)

[264] ———, *A new microstructure noise index*, Quantitative Finance, 11 (2011), pp. 883–899. (Cited on p. 5)

[265] Mathieu Rosenbaum and Jianfei Zhang, *Deep calibration of the quadratic rough Heston model*, Risk.net, September (2022). (Cited on p. 100)

[266] Johannes Ruf, *Hedging under arbitrage*, Mathematical Finance, 23 (2013), pp. 297–317. (Cited on pp. 115, 116)

[267] Stefan G. Samko, Anatoly A. Kilbas, and Oleg I. Marichev, *Fractional Integrals and Derivatives: Theory and Applications*, CRC Press, 1993. (Cited on pp. 75, 82, 242)

[268] Steven E. Shreve, *Stochastic Calculus for Finance* II: *Continuous-Time Models*, Springer Finance 11, Springer-Verlag, New York, 2004. (Cited on p. xxvi)

[269] Carlos A. Sin, *Complications with stochastic volatility models*, Advances in Applied Probability, 30 (1998), pp. 256–268. (Cited on p. 158)

[270] Eric Smith, J. Doyne Farmer, Laszlo Gillemot, and Supriya Krishnamurthy, *Statistical theory of the continuous double auction*, Quantitative Finance, 3 (2003), pp. 481–514. (Cited on p. 61)

[271] Konstantinos Spiliopoulos, *Large deviations and importance sampling for systems of slow-fast motion*, Applied Mathematics & Optimization, 67 (2013), pp. 123–161. (Cited on p. 174)

[272] Bence Tóth, Yves Lemperiere, Cyril Deremble, Joachim De Lataillade, Julien Kockelkoren, and J.-P. Bouchaud, *Anomalous price impact and the critical nature of liquidity in financial markets*, Physical Review X, 1 (2011), 021006. (Cited on pp. 60, 61)

[273] Harry van Zanten, *A multivariate central limit theorem for continuous local martingales*, Statistics & Probability Letters, 50 (2000), pp. 229–235. (Cited on p. 163)

[274] Mark Veraar, *The stochastic Fubini theorem revisited*, Stochastics, 84 (2012), pp. 543–551. (Cited on pp. 75, 107)

[275] Frederi Viens and Jianfeng Zhang, *A martingale approach for fractional Brownian motions and related path dependent PDEs*, The Annals of Applied Probability, 29 (2019), pp. 3489–3540. (Cited on pp. 116, 117)

[276] Nakahiro Yoshida, *Malliavin calculus and asymptotic expansion for martingales*, Probability Theory and Related Fields, 109 (1997), pp. 301–342. (Cited on pp. 164, 167)

[277] ———, *Malliavin calculus and martingale expansion*, Bulletin des Sciences Mathématiques, 125 (2001), pp. 431–456. (Cited on p. 167)

[278] Gilles Zumbach, *Time reversal invariance in finance*, Quantitative Finance, 9 (2009), pp. 505–515. (Cited on p. 98)

[279] ———, *Volatility conditional on price trends*, Quantitative Finance, 10 (2010), pp. 431–442. (Cited on pp. 98, 99)

Index